RAPID CLIMATE CHANGE: PAST EVIDENCE AND FUTURE PROSPECTS

CLIMATE CHANGE AND ITS CAUSES, EFFECTS AND PREDICTION

Additional books in this series can be found on Nova's website under the Series tab.

Additional E-books in this series can be found on Nova's website under the E-book tab.

CLIMATE CHANGE AND ITS CAUSES, EFFECTS AND PREDICTION

RAPID CLIMATE CHANGE: PAST EVIDENCE AND FUTURE PROSPECTS

ALICE E. BENNETT
EDITOR

Nova Science Publishers, Inc.
New York

For permission to use material from this book please contact us:
Telephone 631-231-7269; Fax 631-231-8175
Web Site: http://www.novapublishers.com

NOTICE TO THE READER

The Publisher has taken reasonable care in the preparation of this book, but makes no expressed or implied warranty of any kind and assumes no responsibility for any errors or omissions. No liability is assumed for incidental or consequential damages in connection with or arising out of information contained in this book. The Publisher shall not be liable for any special, consequential, or exemplary damages resulting, in whole or in part, from the readers' use of, or reliance upon, this material. Any parts of this book based on government reports are so indicated and copyright is claimed for those parts to the extent applicable to compilations of such works.

Independent verification should be sought for any data, advice or recommendations contained in this book. In addition, no responsibility is assumed by the publisher for any injury and/or damage to persons or property arising from any methods, products, instructions, ideas or otherwise contained in this publication.

This publication is designed to provide accurate and authoritative information with regard to the subject matter covered herein. It is sold with the clear understanding that the Publisher is not engaged in rendering legal or any other professional services. If legal or any other expert assistance is required, the services of a competent person should be sought. FROM A DECLARATION OF PARTICIPANTS JOINTLY ADOPTED BY A COMMITTEE OF THE AMERICAN BAR ASSOCIATION AND A COMMITTEE OF PUBLISHERS.

Additional color graphics may be available in the e-book version of this book.

LIBRARY OF CONGRESS CATALOGING-IN-PUBLICATION DATA

Rapid climate change : past evidence and future prospects.
   p. cm.
 Includes index.
 ISBN 978-1-60741-422-3 (hardcover)
 1. Climatic changes--History. 2. Climatic changes--Forecasting. 3.
Global environmental change. 4. Environmental policy. I. Climate Change
Science Program (U.S.). Subcommittee on Global Change Research.
 QC903.R37 2010
 551.6--dc22
                                                              2010026964

Published by Nova Science Publishers, Inc. † New York

CONTENTS

PREFACE

This book reports on abrupt climate change which is defined as a large scale change in the climate system that takes place over a few decades or less, persists (or is anticipated to persist) for at least a few decades, and causes substantial disruptions in human and natural systems. Abrupt climate change presents potential risks for society that are poorly understood. An improved ability to understand and model future abrupt climate change is essential to provide decision-makers with the information they need to plan for these potentially significant changes.

Chapter 1 - A large-scale change in the climate system that takes place over a few decades or less, persists (or is anticipated to persist) for at least a few decades, and causes substantial disruptions in human and natural systems.

This chapter considers progress in understanding four types of abrupt change in the paleoclimatic record that stand out as being so rapid and large in their impact that if they were to recur, they would pose clear risks to society in terms of our ability to adapt: (1) rapid change in glaciers, ice sheets, and hence sea level; (2) widespread and sustained changes to the hydrologic cycle; (3) abrupt change in the northward flow of warm, salty water in the upper layers of the Atlantic Ocean associated with the Atlantic Meridional Overturning Circulation (AMOC); and (4) rapid release to the atmosphere of methane trapped in permafrost and on continental margins. While these four types of change pose clear risks to human and natural systems, this chapter does not focus on specific effects on these systems as a result of abrupt change.

Chapter 2 - Scientific conclusions have become more compelling regarding the influence of human activities on the Earth's climate. In 2007, the Intergovernmental Panel on Climate Change declared that evidence of global warming was "unequivocal." It concluded that "[m]ost of the observed increase in globally averaged temperatures since the mid-20th century is very likely due to the observed increase in anthropogenic [human-related] greenhouse gas [GHG] concentrations."

The IPCC concluded that human activities have markedly increased atmospheric concentrations of "greenhouse gases" (GHG), including carbon dioxide (CO_2), methane (CH_4), nitrous oxide (N_2O), and gases (such as chlorofluorocarbons, CFC) that are controlled under the Montreal Protocol to protect the stratospheric ozone layer. From the beginning of the Industrial Revolution, CO_2 has risen from about 280 parts per million (ppm) to about 386 ppm today (up 38%). The concentration of CO_2 is higher now than in at least 800,000 years before present.

Additional human influences on the climate that are not easily compared to GHG emissions could, nonetheless, be managed to moderate regional and global climate change. These include tropospheric ozone pollution (i.e., smog), particulate and aerosol emissions, and land cover change. New chemicals, such as nitrogen trifluoride (NF_3), also may play a small role.

Without radical changes globally from current policies and economic trajectories, experts uniformly expect that GHG emissions will continue to grow and lead to continued warming of the Earth's climate. Experts disagree, however, on the timing, magnitude and patterns of future climate changes. In the absence of concerted climate change mitigation policies, for a wide range of plausible GHG scenarios to 2100, the IPCC projected "best guess" increases in global average temperatures from 1.8°C to 4.0°C (3.2°F to 7.2°F). Although these temperature changes may seem small, they compare to the current global, annual average temperature of around 14°C (57°F). While precipitation overall is expected to increase, its distribution may become more uneven: regions that now are dry are likely to get drier, while regions that now are wet, are likely to get wetter. Extreme precipitation and droughts are expected to become more frequent. Experts project that warming ocean waters will expand, and melting glaciers and ice sheets will further add to sea level rise. The Arctic Ocean could become ice free in summers within a few decades. Ocean salinity is expected to fall, and the Meridional Overturning Circulation in the Atlantic Ocean could slow, reducing ocean productivity and altering regional climates in both North America and Europe. The climate would continue changing for hundreds of years after GHG concentrations were stabilized, according to most models. There are also possibilities of abrupt changes in the state of the climate system, with unpredictable and potentially catastrophic consequences. Much concern is focused now, among scientists and economists, about the likelihoods and implications of exceeding such thresholds of abrupt change, sometimes called "tipping points."

This chapter summarizes highlights of scientific research and assessments related to human- induced climate change. For more extensive explanation of climate change science and analytical methods, see CRS Report RL33849, *Climate Change: Science and Policy Implications.*

In: Rapid Climate Change Past Evidence and Future Prospects   ISBN: 978-1-60741-422-3
Editor: Alice E. Bennett   © 2011 Nova Science Publishers, Inc.

Chapter 1

ABRUPT CLIMATE CHANGE: FINAL REPORT, SYNTHESIS AND ASSESSMENT PRODUCT 3.4

U.S. Climate Change Science Program,Subcommittee on Global Change Research, U.S. Geological Survey, National Oceanic and Atmospheric Administration, and National Science Foundation

EXECUTIVE SUMMARY

Edward, R. Cook,* Lamont-Doherty Earth Observatory, Columbia University, New York,

For this Synthesis and Assessment Report, abrupt climate change is defined as:

> A large-scale change in the climate system that takes place over a few decades or less, persists (or is anticipated to persist) for at least a few decades, and causes substantial disruptions in human and natural systems.

This chapter considers progress in understanding four types of abrupt change in the paleoclimatic record that stand out as being so rapid and large in their impact that if they were to recur, they would pose clear risks to society in terms of our ability to adapt: (1) rapid change in glaciers, ice sheets, and hence sea level; (2) widespread and sustained changes to the hydrologic cycle; (3) abrupt change in the northward flow of warm, salty water in the upper layers of the Atlantic Ocean associated with the Atlantic Meridional Overturning Circulation (AMOC); and (4) rapid release to the atmosphere of methane trapped in permafrost and on continental margins. While these four types of change pose clear risks to human and natural systems, this chapter does not focus on specific effects on these systems as a result of abrupt change.

This chapter reflects the significant progress in understanding abrupt climate change that has been made since the report by the National Research Council in 2002 on this topic, and this chapter provides considerably greater detail and insight on these issues than did the 2007 Intergovernmental Panel on Climate Change (IPCC) Fourth Assessment Report (AR4). New paleoclimatic reconstructions have been developed that provide greater understanding of patterns and mechanisms of past abrupt climate change in the ocean and on land, and new observations are further revealing unanticipated rapid dynamic changes of moderns glaciers, ice sheets, and ice shelves as well as processes that are contributing to these changes. This chapter reviews this progress. A summary and explanation of the main results is presented first, followed by an overview of the types of abrupt climate change considered in this chapter. The subsequent chapters then address each of these types of abrupt climate change, including a synthesis of the current state of knowledge and an assessment of the likelihood that one of these abrupt changes may occur in response to human influences on the climate system. Throughout this chapter we have adopted the IPCC terminology in our expert assessment of the likelihood of a particular outcome or result. The term *virtually certain* implies a >99% probability; *extremely likely*: >95% probability; *very likely*: > 90% probability; *likely*: >65% probability; *more likely than not*: >50% probability; *about as likely as not*: 33%–66% probability; *unlikely*: <33% probability; *very unlikely*: < 10% probability; *extremely unlikely*: <5% probability; *exceptionally unlikely*: <1%.

Based on an assessment of the published scientific literature, the primary conclusions presented in this chapter are:

- Recent rapid changes at the edges of the Greenland and West Antarctic ice sheets show acceleration of flow and thinning, with the velocity of some glaciers increasing more than twofold. Glacier accelerations causing this imbalance have been related to enhanced surface meltwater production penetrating to the bed to lubricate glacier motion, and to ice-shelf removal, ice-front retreat, and glacier ungrounding that reduce resistance to flow. The present generation of models does not capture these processes. It is unclear whether this imbalance is a short- term natural adjustment or a response to recent climate change, but processes causing accelerations are enabled by warming, so these adjustments will very likely become more frequent in a warmer climate. The regions likely to experience future rapid changes in ice volume are those where ice is grounded well below sea level such as the West Antarctic Ice Sheet or large glaciers in Greenland like the Jakobshavn Isbrae that flow into the sea through a deep channel reaching far inland. Inclusion of these processes in models will likely lead to sea-level projections for the end of the 21st century that substantially exceed the projections presented in the IPCC AR4 report (0.28 ± 0.10 m to 0.42 ± 0.16 m rise).

- There is no clear evidence to date of human-induced global climate change on North American precipitation amounts. However, since the IPCC AR4 report, further analysis of climate model scenarios of future hydroclimatic change over North America and the global subtropics indicate that subtropical aridity is likely to intensify and persist due to future greenhouse warming. This projected drying extends poleward into the United States Southwest, potentially increasing the likelihood of severe and persistent drought there in the future. If the model results are correct then this drying may have already begun, but currently cannot be definitively identified amidst the considerable natural variability of hydroclimate in Southwestern North America.

- The AMOC is the northward flow of warm, salty water in the upper layers of the Atlantic, and the southward flow of colder water in the deep Atlantic. It plays an important role in the oceanic transport of heat from low to high latitudes. It is very likely that the strength of the AMOC will decrease over the course of the 21st century in response to increasing greenhouse gases, with a best estimate decrease of 25-30%. However, it is very unlikely that the AMOC will undergo an abrupt transition to a weakened state or collapse during the course of the 21st century, and it is unlikely that the AMOC will collapse beyond the end of the 21st century because of global warming, although the possibility cannot be entirely excluded.

- A dramatic abrupt release of methane (CH_4) to the atmosphere appears very unlikely, but it is very likely that climate change will accelerate the pace of persistent emissions from both hydrate sources and wetlands. Current models suggest that a doubling of northern high latitudes CH_4 emissions could be realized fairly easily. However, since these models do not realistically represent all the processes thought to be relevant to future northern high-latitude CH_4 emissions, much larger (or smaller) increases cannot be discounted. Acceleration of release from hydrate reservoirs is likely, but its magnitude is difficult to estimate.

Major Questions and Related Findings

1. Will there be an abrupt change in sea level?

This question is addressed in Chapter 2 of this chapter, with emphasis on documenting (1) the recent rates and trends in the net glacier and ice-sheet annual gain or loss of ice/snow (known as mass balance) and their contribution to sea level rise (SLR) and (2) the processes responsible for the observed acceleration in ice loss from marginal regions of existing ice sheets. In response to this question, Chapter 2 notes:

1. The record of past changes in ice volume provides important insight to the response of large ice sheets to climate change.
   - Paleorecords demonstrate that there is a strong inverse relation between atmospheric carbon dioxide (CO_2) and global ice volume. Sea level rise associated with the melting of the ice sheets at the end of the last Ice Age ~20,000 years ago averaged 10-20 millimeters per year ($mm\ a^{-1}$) with large "meltwater fluxes" exceeding SLR of 50 $mm\ a^{-1}$ and lasting several centuries, clearly demonstrating the potential for ice sheets to cause rapid and large sea level changes.

2. Sea level rise from glaciers and ice sheets has accelerated.
   - Observations demonstrate that it is extremely likely that the Greenland Ice Sheet is losing mass and that this has very likely been accelerating since the mid-1990s. Greenland has been thickening at high elevations because of the increase in snowfall that is consistent with high-latitude warming, but this gain is more than offset by an accelerating mass loss, with a large component from rapidly thinning and accelerating outlet glaciers. The balance between gains and losses of mass decreased from near-zero in the early 1990s to net losses of 100

gigatonnes per year (Gt a^{-1}) to more than 200 Gt a^{-1} for the most recent observations in 2006.

- The mass balance for Antarctica is a net loss of about 80 Gt a^{-1} in the mid 1990s, increasing to almost 130 Gt a^{-1} in the mid 2000s. Observations show that while some higher elevation regions are thickening, substantial ice losses from West Antarctica and the Antarctic Peninsula are very likely caused by changing ice dynamics.
- The best estimate of the current (2007) mass balance of small glaciers and ice caps is a loss that is at least three times greater (380 to 400 Gt a^{-1}) than the net loss that has been characteristic since the mid-19th century.

3. Recent observations of the ice sheets have shown that changes in ice dynamics can occur far more rapidly than previously suspected.
- Recent observations show a high correlation between periods of heavy surface melting and increase in glacier velocity. A possible cause is rapid meltwater drainage to the base of the glacier, where it enhances basal sliding. An increase in meltwater production in a warmer climate will likely have major consequences on ice-flow rate and mass loss.
- Recent rapid changes in marginal regions of the Greenland and West Antarctic ice sheets show mainly acceleration and thinning, with some glacier velocities increasing more than twofold. Many of these glacier accelerations closely followed reduction or loss of their floating extensions known as ice shelves. Significant changes in ice-shelf thickness are most readily caused by changes in basal melting induced by oceanic warming. The interaction of warm waters with the periphery of the large ice sheets represents one of the most significant possibilities for abrupt change in the climate system. The likely sensitive regions for future rapid changes in ice volume by this process are those where ice is grounded well below sea level, such as the West Antarctic Ice Sheet or large outlet glaciers in Greenland like the Jakobshavn Isbrae that flow through a deep channel that extends far inland.
- Although no ice-sheet model is currently capable of capturing the glacier speedups in Antarctica or Greenland that have been observed over the last decade, including these processes in models will very likely show that IPCC AR4 projected sea level rises for the end of the 21st century are too low.

2. Will there be an abrupt change in land hydrology?

This question is addressed in Chapter 3 of this chapter. In general, variations in water supply and in particular protracted droughts are among the greatest natural hazards facing the United States and the globe today and in the foreseeable future. In contrast to floods, which reflect both previous conditions and current meteorological events, and which are consequently more localized in time and space, droughts occur on subcontinental to continental scales and can persist for decades and even centuries.

On interannual to decadal time scales, droughts can develop faster than human societies can adapt to the change. Thus, a severe drought lasting several years can be regarded as an abrupt change, although it may not reflect a permanent change in the state of the climate system.

Empirical studies and climate model experiments conclusively show that droughts over North America and around the world are significantly influenced by the state of tropical sea-surface temperatures (SSTs), with cool La Niña-like SSTs in the eastern equatorial Pacific being especially responsible for the development of droughts over the southwestern United States and northern Mexico. Warm subtropical North Atlantic SSTs played a role in forcing the 1930s Dust Bowl and 1950s droughts as well. Unusually warm Indo-Pacific SSTs have also been strongly implicated in the development of global patterns of drought observed in recent years.

Historic droughts over North America have been severe, but not nearly as prolonged as a series of "megadroughts" reconstructed from tree rings from about A.D. 900 up to about A.D. 1600. These megadroughts are significant because they occurred in a climate system that was not being perturbed in a major way by human activity (i.e., the ongoing anthropogenic changes in greenhouse gas concentrations, atmospheric dust loadings, and land-cover changes). Modeling experiments indicate that these megadroughts may have occurred in response to cold tropical Pacific SSTs and warm subtropical North Atlantic SSTs externally forced by high irradiance and weak volcanic activity. However, this result is tentative, and the exceptional duration of the droughts has not been adequately explained, nor whether they also involved forcing from SST changes in other ocean basins.

Even larger and more persistent changes in hydroclimatic variability worldwide are indicated over the last 10,000 years by a diverse set of paleoclimatic indicators. The climate conditions associated with those changes were quite different from those of the past millennium and today, but they show the additional range of natural variability and truly abrupt hydroclimatic change that can be expressed by the climate system.

With respect to this question, Chapter 3 concludes:

- There is no clear evidence to date of human-induced global climate change on North American precipitation amounts. However, since the IPCC AR4 report, further analysis of climate model scenarios of future hydroclimatic change over North America and the global subtropics indicate that subtropical aridity is likely to intensify and persist due to future greenhouse warming. This projected drying extends poleward into the United States Southwest, potentially increasing the likelihood of severe and persistent drought there in the future. If the model results are correct then this drying may have already begun, but currently cannot be definitively identified amidst the considerable natural variability of hydroclimate in Southwestern North America.

- The cause of model-projected subtropical drying is an overall widespread warming of the ocean and atmosphere, in contrast to the causes of historic droughts, and the likely causes of Medieval megadroughts, which were related to changes in the patterns of SSTs. However, systematic biases within current coupled atmosphere-ocean models raise concerns as to whether they correctly represent the response of the tropical climate system to radiative forcing and whether greenhouse forcing will actually induce El Nino/Southern Oscillation-like patterns of tropical SST change that will create impacts on global hydroclimate in addition to those caused by overall warming.

3. Do we expect an abrupt change in the Atlantic Meridional Overturning Circulation?

This question is addressed in Chapter 4 of this chapter. The Atlantic Meridional Overturning Circulation (AMOC) is an important component of the Earth's climate system, characterized by a northward flow of warm, salty water in the upper layers of the Atlantic, and a southward flow of colder water in the deep Atlantic. This ocean current system transports a substantial amount of heat from the Tropics and Southern Hemisphere toward the North Atlantic, where the heat is transferred to the atmosphere. Changes in this ocean circulation could have a profound impact on many aspects of the global climate system.

There is growing evidence that fluctuations in Atlantic sea surface temperatures, hypothesized to be related to fluctuations in the AMOC, have played a prominent role in significant climate fluctuations around the globe on a variety of time scales. Evidence from the instrumental record shows pronounced, multidecadal swings in widespread Atlantic temperature that may be at least partly due to fluctuations in the AMOC. Evidence from paleorecords suggests that there have been large, decadal-scale changes in the AMOC, particularly during glacial times. These abrupt changes have had a profound impact on climate, both locally in the Atlantic and in remote locations around the globe.

At its northern boundary, the AMOC interacts with the circulation of the Arctic Ocean. The summer arctic sea ice cover has undergone dramatic retreat since satellite records began in 1979, amounting to a loss of almost 30% of the September ice cover in 29 years. The late summer ice extent in 2007 was particularly startling and broke the previous record minimum with an extent that was three standard deviations below the linear trend. Conditions over the 2007-200 8 winter promoted further loss of multiyear ice due to anomalous transport through Fram Strait, raising the possibility that rapid and sustained ice loss could result. Climate model simulations suggest that rapid and sustained September arctic ice loss is likely in future 21st century climate projections.

In response to the question of an abrupt change in the AMOC, Chapter 4 notes:

- It is very likely that the strength of the AMOC will decrease over the course of the 21st century in response to increasing greenhouse gases, with a best estimate decrease of 25-30%.
- Even with the projected moderate AMOC weakening, it is still very likely that on multidecadal to century time scales a warming trend will occur over most of the European region downstream of the North Atlantic Current in response to increasing greenhouse gases, as well as over North America.
- It is very unlikely that the AMOC will undergo a collapse or an abrupt transition to a weakened state during the 21st century.
- It is also unlikely that the AMOC will collapse beyond the end of the 21st century because of global warming, although the possibility cannot be entirely excluded.
- Although it is very unlikely that the AMOC will collapse in the 21st century, the potential consequences of this event could be severe. These might include a southward shift of the tropical rainfall belts, additional sea level rise around the North Atlantic, and disruptions to marine ecosystems.

4. What is the potential for abrupt changes in atmospheric methane?

This question is addressed in Chapter 5 of this chapter. The main concerns about abrupt changes in atmospheric methane stem from (1) the large quantity of methane believed to be stored in clathrate hydrates in the sea floor and to a lesser extent in permafrost soils and (2) climate-driven changes in emissions from northern high-latitude and tropical wetlands. The size of the hydrate reservoir is uncertain, perhaps by up to a factor of 10. Because the size of the reservoir is directly related to the perceived risks, it is difficult to make certain judgment about those risks.

Observations show that there have not yet been significant increases in methane emissions from northern high-latitude hydrates and wetlands resulting from increasing arctic temperatures. Although there are a number of suggestions in the literature about the possibility of a dramatic abrupt release of methane to the atmosphere, modeling and

isotopic fingerprinting of ice-core methane do not support such a release to the atmosphere over the last 100,000 years or in the near future. Previous suggestions of a large release of methane at the Paleocene-Eeocene boundary (about 55 million years ago) face a number of objections, but may still be viable.

In response to the question of an abrupt increase in atmospheric methane, Chapter 5 notes:

- While the risk of catastrophic release of methane to the atmosphere in the next century appears very unlikely, it is very likely that climate change will accelerate the pace of persistent emissions from both hydrate sources and wetlands. Current models suggest that wetland emissions could double in the next century. However, since these models do not realistically represent all the processes thought to be relevant to future northern high-latitude CH_4 emissions, much larger (or smaller) increases cannot be discounted. Acceleration of persistent release from hydrate reservoirs is likely, but its magnitude is difficult to estimate.

Recommendations

How can the understanding of the potential for abrupt changes be improved?

We answer this question with nine primary recommendations that are required to substantially improve our understanding of the likelihood of an abrupt change occurring in the future. An overarching recommendation is the urgent need for committed and sustained monitoring of those components of the climate system identified in this chapter that are particularly vulnerable to abrupt climate change. The nine primary recommendations are:

1. Efforts should be made to (i) reduce uncertainties in estimates of mass balance and (ii) derive better measurements of glacier and ice-sheet topography and velocity through improved observation of glaciers and ice sheets. This includes continuing mass-balance measurements on small glaciers and completing the World Glacier Inventory. This further includes observing flow rates of glaciers and ice sheets from satellites, and sustaining aircraft observations of surface elevation and ice thickness

to ensure that such information is acquired at the high spatial resolution that cannot be obtained from satellites.

2. Address shortcomings in ice-sheet models currently lacking proper representation of the physics of the processes likely to be most important in potentially causing an abrupt loss of ice and resulting sea level rise. This will significantly improve the prediction of future sea level rise.

3. Research is needed to improve existing capabilities to forecast short- and longterm drought conditions and to make this information more useful and timely for decision making to reduce drought impacts. In the future, drought forecasts should be based on an objective multimodel ensemble prediction system to enhance their reliability and the types of information should be expanded to include soil moisture, runoff, and hydrological variables.

4. Improved understanding of the dynamic causes of long-term changes in oceanic conditions, the atmospheric responses to these ocean conditions, and the role of soil moisture feedbacks are needed to advance drought prediction capabilities. Ensemble drought prediction is needed to maximize forecast skill, and "downscaling" is needed to bring coarse-resolution drought forecasts from General Circulation Models down to the resolution of a watershed.

5. Efforts should be made to improve the theoretical understanding of the processes controlling the AMOC, including its inherent variability and stability, especially with respect to climate change. This will likely be accomplished through synthesis studies combining models and observational results.

6. Improve long-term monitoring of the AMOC. Parallel efforts should be made to more confidently predict the future behavior of the AMOC and the risk of an abrupt change. Such a prediction system should include advanced computer models, systems to start model predictions from the observed climate state, and projections of future changes in greenhouse gases and other agents that affect the Earth's energy balance.

7. Prioritize the monitoring of atmospheric methane abundance and its isotopic composition with spatial density sufficient to allow detection of any change in net emissions from northern and tropical wetland regions. The feasibility of monitoring methane in the ocean water column or in the atmosphere to detect emissions from the hydrate reservoir should be investigated. Efforts are needed to reduce uncertainties in the size of the global methane hydrate reservoir in marine and terrestrial environments and to identify the size and location of hydrate reservoirs that are most vulnerable to climate change.

8. Additional modeling efforts should be focused on (i) processes involved in releasing methane from the hydrate reservoir and (ii) the current and future climate-driven acceleration of release of methane from wetlands and terrestrial hydrate deposits.

9. Improve understanding of past abrupt changes through the collection and analysis of those proxy records that most effectively document past abrupt changes in sea level, ice-sheet and glacier extent, distribution of drought, the AMOC, and methane, and their impacts.

1. Introduction: Abrupt Changes in the Earth's Climate System

1. Background

Ongoing and projected growth in global population and its attendant demand for carbon-based energy is placing human societies and natural ecosystems at ever-increasing risk to climate change (*IPCC, 2007*). In order to mitigate this risk, the United Nations Framework Convention on Climate Change (UNFCCC) would stabilize greenhouse gas (GHG) concentrations in the atmosphere at a level that would prevent "dangerous anthropogenic interference" with the climate system (*UNFCCC, 1992*, Article 2). Successful implementation of this objective requires that such a level be achieved "within a time frame sufficient to allow ecosystems to adapt naturally to climate change, to ensure that food production is not threatened and to enable economic development to proceed in a sustainable manner" (*UNFCCC, 1992*, Article 2).

Among the various aspects of the climate change problem, the rate of climate change is clearly important in determining whether proposed implementation measures to stabilize GHG concentrations are adequate to allow sufficient time for mitigation and adaptation. In particular, the notion of adaptation and vulnerability takes on a new meaning when considering the possibility that the response of the climate system to radiative forcing[1] from increased GHG concentrations may be abrupt. Because the societal, economic, and ecological impacts of such an abrupt climate change would be far greater than for the case of a gradual change, assessing the likelihood of an abrupt, or nonlinear, climate response becomes critical to evaluating what constitutes dangerous human interference (*Alley et al., 2003*).

Studies of past climate demonstrate that abrupt changes have occurred frequently in Earth history, even in the absence of radiative forcing. Although geologic records of abrupt change have been available for decades, the decisive evidence that triggered widespread scientific and public interest in this behavior of the climate system came in the early 1990s with the publication of climate records from long ice cores from the Greenland Ice Sheet (Figure 1.1). Subsequent development of marine and terrestrial records (Figure 1.1) that also resolve changes on these short time scales has yielded a wide variety of climate signals from highly resolved and well-dated records from which the following generalizations can be drawn:

- abrupt climate change is a fundamental characteristic of the climate system;
- some past changes were subcontinental to global in extent;
- the largest of these changes occurred during times of greater-than-present global ice volume;
- all components of the Earth's climate system (ocean, atmosphere, cyrosphere, biosphere) were involved in the largest changes, indicating a closely coupled system response with important feedbacks.
- many past changes can be linked to forcings associated with changes in sea-surface temperatures or increased freshwater fluxes from former ice sheets.

These developments have led to an intensive effort by climate scientists to understand the possible mechanisms of abrupt climate change. This effort is motivated by the fact that if such large changes were to recur, they would have a potentially devastating impact on human society and natural ecosystems because of the inability of either to adapt on such short time scales. While past abrupt changes occurred in response to natural forcings, or were unforced, the prospect that human influences on the climate system may trigger similar abrupt changes in the near future (*Broecker, 1997*) adds further urgency to the topic.

Significant progress has been made since the report on abrupt climate change by the National Research Council (NRC) in 2002 (*NRC, 2002*), and this chapter provides considerably greater detail and insight on many of these issues than was provided in the 2007 Intergovernmental Panel on Climate Change (IPCC) Fourth Assessment Report (AR4) (*IPCC, 2007*). New paleoclimate reconstructions have been developed that provide greater understanding of patterns and mechanisms of past abrupt climate change in the ocean and on land, and new observations are further revealing unanticipated rapid dynamical changes of modern glaciers, ice sheets, and ice shelves as well as processes that are contributing to these changes. Finally, improvements in modeling of the climate system have further reduced uncertainties in assessing the likelihood of an abrupt change. The present report reviews this progress.

2. Definition of Abrupt Climate Change

What is meant by abrupt climate change? Several definitions exist, with subtle but important differences. *Clark et al. (2002)* defined abrupt climate change as "a persistent transition of climate (over subcontinental scale) that occurs on the time scale of decades." The NRC report "Abrupt Climate Change" (*NRC, 2002*) offered two definitions of abrupt climate change. A mechanistic definition defines abrupt climate change as occurring when "the climate system is forced to cross some threshold, triggering a transition to a new state at a rate determined by the climate system itself and faster than the cause." This definition implies that abrupt climate changes involve a threshold or nonlinear feedback within the climate system from one steady state to another, but is not restrictive to the short time scale (1-100 years) that has clear societal and ecological implications. Accordingly, the NRC report also provided an impacts-based definition of abrupt climate change as "one that takes place so rapidly and unexpectedly that human or natural systems have difficulty adapting to it." Finally, *Overpeck and Cole (2006)* defined abrupt climate change as "a transition in the climate system whose duration is fast relative to the duration of the preceding or subsequent state." Similar to the NRC's mechanistic definition, this definition transcends many possible time scales, and thus includes many different behaviors of the climate system that would have little or no detrimental impact on human (economic, social) systems and ecosystems.

For this chapter, we have modified and combined these definitions into one that emphasizes both the short time scale and the impact on ecosystems. In what follows we define abrupt climate change as:

A large-scale change in the climate system that takes place over a few decades or less, persists (or is anticipated to persist) for at least a few decades, and causes substantial disruptions in human and natural systems.

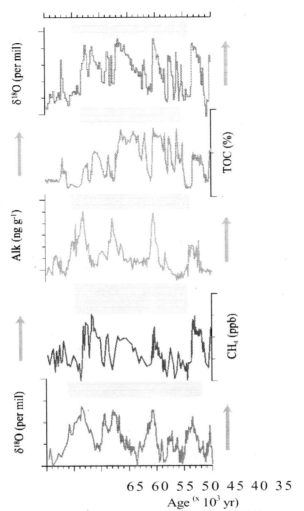

Figure 1.1. Records of climate change from the time period 35,000 to 65,000 years ago, illustrating how many aspects of the Earth's climate system have changed abruptly in the past. In all panels, the upward-directed gray arrows indicate the direction of increase in the climate variable recorded in these geologic archives (i.e., increase in temperature, increase in monsoon strength, etc.). The upper panel shows changes in the oxygen- isotopic composition of ice (δ^{18}O) from the GISP2 Greenland ice core (*Grootes et al., 1993*). Isotopic variations record changes in temperature of the high northern latitudes, with intervals of cold climate (more negative values) abruptly switching to intervals of warm climate (more positive values), representing temperature increases of 8°C to 15°C typically occurring within decades (*Huber et al., 2006*). The next panel down shows a record of strength of the Indian monsoon, with increasing values of total organic content (TOC) indicating an increase in monsoon strength (*Schulz et al., 1998*). This record indicates that changes in monsoon strength occurred at the same time as, and at similar rates as, changes in high northern-latitude temperatures. The next panel down shows a record of the biological productivity of the surface waters in the southwest Pacific Ocean east of New Zealand, as recorded by the concentration of alkenones in marine sediments (*Sachs and Anderson, 2005*). This record indicates that large increases in biological productivity of these surface waters occurred at the same time as cold temperatures in high-northern latitudes and weakened Indian monsoon strength. The next panel down is a record of changes in the concentration of atmospheric methane (CH4) from the GISP2 ice core (*Brook et al., 1996*). As discussed in Chapter 5 of this chapter, methane is a powerful greenhouse gas, but the variations recorded were not large enough to have a significant effect on radiative forcing. However, these variations are important in that they are thought to reflect changes in the tropical water balance that controls the distribution of methane-producing wetlands. Times of high-

atmospheric methane concentrations would thus correspond to a greater distribution of wetlands, which generally correspond to warm high northern latitudes and a stronger Indian monsoon. The bottom panel is an oxygen-isotopic ($\delta^{18}O$) record of air temperature changes over the Antarctic continent (*Blunier and Brook, 2001*). In this case, warm temperatures over Antarctica correspond to cold high northern latitudes, weakened Indian monsoon and drier tropics, and great biological productivity of the southwestern Pacific Ocean.

3. Organization of Report

Synthesis and Assessment Product 3.4 considers four types of change documented in the paleoclimate record that stand out as being so rapid and large in their impact that they pose clear risks to the ability of society and ecosystems to adapt. These changes are (i) rapid decrease in ice sheet mass with resulting global sea level rise; (ii) widespread and sustained changes to the hydrologic cycle that induces drought; (iii) changes in the Atlantic meridional overturning circulation (AMOC); and (iv) rapid release to the atmosphere of the potent greenhouse gas methane, which is trapped in permafrost and on continental slopes. Based on the published scientific literature, each chapter examines one of these types of change (sea level, drought, AMOC, and methane), providing a detailed assessment of the likelihood of future abrupt change as derived from reconstructions of past changes, observations and modeling of the present physical systems that are subject to abrupt change, and where possible, climate model simulations of future behavior of changes in response to increased GHG concentrations. In providing this assessment, we adopt the IPCC AR4 standard terms used to define the likelihood of an outcome or result where this can be determined probabilistically (Box 1.1).

BOX 1.1. TREATMENT OF UNCERTAINTIES IN THE SAP 3.4 ASSESSMENT

This chapter follows the 2007 Intergovernmental Panel on Climate Change (IPCC) Fourth Assessment Report (AR4) (*IPCC, 2007*) in the treatment of uncertainty, whereby the following standard terms are used to define the likelihood of an outcome or result where this can be estimated probabilistically based on expert judgment about the state of that knowledge:

Likelihood terminology	Likelihood of occurrence/outcome
Virtually certain	>99% probability
Extremely likely	>95% probability
Very likely	>90% probability
Likely	>66% probability
More likely than not	>50% probability
About as likely as not	33 to 66% probability
Unlikely	<33% probability
Very unlikely	<10% probability
Extremely unlikely	<5% probability
Exceptionally unlikely	<1% probability

4. Abrupt Change in Sea Level

Population densities in coastal regions and on islands are about three times higher than the global average, with approximately 23% of the world's population living within 100 kilometers (km) distance of the coast and <100 meters (m) above sea level (*Nicholls et al., 2007*). This allows even small sea level rise to have significant societal and economic impacts through coastal erosion, increased susceptibility to storm surges and resulting flooding, ground-water contamination by salt intrusion, loss of coastal wetlands, and other issues (Figure 1.2).

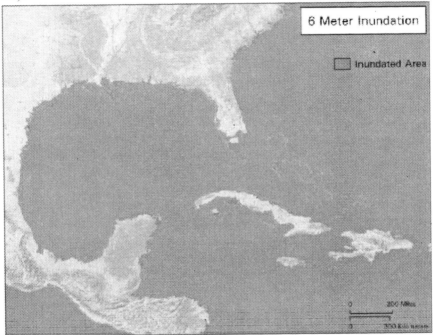

Figure 1.2. Portions (shown in red) of the southeastern United States, Central America, and the Caribbean surrounding the Gulf of Mexico that would be inundated by a 6-meter sea level rise (from *Rowley et al., 2007*). Note that additional changes in the position of the coastline may occur in response to erosion from the rising sea level.

An increase in global sea level largely reflects a contribution from water expansion from warming, and from the melting of land ice which dominates the actual addition of water to the oceans. Over the last century, the global average sea level rose at a rate of ~1 .7 ± 0.5 millimeters per year (mm yr^{-1}). However, the rate of global sea level rise for the period 1993 to 2003 accelerated to 3.1 ± 0.7 mm yr^{-1}, reflecting either variability on decadal time scales or an increase in the longer term trend. Relative to the period 1961- 2003, estimates of the contributions from thermal expansion and from glaciers and ice sheets indicate that increases in both of these sources contributed to the acceleration in global sea level rise that characterized the 1992-2003 period (*Bindoff et al., 2007*).

By the end of the 21st century, and in the absence of ice-dynamical contributions, the IPCC AR4 projects sea level to rise by 0.28 ± 0.10 m to 0.42 ± 0.16 m in response to additional global warming, with the contribution from thermal expansion accounting for 70-75% of this rise (*Meehl et al., 2007*). Projections for contributions from ice sheets are based on models that emphasize accumulation and surface melting in controlling the amount of mass gained and lost by ice sheets (mass balance), with different relative contributions for the Greenland and Antarctic ice sheets. Because the increase in mass loss (ablation) is greater than the increase in mass gain (accumulation), the Greenland Ice Sheet is projected to contribute to a positive sea level rise and may melt entirely from future global warming (*Ridley et al., 2005*). In contrast, the Antarctic Ice Sheet is projected to grow through increased accumulation relative to ablation and thus contribute to a negative sea level rise. The net projected effect on global sea level from these two differing ice-sheet responses to global warming over the remainder of this century is to nearly cancel each other out. Accordingly, the primary contribution to sea level rise from projected mass changes in the IPCC AR4 is associated with retreat of glaciers and ice caps (*Meehl et al., 2007*).

Rahmstorf (2007) used the relation between 20th century sea level rise and global mean surface temperature increase to predict a sea level rise of 0.5 to 1.4 m above the 1990 level by the end of the 21st century, considerably higher than the projections by the IPCC AR4 (*Meehl et al., 2007*). Insofar as the contribution to 20th century sea level rise from melting land ice is thought to have been dominated by glaciers and ice caps (*Bindoff et al., 2007*), the *Rahmstorf (2007)* projection does not include the possible contribution to sea level rise from ice sheets.

Recent observations of startling changes at the margins of the Greenland and Antarctic ice sheets indicate that dynamic responses to warming may play a much greater role in the future mass balance of ice sheets than considered in current numerical projections of sea level rise. Ice-sheet models used as the basis for the IPCC AR4 numerical projections did not include the physical processes that may be governing these dynamical responses, but if they prove to be significant to the long-term mass balance of the ice sheets, sea level projections will likely need to be revised upwards substantially. By implicitly excluding the potential contribution from ice sheets, the *Rahmstorf (2007)* estimate will also likely need to be revised upwards if dynamical processes cause future ice-sheet mass balance to become more negative.

Chapter 2 of this chapter summarizes the available evidence for recent changes in the mass of glaciers and ice sheets. The Greenland Ice Sheet is losing mass and very likely on an accelerated path since the mid-1990s. Observations show that Greenland is thickening at high elevations, because of an increase in snowfall, but that this gain is more than offset by an accelerating mass loss at the coastal margins, with a large component from rapidly thinning and accelerating outlet glaciers. The mass balance of the Greenland Ice Sheet during the period with good observations indicates that the loss increased from 100 gigatonnes per year (Gt a^{-1}) (where 360 Gt of ice = 1 mm of sea level) in the late 1990s to more than 200 Gt a^{-1} for the most recent observations in 2006.

Determination of the mass budget of the Antarctic ice sheet is not as advanced as that for Greenland. The mass balance for Antarctica as a whole has likely experienced a net loss since 2000 at rates of a few tens of Gt a^{-1} that are increasing with time, but with uncertainty of a similar magnitude to the estimated amount. There is little surface melting in Antarctica, but substantial ice losses are occurring from West Antarctica and the Antarctic Peninsula primarily in response to changing ice dynamics.

The record of past changes provides important insight to the behavior of large ice sheets during warming. At the last glacial maximum about 21,000 years ago, ice volume and area were about 2.5 times modern. Deglaciation was forced by warming from changes in the

Earth's orbital parameters, increasing greenhouse gas concentrations, and attendant feedbacks. Deglacial sea level rise averaged 10 mm a^{-1}, but with variations including two extraordinary episodes at 19 thousand years ago (ka) and 14.5 ka when peak rates potentially exceeded 50 mm a^{-1} (*Fairbanks, 1989; Yokoyama et al., 2000*). Each of these "meltwater pulses" added the equivalent of 1.5 to 3 Greenland ice sheets (~1 0-20 m) to the oceans over a one- to five-century period, clearly demonstrating the potential for ice sheets to cause rapid and large sea level changes.

The primary factor that raises concerns about the potential of future abrupt changes in sea level is that large areas of modern ice sheets are currently grounded below sea level. Where it exists, it is this condition that lends itself to many of the processes that can lead to rapid ice-sheet changes, especially with regard to atmosphere-ocean-ice interactions that may affect ice shelves and calving fronts of glaciers terminating in water (tidewater glaciers). An important aspect of these marine-based ice sheets is that the beds of ice sheets grounded below sea level tend to deepen inland. The grounding line is the critical juncture that separates ice that is thick enough to remain grounded from either an ice shelf or a calving front. In the absence of stabilizing factors, this configuration indicates that marine ice sheets are inherently unstable, whereby small changes in climate could trigger irreversible retreat of the grounding line.

The amount of retreat clearly depends on how far inland glaciers remain below sea level. Of greatest concern is the West Antarctic Ice Sheet, with 5 to 6 m sea level equivalent, where much of the base of the ice sheet is grounded well below sea level, with deeper trenches lying well inland of their grounding lines. A similar situation applies to the entire Wilkes Land sector of East Antarctica. In Greenland, a number of outlet glaciers remain below sea level, indicating that glacier retreat by this process will continue for some time. A notable example is Greenland's largest outlet glacier, Jakobshavn Isbrae, which appears to tap into the central region of Greenland that is below sea level. Accelerated ice discharge is possible through such outlet glaciers, but we consider the potential for destabilization of the Greenland Ice Sheet by this mechanism to be very unlikely.

The key requirement for stabilizing grounding lines of marine-based ice sheets appears to be the presence of an extension of floating ice beyond the grounding line, referred to as an ice shelf. A thinning ice shelf results in ice-sheet ungrounding, which is the main cause of the ice acceleration because it has a large effect on the force balance near the ice front. Recent rapid changes in marginal regions of both ice sheets are characterized mainly by acceleration and thinning, with some glacier velocities increasing more than twofold. Many of these glacier accelerations closely followed reduction or loss of ice shelves. If glacier acceleration caused by thinning ice shelves can be sustained over many centuries, sea level will rise more rapidly than currently estimated.

Such behavior was predicted almost 30 years ago by *Mercer (1978)* but was discounted as recently as the IPCC Third Assessment Report (*Church et al., 2001*) by most of the glaciological community based largely on results from prevailing model simulations. Considerable effort is now underway to improve the models, but it is far from complete, leaving us unable to make reliable predictions of ice-sheet responses to a warming climate if such glacier accelerations were to increase in size and frequency.

A nonlinear response of ice-shelf melting to increasing ocean temperatures is a central tenet in the scenario for abrupt sea-level rise arising from ocean – ice-shelf interactions. Significant changes in ice-shelf thickness are most readily caused by changes in basal melting. The susceptibility of ice shelves to high melt rates and to collapse is a function of the presence of warm waters entering the cavities beneath ice shelves. Future changes in ocean circulation and ocean temperatures will produce changes in basal melting, but the magnitude of these changes is currently neither modeled nor predicted.

Another mechanism that can potentially increase the sensitivity of ice sheets to climate change involves enhanced flow of the ice over its bed due to the presence of pressurized water, a process known as sliding. Where such basal flow is enabled, total ice flow rates may increase by 1 to 10 orders of magnitude, significantly decreasing the response time of an ice sheet to a climate or ice-marginal perturbation.

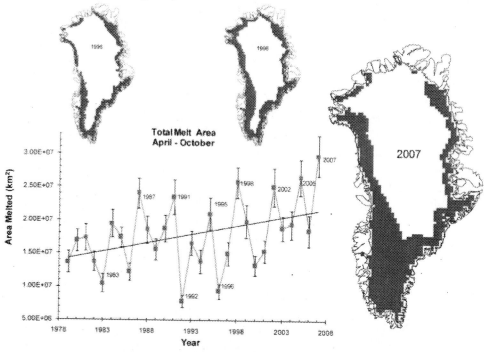

Figure 1.3. The graph shows the total melt area 1979 to 2007 for the Greenland ice sheet derived from passive microwave satellite data. Error bars represent the 95% confidence interval. The map inserts display the area of melt for 1996, 1998, and the record year 2007 (from K. Steffen, CIRES, University of Colorado)

Recent data from Greenland show a high correlation between periods of heavy surface melting and an increase in glacier velocity (*Zwally et al., 2002*). A possible cause for this relation is rapid drainage of surface meltwater to the glacier bed, where it enhances lubrication and basal sliding. There has been a significant increase in meltwater runoff from the Greenland Ice Sheet for the 1998-2007 period compared to the previous three decades (Figure 1.3). Total melt area is continuing to increase during the melt season and has already reached up to 50% of the Greenland Ice Sheet; further increase in Arctic temperatures will very likely continue this process and will add additional runoff. Because water represents such an important control on glacier flow, an increase in meltwater production in a warmer climate will likely have major consequences on flow rate and mass loss.

Because sites of global deep water formation occur immediately adjacent to the Greenland and Antarctic ice sheets, any significant increase in freshwater fluxes from these ice sheets may induce changes in ocean heat transport and thus climate. This topic is addressed in Chapter 4 of this chapter.

Summary

The Greenland and Antarctic Ice Sheets are losing mass, likely at an accelerating rate. Much of the loss from Greenland is by increased summer melting as temperatures rise, but an increasing proportion of the combined mass loss is caused by increasing ice discharge from the ice-sheet margins, indicating that dynamical responses to warming may play a much greater role in the future mass balance of ice sheets than previously considered. The interaction of warm waters with the periphery of the ice sheets is very likely one of the most significant mechanisms to trigger an abrupt rise in global sea level. The potentially sensitive regions for rapid changes in ice volume are thus likely those ice masses grounded below sea level such as the West Antarctic Ice Sheet or large glaciers in Greenland like the Jakobshavn Isbrae with an over-deepened channel (channel below sea level, see Chapter 2, Figure 2.10) reaching far inland. Ice-sheet models currently do not include the physical processes that may be governing these dynamical responses, so quantitative assessment of their possible contribution to sea level rise is not yet possible. If these processes prove to be significant to the long-term mass balance of the ice sheets, however, current sea level projections based on present-generation numerical models will likely need to be revised substantially upwards.

5. Abrupt Change in Land Hydrology

Much of the research on the climate response to increased GHG concentrations, and most of the public's understanding of that work, has been concerned with global warming. Accompanying this projected globally uniform increase in temperature, however, are spatially heterogeneous changes in water exchange between the atmosphere and the Earth's surface that are expected to vary much like the current daily mean values of precipitation and evaporation (*IPCC, 2007*). Although projected spatial patterns of hydroclimate change are complex, these projections suggest that many already wet areas are likely to get wetter and already dry areas are likely to get drier, while some intermediate regions on the poleward flanks of the current subtropical dry zones are likely to become increasingly arid.

These anticipated changes will increase problems at both extremes of the water cycle, stressing water supplies in many arid and semi-arid regions while worsening flood hazards and erosion in many wet areas. Moreover, the instrumental, historical, and prehistorical record of hydrological variations indicates that transitions between extremes can occur rapidly relative to the time span under consideration. Over the course of several decades, for example, transitions between wet conditions and dry conditions may occur within a year and can persist for several years.

Abrupt changes or shifts in climate that lead to drought have had major impacts on societies in the past. Paleoclimatic data document rapid shifts to dry conditions that coincided with downfall of advanced and complex societies. The history of the rise and fall of several empires and societies in the Middle East between 7000 and 2000 B.C. have been linked to abrupt shifts to persistent drought conditions (*Weiss and Bradley, 2001*). Severe drought leading to crop failure and famine in the mid-8th century has been suggested as cause for the decline and collapse of the Tang Dynasty (*Yancheva et al., 2007*) and the Classic Maya (*Hodell et al., 1995*). A more recent example of the impact of severe and persistent drought on society is the 193 0s Dust Bowl in the Central United States (Figure 1.4), which led to a large-scale migration of farmers from the Great Plains to the Western United States. Societies in many parts of the world today may now be more insulated to the impacts of abrupt climate shifts in the form of drought through managed water resources and reservoir systems. Nevertheless, population growth and over- allocation of scarce water supplies in a number of regions have made societies even more vulnerable to the impacts of abrupt climate change involving drought.

Figure 1.4. Photograph showing a dust storm approaching Stratford, Texas, during the 1930's Dust Bowl. (NOAA Photo Library, Historic NWS collection).

Variations in water supply in general, and protracted droughts in particular, are among the greatest natural hazards facing the United States and the globe today and in the foreseeable future. According to the National Climatic Data Center, National Oceanic and Atmospheric Administration (NCDC, NOAA), over the period from 1980 to 2006 droughts and heat waves were the second most expensive natural disaster in the United States behind tropical storms. The annual cost of drought to the United States is estimated to be in the billions of dollars. Although there is much uncertainty in these figures, it is clear that drought leads to (1) crop losses, which result in a loss of farm income and an increase in Federal disaster relief funds and food prices, (2) disruption of recreation and tourism, (3) increased fire risk and loss of life and property, (4) reduced hydroelectric energy generation, and (5) enforced water conservation to preserve essential municipal water supplies and aquatic ecosystems (Changnon et al., 2000; Pielke and Landsea, 1998; Ross and Lott, 2003).

5.1. History of North American Drought

In Chapter 3 of this chapter, we examine North American drought and its causes from the perspective of the historical record and, based on paleoclimate records, the last 1,000 years and the last 10,000 years. This longer temporal perspective relative to the historical record allows us to evaluate the natural range of drought variability under a diverse range of mean climatic conditions, including those similar to the present.

Instrumental precipitation and temperature data and tree-ring analyses provide sufficient information to identify six serious multiyear droughts in western North America since 1856. Of these, the most famous is the 'Dust Bowl' drought that included most of the 1930s decade (Figure 1.4). The other two in the 20th century are the severe drought in the Southwest from that late 1940s to the late 1950s and the drought that began in 1998 and is ongoing. Three droughts in the middle to late 19th century occurred (with approximate dates) from 1856 to 1865, from 1870 to 1876, and from 1890 to 1896.

Is the 1930s Dust Bowl drought the worst that can conceivably occur over North America? The instrumental and historical data only go back about 130 years with an acceptable degree of spatial completeness over the United States, which does not provide us with enough time to characterize the full range of hydroclimatic variability that has happened in the past and could conceivably happen in the future independent of any added effects due to greenhouse warming. To do so, we must look beyond the historical data to longer natural archives of past climate information to gain a better understanding of the past occurrence of drought and its natural range of variability.

Much of what we have learned about the history of North American drought over the past 1,000 years is based on annual ring-width patterns of long-lived trees that are used to reconstruct summer drought based on the Palmer Drought Severity Index (PD SI). This information and other paleoclimate data have identified a period of elevated aridity during the "Medieval Climate Anomaly" (MCA) period (A.D. 900-1300) that included four particularly severe multi-decadal megadroughts (Figure 1.5) (*Cook et al., 2004*). The range of annual drought variability during this period was not any larger than that seen after 1470, suggesting that the climate conditions responsible for these early droughts each year were apparently no more extreme than those conditions responsible for droughts during more recent times. This can be appreciated by noting that only 1 year of drought during the MCA was marginally more severe than the 1934 Dust Bowl year. This suggests that the 1934 event may be used as a worst-case scenario for how severe a given year of drought can get over the West. What sets these MCA megadroughts apart from droughts of more modern times, however, is their duration, with droughts during the MCA lasting much longer than historic droughts in the Western United States.

The emphasis up to now has been on the semi-arid to arid Western United States because that is where the late-20th century drought began and has largely persisted up to the present time. Yet, previous studies indicate that megadroughts have also occurred in the important crop-producing states in the Midwest and Great Plains as well (*Stahle et al., 2007*). In particular, a tree-ring PDSI reconstruction for the Great Plains shows the MCA period with even more persistent drought than the Southwest, but now on a centennial time scale.

Examination of drought history over the last 11,500 years (referred to as the Holocene Epoch) is motivated by noting that the projected changes in both the radiative forcing and the resulting climate of the 21st century far exceed those registered by either the instrumental records of the past century or by geologic archives that can be calibrated to derive climate (proxy records) of the past few millennia. In other words, all of the variations in climate over the instrumental period and over the past millennia reviewed above have occurred in a climate system whose controls have not differed much from those of the 20th century. Consequently, a longer term perspective is required to describe the behavior of the climate system under controls as different from those at present as those of the 21st century will be, and to assess the potential for abrupt climate changes to occur in response to gradual changes in large-scale forcing.

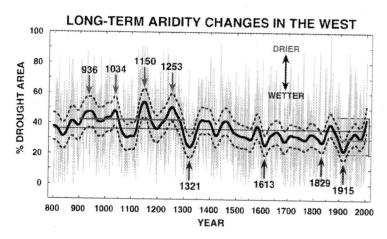

Figure 1.5. Percent area affected by drought (PDSI<-1) in the area defined as the West (see Chapter 3 of this chapter) (from *Cook et al., 2004*). Annual data are in gray and a 60- year low-pass filtered version is indicated by the thick smooth curve. Dashed blue lines are 2-tailed 95% confidence limits based on bootstrap resampling. The modern (mostly 20th century) era is highlighted in yellow for comparison to an increase in aridity prior to about A.D. 1300

It is important to emphasize that the controls of climate during the 21st century and during the Holocene differ from one another, and from those of the 20th century, in important ways. The major difference in controls of climate between the early 20th, late 20th, and 21st century is in atmospheric composition (with an additional component of land-cover change). In contrast, the major difference between the controls in the 20th and 21st centuries and those in the early to middle Holocene is in the latitudinal and seasonal distribution of solar radiation. Accordingly, climatic variations during the Holocene should not be thought of either as analogs for future climates or as examples of what might be observable under present-day climate forcing if records were longer, but instead should be thought of as the result of a natural experiment within the climate system that features large perturbations of the controls of climate.

The paleoclimatic record from North America indicates that drier conditions than present commenced in the mid-continent between 10 and 8 thousand years ago (ka) (*Webb et al., 1993*), and ended after 4 ka. The variety of paleoenvironmental indicators reflect the spatial extent and timing of these moisture variations, and in general suggest that the dry conditions increased in their intensity during the interval from 11 ka to 8 ka, and then gave way to increased moisture after 4 ka. During the middle of this interval (around 6 ka) dry conditions were widespread. Lake-status indicators at 6 ka indicate lower-thanpresent levels (and hence drier-than-present conditions) across most of the continent, and quantitative interpretation of pollen data shows a similar pattern of overall aridity, but again with some regional and local variability, such as moister-than-present conditions in the Southwestern United States (*Williams et al., 2004*). Although the region of drier-thanpresent conditions extends into the Northeastern United States and eastern Canada, most of the evidence for mid-Holocene dryness is focused on the mid-continent, in particular the Great Plains and Midwest, where the evidence for aridity is particularly clear.

5.2. Causes of North American Drought

Empirical studies and climate model experiments show that droughts over North America and globally are significantly influenced by the state of tropical sea surface temperatures (SSTs), with cool, persistent La Niña-like SSTs in the eastern equatorial Pacific frequently causing development of droughts over the southwestern United States and northern Mexico. Climate models that have evaluated this linkage need only prescribe small changes in SSTs, no more than a fraction of a degree Celsius, to result in reductions in precipitation. It is the persistence of the SST anomalies and associated moisture deficits that creates serious drought conditions. In the Pacific, the SST anomalies presumably arise naturally from dynamics similar to those associated with the El Niño Southern Oscillation (ENSO) on time scales of a year to a decade (*Newman et al., 2003*). On long time scales, the dynamics that link tropical Pacific SST anomalies to North American hydroclimate appear as analogs of higher frequency phenomena associated with ENSO (*Shin et al., 2006*). In general, the atmospheric response to La Niña-like conditions forces descent of air over western North America that suppresses precipitation. In addition to the ocean influence, some modeling and observational estimates indicate that soil-moisture feedbacks also influence precipitation variability.

The causes of the MCA megadroughts appear to have similar origin to the causes of modern droughts, which is consistent with the similar spatial patterns expressed by MCA and modern droughts (*Herweijer et al., 2007*). In particular, modeling experiments indicate that these megadroughts may have occurred in response to cold tropical Pacific SSTs and warm subtropical North Atlantic SSTs externally forced by high irradiance and weak volcanic activity (*Mann et al., 2005; Emile-Geay et al., 2007*). However, this result is tentative, and the exceptional duration of the droughts has not been adequately explained, nor whether they also involved forcing from SST changes in other ocean basins.

Over longer time spans, the paleoclimatic record indicates that even larger hydrological changes have taken place in response to past changes in the controls of climate that rival in magnitude those predicted for the next several decades and centuries. These changes were driven ultimately by variations in the Earth's orbit that altered the seasonal and latitudinal distribution of incoming solar radiation. The climate boundary conditions associated with those changes were quite different from those of the past millennium and today, but they show the additional range of natural variability and truly abrupt hydroclimatic change that can be expressed by the climate system.

Summary

The paleoclimatic record reveals dramatic changes in North American hydroclimate over the last millennium that were not associated with changes in greenhouse gases and human-induced global warming. Accordingly, one important implication of these results is that because these megadroughts occurred under conditions not too unlike today's, the United States still has the capacity to enter into a prolonged state of dryness even in the absence of increased greenhouse-gas forcing.

In response to increased concentration of GHGs, the semi-arid regions of the Southwest are projected to dry in the 21st century, with the model results suggesting, if they are correct, that the transition may already be underway (*Seager et al., 2007*). The drying in the Southwest is a matter of great concern because water resources in this region are already stretched, new development of resources will be extremely difficult, and the population and thus demand for water) continues to grow rapidly. Other subtropical regions of the world are also expected to dry in the near future, turning this feature of global hydroclimatic change into an international issue with potential impacts on migration and social stability. The midcontinental U.S. Great Plains could also experience changes in water supply impacting agricultural practices, grain exports, and biofuel production.

6. Abrupt Change in the Atlantic Meridional Overturning Circulation

The Atlantic Meridional Overturning Circulation (AMOC) is an important component of the Earth's climate system, characterized by a northward flow of warm, salty water in the upper layers of the Atlantic, a transformation of water mass properties at higher northern latitudes of the Atlantic in the Nordic and Labrador Seas that induces sinking of surface waters to form deep water, and a southward flow of colder water in the deep Atlantic (Figure 1.6). There is also an interhemispheric transport of heat associated with this circulation, with heat transported from the Southern Hemisphere to the Northern Hemisphere. This ocean current system thus transports a substantial amount of heat from the Tropics and Southern Hemisphere toward the North Atlantic, where the heat is released to the atmosphere (Figure 1.7).

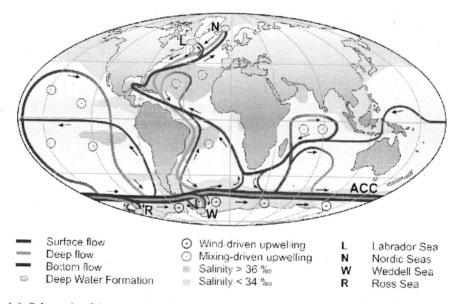

▬	Surface flow	⊙	Wind-driven upwelling	**L**	Labrador Sea
▬	Deep flow	○	Mixing-driven upwelling	**N**	Nordic Seas
▬	Bottom flow		Salinity > 36 ‰	**W**	Weddell Sea
⊙	Deep Water Formation		Salinity < 34 ‰	**R**	Ross Sea

Figure 1.6. Schematic of the ocean circulation (from *Kuhlbrodt et al., 2007*) associated with the global Meridional Overturning Circulation (MOC), with special focus on the Atlantic section of the flow (AMOC). The red curves in the Atlantic indicate the northward flow of water in the upper layers. The filled orange circles in the Nordic and Labrador Seas indicate regions where near-surface water cools and becomes denser, causing the water to sink to deeper layers of the Atlantic. The light blue curve denotes the southward flow of cold water at depth. See Chapter 4 of this chapter for further explanation

Changes in the AMOC have a profound impact on many aspects of the global climate system. There is growing evidence that fluctuations in Atlantic sea surface temperatures, hypothesized to be related to fluctuations in the AMOC, have played a prominent role in significant climate fluctuations around the globe on a variety of time scales. Evidence from the instrumental record (based on the last ~130 years) shows pronounced, multidecadal swings in large-scale Atlantic temperature that may be at least partly a consequence of fluctuations in the AMOC. Recent modeling and observational analyses have shown that these multidecadal shifts in Atlantic temperature exert a substantial influence on the climate system ranging from modulating African and Indian monsoonal rainfall to tropical Atlantic atmospheric circulation conditions of relevance for hurricanes. Atlantic SSTs also influence summer climate conditions over North America and Western Europe.

Figure 1.7. Palm trees on Mullaghmore Head, County Sligo, Ireland, which are symbolic of the relatively balmy climates of Ireland provided in part by the heat supplied from the Atlantic Meridional Overturning Circulation.

Evidence from paleorecords suggests that there have been large, decadal-scale changes in the AMOC, particularly during glacial times. These abrupt change events have had a profound impact on climate, both locally in the Atlantic and in remote locations around the globe (Figure 1.1). Research suggests that these abrupt events were related to discharges of freshwater into the North Atlantic from surrounding land-based ice sheets. Subpolar North Atlantic air temperature changes of more than 10°C on time scales of a decade or two have been attributed to these abrupt change events.

6.1. Uncertainties in Modeling the AMOC

As with any projection of future behavior of the climate system, our understanding of the AMOC in the 21st century and beyond relies on numerical models that simulate the important physical processes governing the overturning circulation. An important test of model skill is to conduct transient simulations of the AMOC in response to the addition of freshwater and compare with paleoclimatic data. Such a test requires accurate, quantitative reconstructions of the freshwater forcing, including its volume, duration, and location, plus the magnitude and duration of the resulting reduction in the AMOC. This information is not easy to obtain; coupled general circulation model (GCM) simulations of most events have been forced with idealized freshwater pulses and compared with qualitative reconstructions of the AMOC (e.g., *Hewitt et al., 2006; Peltier et al., 2006;* see also *Stouffer et al., 2006*). There is somewhat more information about the freshwater pulse associated with an event 8200 years ago, but important uncertainties remain (*Clarke et al., 2004; Meissner and Clark, 2006*). Thus, simulations of such paleoclimatic events provide important qualitative perspectives on the ability of models to simulate the response of the AMOC to forcing changes, but their ability to provide quantitative assessments is limited. Improvements in this area would be an important advance, but the difficulty in measuring even the current AMOC makes this task daunting.

Although numerical models show good skill in reproducing the main features of the AMOC, there are known errors that introduce uncertainty in model results. Some of these model errors, particularly in temperature and heat transport, are related to the representation of western boundary currents and deep-water overflow across the Greenland-Iceland-Scotland ridge. Increasing the resolution of current coupled ocean- atmosphere models to better address these errors will require an increase in computing power by an order of magnitude. Such higher resolution offers the potential of more realistic and robust treatment of key physical processes, including the representation of deep-water overflows. Efforts are being made to improve this model deficiency (*Willebrand et al., 2001; Thorpe et al., 2004; Tang and Roberts, 2005*). Nevertheless, recent work by *Spence et al. (2008)* using an Earth-system model of intermediate complexity (EMIC) found that the duration and maximum amplitude of their coupled model response to freshwater forcing showed little sensitivity to increasing resolution. They concluded that the coarse-resolution model response to boundary layer freshwater forcing remained robust at finer horizontal resolutions.

6.2. Future Changes in the AMOC

A particular focus on the AMOC in Chapter 4 of this chapter is to address the widespread notion, both in the scientific and popular literature, that a major weakening or even complete shutdown of the AMOC may occur in response to global warming. This discussion is driven in part by model results indicating that global warming tends to weaken the AMOC both by warming the upper ocean in the subpolar North Atlantic and through increasing the freshwater input (by more precipitation, more river runoff, and melting inland ice) into the Arctic and North Atlantic. Both processes reduce the density of the upper ocean in the North Atlantic, thereby stabilizing the water column and weakening the AMOC.

It has been theorized that these processes could cause a weakening or shutdown of the AMOC that could significantly reduce the poleward transport of heat in the Atlantic, thereby possibly leading to regional cooling in the Atlantic and surrounding continental regions, particularly Western Europe. This mechanism can be inferred from paleodata and is reproduced at least qualitatively in the vast majority of climate models (*Stouffer et al., 2006*). One of the most misunderstood issues concerning the future of the AMOC under anthropogenic climate change, however, is its often-cited potential to cause the onset of the next ice age. As discussed by *Berger and Loutre (2002)* and *Weaver and Hillaire-Marcel (2004)*, it is not possible for global warming to cause an ice age by this mechanism.

In the past, there was disagreement in determining which of the two processes governing upper-ocean density will dominate under increasing GHG concentrations, but a recent 11-model intercomparison project found that an MOC reduction in response to increasing GHG concentrations was caused more by changes in surface heat flux than by changes in surface freshwater flux (*Gregory et al., 2005*). Nevertheless, different climate models show different sensitivities toward an imposed freshwater flux (*Gregory et al., 2005*). It is therefore not fully clear to what degree salinity changes will affect the total overturning rate of the AMOC. In addition, by today's knowledge, it is hard to assess how large future freshwater fluxes into the North Atlantic might be. This is due to uncertainties in modeling the hydrological cycle in the atmosphere, in modeling the sea-ice dynamics in the Arctic, as well as in estimating the melting rate of the Greenland ice sheet (see Chapter 2 of this chapter).

It is important to distinguish between an AMOC weakening and an AMOC collapse. Historically, coupled models that eventually lead to a collapse of the AMOC under global warming scenarios have fallen into two categories: (1) coupled atmosphere-ocean general circulation models (AOGCMs) that required ad hoc adjustments in heat or moisture fluxes to prevent them from drifting away from observations, and (2) intermediate-complexity models with longitudinally averaged ocean components. Current AOGCMs used in the IPCC AR4 assessment typically do not use flux adjustments and incorporate improved physics and resolution. When forced with plausible estimates of future changes in greenhouse gases and aerosols, these newer models project a gradual 25-30% weakening of the AMOC, but not an abrupt change or collapse. Although a transient collapse with climatic impacts on the global scale can always be triggered in models by a large enough freshwater input (e.g., *Vellinga and Wood, 2007*), the magnitude of the required freshwater forcing is not currently viewed as a plausible estimate of the future. In addition, many experiments have been conducted with idealized forcing changes, in which atmospheric CO_2 concentration is increased at a rate of 1%/year to either two times or four times the preindustrial levels and held fixed thereafter. In virtually every simulation, the AMOC reduces but recovers to its initial strength when the radiative forcing is stabilized at two times or four times the preindustrial levels.

Perhaps more important for 21st century climate change is the possibility for a rapid transition to seasonally ice-free Arctic conditions. In one climate model simulation, a transition from conditions similar to pre-2007 levels to a near-ice-free September extent occurred in a decade (*Holland et al., 2006*). Increasing ocean heat transport was implicated in this simulated rapid ice loss, which ultimately resulted from the interaction of large, intrinsic variability and anthropogenically forced change. It is notable that climate models are generally conservative in the modeled rate of Arctic ice loss as compared to observations (Stroeve et al., 2007; Figure 1-3), suggesting that future ice retreat could occur even more abruptly than simulated.

This nonlinear response occurs because sea ice has a strong inherent threshold in that its existence depends on the freezing temperature of seawater. Additionally, strong positive feedbacks associated with sea ice act to accelerate its change. The most notable of these is the positive surface albedo feedback in which changes in ice cover and surface properties modify the surface reflection of solar radiation. For example, in a warming climate, reductions in ice cover expose the dark underlying ocean, allowing more solar radiation to be absorbed. This enhances the warming and leads to further ice melt. Because the AMOC interacts with the circulation of the Arctic Ocean at its northern boundary, future changes in the AMOC and its attendant heat transport thus have the potential to further influence the future of sea ice.

Summary

Our analysis indicates that it is very likely that the strength of the AMOC will decrease over the course of the 21st century. In models where the AMOC weakens, warming still occurs downstream over Europe due to the radiative forcing associated with increasing greenhouse gases. No model under plausible estimates of future forcing exhibits an abrupt collapse of the MOC during the 21st century, even accounting for estimates of accelerated Greenland ice sheet melting. We conclude that it is very unlikely that the AMOC will abruptly weaken or collapse during the course of the 21st century. Based on available model simulations and sensitivity analyses, estimates of maximum Greenland ice sheet melting rates, and our understanding of mechanisms of abrupt climate change from the paleoclimatic record, we further conclude that it is unlikely that the AMOC will collapse beyond the end of the 21st century as a consequence of global warming, although the possibility cannot be entirely excluded.

The above conclusions depend upon our understanding of the climate system and on the ability of current models to simulate the climate system. An abrupt collapse of the AMOC in the 21st century would require either a sensitivity of the AMOC to forcing that is far greater than current models suggest or a forcing that greatly exceeds even the most aggressive of current projections (such as extremely rapid melting of the Greenland ice sheet). While we view these as very unlikely, we cannot exclude either possibility. Further, even if a collapse of the AMOC is very unlikely, the large climatic impacts of such an event, coupled with the significant climate impacts that even decadal scale AMOC fluctuations induce, argue for a strong research effort to develop the observations, understanding, and models required to predict more confidently the future evolution of the AMOC.

7. Abrupt Change in Atmospheric Methane Concentration

After carbon dioxide (CO_2), methane (CH_4) is the next most important greenhouse gas that humans directly influence. Methane is a potent greenhouse gas because it strongly absorbs terrestrial infrared (IR) radiation. Methane's atmospheric abundance has more than doubled since the start of the Industrial Revolution (*Etheridge et al., 1998; MacFarling-Meure et al., 2006*), amounting to a total contribution to radiative forcing over this time of ~0.7 watts per square meter (W m^{-2}), or nearly half of that resulting from parallel increase in the atmospheric concentration of CO_2 (*Hansen and Sato, 2001*). Additionally, CO_2 produced by CH_4 oxidation is equivalent to ~6% of CO_2 emissions from fossil fuel combustion. Over a 100-year time horizon, the direct and indirect effects on radiative forcing from emission of 1 kg CH_4 are 25 times greater than for emission of 1 kg CO_2 (*IPCC, 2007*). On shorter time scales, methane's impact on radiative forcing is higher.

The primary geological reservoirs of methane that could be released abruptly to the atmosphere are found in ocean sediments and terrestrial soils as methane hydrate. Methane hydrate is a solid in which methane molecules are trapped in a lattice of water molecules (Figure 1.8). On Earth, methane hydrate forms under high pressure – low temperature conditions in the presence of sufficient methane. These conditions are most often found in relatively shallow marine sediments on continental margins but also in some high-latitude soils (*Kvenvolden, 1993*). Estimates of the total amount of methane hydrate vary widely, from 500 to 10,000 gigatons of carbon (GtC) total stored as methane in hydrates in marine sediments, and 7.5-400 GtC in permafrost (both figures are uncertain). The total amount of carbon in the modern atmosphere is ~8 10 GtC, but the total methane content of the atmosphere is only ~4 GtC (*Dlugokencky et al., 1998*). Therefore, even a release of a small portion of the methane hydrate reservoir to the atmosphere could have a substantial impact on radiative forcing.

There is little evidence to support massive releases of methane from marine or terrestrial hydrates in the past. Evidence from the ice core record indicates that abrupt shifts in methane concentration have occurred in the past 110,000 years (*Brook et al., 1996*), but the concentration changes during these events were relatively small. Farther back in geologic time, an abrupt warming at the Paleocene-Eocene boundary about 55 million years ago has been attributed by some to a large release of methane to the atmosphere.

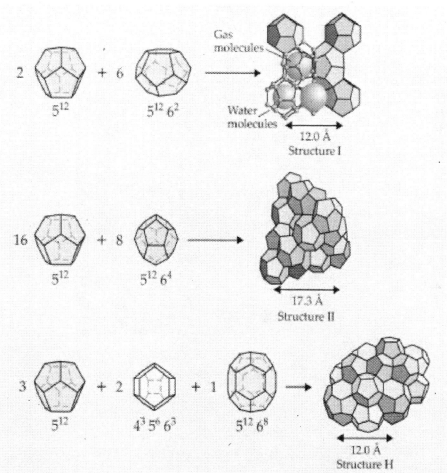

Figure 1.8. Clathrate hydrates are inclusion compounds in which a hydrogen-bonded water framework—the host lattice—traps "guest" molecules (typically gases) within ice cages. The gas and water don't chemically bond, but interact through weak van der Waals forces, with each gas molecule—or cluster of molecules in some cases—confined to a single cage. Clathrates typically crystallize into one of the three main structures illustrated here. As an example, structure I is composed of two types of cages: dodecahedra, 20 water molecules arranged to form 12 pentagonal faces (designated 5^{12}), and tetrakaidecahedra, 24 water molecules that form 12 pentagonal faces and two hexagonal ones ($5^{12}6^2$). Two 5^{12} cages and six $5^{12}6^2$ cages combine to form the unit cell. The pictured structure I illustrates the water framework and trapped gas molecules (from *Mao et al., 2007*). See Chapter 5 of this chapter for further explanation

Concern about future abrupt release in atmospheric methane stems largely from the possibility that the massive amounts of methane present as solid methane hydrate in ocean sediments and terrestrial soils may become unstable in the face of global warming. Warming or release of pressure can destabilize methane hydrate, forming free gas that may ultimately be released to the atmosphere (Figure 1.9).

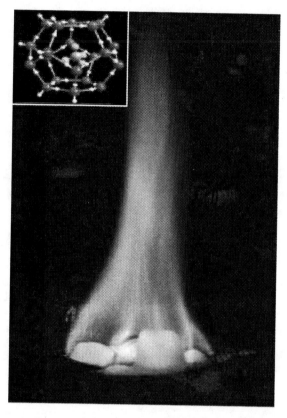

Figure 1.9. A piece of methane clathrate displays its potential as an energy source. As the compound melts, released gas feeds the flame and the ice framework drips off as liquid water. Inlay shows the clathrate structure. Source: U.S. Geological Survey

The processes controlling hydrate stability and gas transport are complex, and only partly understood. In Chapter 5 of this chapter, three categories of mechanisms are considered as potential causes of abrupt increases in atmospheric methane concentration in the near future. These are summarized in the following.

7.1. Destabilization of Marine Methane Hydrates

This issue is probably the most well known due to extensive research on the occurrence of methane hydrates in marine sediments, and the large quantities of methane apparently present in this solid phase in primarily continental margin marine sediments. Destabilization of this solid phase requires mechanisms for warming the deposits and/or reducing pressure on the appropriate time scale, transport of free methane gas to the sediment-water interface, and transport through the water column to the atmosphere (*Archer, 2007*). Warming of bottom waters, slope failure, and their interaction are the most commonly discussed mechanisms for abrupt release. However, bacteria are efficient at consuming methane in oxygen-rich sediments and the ocean water column, and there are a number of physical impediments to abrupt release from marine sediments.

On the time scale of the coming century, it is likely that most of the marine hydrate reservoir will be insulated from anthropogenic climate change. The exception is in shallow ocean sediments where methane gas is focused by subsurface migration. These deposits will very likely respond to anthropogenic climate change with an increased background rate of sustained methane release, rather than an abrupt release.

7.2. Destabilization of Permafrost Hydrates

Hydrate deposits at depth in permafrost soils are known to exist, and although their extent is uncertain, the total amount of methane in permafrost hydrates appears to be much smaller than in marine sediments. Surface warming eventually would increase melting rates of permafrost hydrates. Inundation of some deposits by warmer seawater and lateral invasion of the coastline are also concerns and may be mechanisms for more rapid change.

Destabilization of hydrates in permafrost by global warming is unlikely over the next few centuries (*Harvey and Huang, 1995*). No mechanisms have been proposed for the abrupt release of significant quantities of methane from terrestrial hydrates (*Archer, 2007*). Slow and perhaps sustained release from permafrost regions may occur over decades to centuries from mining extraction of methane from terrestrial hydrates in the Arctic (*Boswell, 2007*), over decades to centuries from continued erosion of coastal permafrost in Eurasia (*Shakova et al., 2005*), and over centuries to millennia from the propagation of any warming 100 to 1,000 meters down into permafrost hydrates (*Harvey and Huang, 1995*).

7.3. Changes in Wetland Extent and Methane Productivity

Although a destabilization of either the marine or terrestrial methane hydrate reservoirs is the most likely pathway for an abrupt increase in atmospheric methane concentration, the potential exists for a more gradual, but substantial, increase in natural methane emissions in association with projected changes in climate. The most likely region to experience a dramatic change in natural methane emission is the northern high latitudes, where there is increasing evidence for accelerated warming, enhanced precipitation, and widespread permafrost thaw which could lead to an expansion of wetland areas into organic-rich soils that, given the right environmental conditions, would be fertile areas for methane production (*Jorgenson et al., 2001, 2006*).

Tropical wetlands are a stronger methane source than boreal and arctic wetlands and will likely continue to be over the next century, during which fluxes from both regions are expected to increase. However, several factors that differentiate northern wetlands from tropical wetlands make them more likely to experience a larger increase in fluxes.

The balance of evidence suggests that anticipated changes to northern wetlands in response to large-scale permafrost degradation, thermokarst development, a positive trend in water balance in combination with substantial soil warming, enhanced vegetation productivity, and an abundant source of organic matter will very likely drive a sustained increase in CH4 emissions from the northern latitudes during the 21st century. A doubling of northern CH4 emissions could be realized fairly easily. Much larger increases cannot be discounted.

Summary

The prospect of a catastrophic release of methane to the atmosphere as a result of anthropogenic climate change appears very unlikely. However, the carbon stored as methane hydrate and as potential methane in the organic carbon pool of northern (and tropical) wetland soils is likely to play a role in future climate change. Changes in climate, including warmer temperatures and more precipitation in some regions, particularly the arctic, will very likely gradually increase emission of methane from both melting hydrates and natural wetlands. The magnitude of this effect cannot be predicted with great accuracy yet, but is likely to be at least equivalent to the current magnitude of many anthropogenic sources.

2. RAPID CHANGES IN GLACIERS AND ICE SHEETS AND THEIR IMPACTS ON SEA LEVEL

Key Findings

- Since the mid-19th century, small glaciers (sometimes called "glaciers and ice caps"; see Box 2.1 for definitions) have been losing mass at an average rate equivalent to 0.3 to 0.4 millimeters per year of sea level rise.
- The best estimate of the current (2007) mass balance of small glaciers is about -400 gigatonnes per year (Gt a^{-1}), or nearly 1.1 millimeters sea level equivalent per year.
- The mass balance loss of the Greenland Ice Sheet during the period with good observations increased from 100 Gt a^{-1} in the mid-1990s to more than 200 Gt a-1 for the most recent observations in 2006. Much of the loss is by increased summer melting as temperatures rise, but an increasing proportion is by enhanced ice discharge down accelerating glaciers.
- The mass balance for Antarctica is a net loss of about 80 Gt a^{-1} in the mid 1 990s, increasing to almost 130 Gt a^{-1} in the mid 2000s. There is little surface melting in Antarctica, and the substantial ice losses from West Antarctica and the Antarctic Peninsula are very likely caused by increasing ice discharge as glacier velocities increase.
- During the last interglacial period (~120 thousand years ago) with similar carbon dioxide levels to pre-industrial values and Arctic summer temperatures up to 4° C warmer than today, sea level was 4-6 meters above present. The temperature increase during the Eamian was the result of orbital changes of the sun. During the last two deglaciations sea level rise averaged 10-20 millimeters per year with large "meltwater fluxes" exceeding sea level rise of 50 millimeters per year lasting several centuries.
- The potentially sensitive regions for rapid changes in ice volume are those with ice masses grounded below sea level such as the West Antarctic Ice Sheet, with 5 to 6 meters sea level equivalent, or large glaciers in Greenland like the Jakobshavn Isbræ,

also known as Jakobshavn Glacier and Sermeq Kujalleq (in Greenlandic), with an over-deepened channel (channel below sea level, see Figure 2.10) reaching far inland; total breakup of Jakobshavn Isbræ ice tongue in Greenland, as well as other tidewater glaciers and ice cap outlets, was preceded by its very rapid thinning.

- Several ice shelves in Antarctica are thinning, and their area declined by more than 13,500 square kilometers in the last 3 decades of the 20th century, punctuated by the collapse of the Larsen A and Larsen B ice shelves, soon followed by several-fold increases in velocities of their tributary glaciers.

- The interaction of warm waters with the periphery of the large ice sheets represents a strong potential cause of abrupt change in the big ice sheets, and future changes in ocean circulation and ocean temperatures will very likely produce changes in ice-shelf basal melting, but the magnitude of these changes cannot currently be modeled or predicted. Moreover, calving, which can originate in fractures far back from the ice front, and ice-shelf breakup, are very poorly understood.

- Existing models suggest that climate warming would result in increased melting from coastal regions in Greenland and an overall increase in snowfall. However, they are incapable of realistically simulating the outlet glaciers that discharge ice into the ocean and cannot predict the substantial acceleration of some outlet glaciers that we are already observing.

Recommendations

- Reduce uncertainties in estimates of mass balance. This includes continuing mass-balance measurements on small glaciers and completing the World Glacier Inventory.

- Maintain climate networks on ice sheets to detect regional climate change and calibrate climate models.

- Derive better measurements of glacier and ice-sheet topography and velocity through improved observation of glaciers and ice sheets. This includes utilizing existing satellite interferometric synthetic aperture radar (InSAR) data to measure ice velocity.

- Use observations of the time-varying gravity field from satellites to estimate changes in ice sheet mass.

- Survey changes in ice sheet topography using tools such as satellite radar (e.g., Envisat and Cryosat-2), laser (e.g., ICESat-1/2), and wide-swath altimeters.

- Monitor the polar regions with numerous satellites at various wavelengths to detect change and to understand processes responsible for the accelerated ice loss of ice sheets, the disintegration of ice shelves, and the reduction of sea ice. It is the integrated satellite data evaluation that provides the tools and understanding to model the future response of cryospheric processes to climate change.

- Utilize aircraft observations of surface elevation, ice thickness, and basal characteristics to ensure that such information is acquired at high spatial resolution along specific routes, such as glacier flow lines, and along transects close to the grounding lines.

- Improve coverage of longer term (centennial to millennial) records of ice sheet and ocean history from geological observations.
- Support field, theoretical, and computational investigations of physical processes beneath and along ice shelves and beneath glaciers, especially near to the grounding lines of the latter, with the goal of understanding recent increases in mass loss.
- Develop ice-sheet models on a par with current models of the atmosphere and ocean. Particular effort is needed with respect to the modeling of ocean/ice-shelf interactions and physical processes, of surface mass balance from climatic information, and of all (rather than just some, as now) of the forces which drive the motion of the ice.

BOX 2.1. GLACIERS: SOME DEFINITIONS

Glaciers are bodies of ice resting on the Earth's solid surface (Box 2.1 Figure 1). We distinguish between *ice sheets* (Box 2.1 Figure 2), which are glaciers of near-continental extent and of which there are at present two, the Antarctic Ice Sheet and the Greenland Ice Sheet, and *small glaciers*, sometimes also referred to as *glaciers and ice caps* (Box 2.1 Figure 2). There are several hundred thousand small glaciers. They are typically a few hundred meters to a few tens of kilometers long, while the ice sheets are drained by ice streams many tens to hundreds of kilometers long. In terms of volume, the ice sheets dwarf the small glaciers. If they all melted, the equivalent sea level rise would be 57 m from Antarctica and 7 m from Greenland but only 0.5 m from the small glaciers. Of the Antarctic total, about 7 m would come from West Antarctica, which may be especially vulnerable to abrupt changes.

Ice at the Earth's surface is a soft solid because it is either at or not far below its melting point. It therefore deforms readily under stress, spreading under its own weight until a balance is achieved between mass gains, mainly as snowfall, in the cold interior or upper parts of the glacier, and mass loss in the lower parts by melting or right at sea level by the calving of icebergs. The glacier may, however, keep spreading when it reaches sea level, and in this case it has a floating tongue or, when several glaciers are involved, a buttressing *ice shelf* (Box 2.1 Figure 3), the weight of which is supported not by the solid earth but by the ocean. A glacier which reaches sea level is called a *tidewater glacier*.

Ice shelves, which are mostly confined to Antarctica, are typically a few hundred meters thick and must not be confused with sea ice, typically a few meters thick. They are a critical part of the picture because they can lose mass not just by melting at their surfaces and by calving but also by melting at their bases. Increased basal melting, due for example to the arrival of warmer seawater, can "pull" more ice across the grounding line.

Box 2.1 Figure 1. Glaciers are slow-moving rivers of ice, formed from compacted layers of snow, that slowly deform and flow in response to gravity. Glacier ice is the largest reservoir of freshwater, and second only to oceans the largest reservoir of total water. Glaciers cover vast areas of polar regions and are restricted to the mountains in mid-latitudes. Glaciers are typically a few hundred meters to a few tens of kilometers long; most of the glaciers in mid-latitudes have been retreating in the last two centuries (Rhône Glacier, Switzerland, photograph courtesy of K. Steffen, CIRES, University of Colorado at Boulder.)

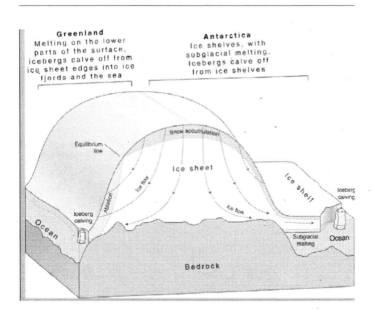

Box 2.1 Figure 2. The ice cover in Greenland and Antarctica has two components – thick, grounded, inland ice that rests on a more or less solid bed, and thinner floating ice shelves and glacier tongues. An ice sheet is actually a giant glacier, and like most glaciers it is nourished by the continual accumulation of snow on its surface. As successive layers of snow build up, the layers beneath are gradually compressed into solid ice. Snow input is balanced by glacial outflow, so the height of the ice sheet stays approximately constant through time. The ice is driven by gravity to slide and to flow downhill from the highest points of the interior to the coast. There it either melts or is carried away as icebergs which also eventually melt, thus returning the water to the ocean whence it came. Outflow from the inland ice is organized into a series of drainage basins separated by ice divides that concentrate the flow of ice into either narrow mountain-bounded outlet glaciers

or fast-moving ice streams surrounded by slow-moving ice rather than rock walls. In Antarctica, much of this flowing ice has reached the coast and has spread over the surface of the ocean to form ice shelves that are floating on the sea but are attached to ice on land. There are ice shelves along more than half of Antarctica's coast, but very few in Greenland (UNEP Maps and Graphs; K. Steffen, CIRES, University of Colorado at Boulder.).

The *grounding line* separates the grounded inland ice from the floating shelf or tongue ice. It is also where the ice makes its contribution to sea level change. When it begins to float, it displaces seawater whether or not it becomes an iceberg.

There is another crucial role for ice shelves, for they appear to be thermally unstable – there are no ice shelves where the annual average temperature is higher than about minus 5 C. Recently several "warm" ice shelves have collapsed dramatically, and their disintegration has been followed by equally dramatic acceleration of tributary glaciers across what was once the grounding line, where the grounded ice calves directly into the ocean at a far greater rate than before ice-shelf breakup.

Ice streams are rapid flows of ice with walls of slower ice, and are the principal means by which ice is evacuated from the interiors of the ice sheets and supplied to the larger ice shelves. Similar flows with walls of rock are called *outlet glaciers*, although this term is sometimes used quite loosely.

Box 2.1 Figure 3. An ice shelf is a thick, floating platform of ice that forms where a glacier or ice sheet flows down to a coastline and onto the ocean surface. Ice shelves are found in Antarctica, Greenland, and Canada. The boundary between the floating ice shelf and the grounded (resting on bedrock) ice that feeds it is called the grounding line. The thickness of modern-day ice shelves ranges from about 100 to 1,000 meters. The density contrast between solid ice and liquid water means that only about 1/9 of the floating ice is above the ocean surface. The picture shows the ice shelf of Petermann Glacier in northwestern Greenland (right side of picture) with a floating ice tongue of 60 km in length and 20 km wide. Glaciers from the left are merging with the ice shelf. (Petermann Glacier, northwest Greenland, photograph courtesy of K. Steffen, CIRES, University of Colorado at Boulder.)

1. Summary

1.1. Paleorecord

The most recent time with no appreciable ice on the globe was 35 million years ago during a period when the atmospheric carbon dioxide (CO_2) was $1,250 \pm 250$ parts per million by volume (ppmV) and a sea level 73 meters (m) higher than today. During the last interglacial period (~120 thousand years ago, ka) with similar CO_2 levels to preindustrial values and arctic summer temperatures warmer than today, sea level was 4-6 m above present. Most of that sea level rise (SLR) is believed to have originated from the Greenland Ice Sheet, but the rate of SLR is unknown Sea level rise averaged 10-20 millimeters per year (mm a^{-1}) during the last two deglaciation periods (130-116 ka and 21 – 14 ka, respectively), with large "meltwater fluxes" with rates of SLR exceeding 50 mm a-1 lasting several centuries *(Fairbanks, 1989; Rohling et al., 2008)*. Each of these meltwater fluxes added 1.5–3 times the volume of the current Greenland Ice Sheet (7 m) to the oceans. The cause, ice-sheet source, and mechanism of the meltwater fluxes is not well understood, yet the rapid loss of ice must have had an effect on ocean circulation resulting in a forcing of the global climate.

1.2. Ice Sheets

Rapid changes in ice-sheet mass have surely contributed to abrupt changes in climate and sea level in the past. The mass balance loss of the Greenland Ice Sheet increased in the late 1990s to 100 gigatonnes per year (Gt a^{-1}) or even more than 200 Gt a^{-1} for the most recent observations in 2006. It is extremely likely that the Greenland Ice Sheet is losing mass and very likely on an accelerated path since the mid-1990s. The mass balance for Antarctica as a whole is close to balance, but with a likely net loss since 2000 at rates of a few tens of gigatonnes per year. The largest losses are concentrated along the Amundsen and Bellinghausen sectors of West Antarctica and the northern tip of the Antarctic Peninsula. The potentially sensitive regions for rapid changes in ice volume are those with ice masses grounded below sea level such as the West Antarctic Ice Sheet, with 7 m sea level equivalent (SLE), or large glaciers in Greenland like the Jakobshavn, also known as Jakobshavn Isbræ and Sermeq Kujalleq (in Greenlandic), with an over- deepened channel reaching far inland. There are large mass-budget uncertainties from errors in both snow accumulation and calculated ice losses for Antarctica (~±160 Gt a^{-1}) and for Greenland (~±35 Gt a^{-1}). Mass-budget uncertainties from aircraft or satellite observations (i.e., radar altimeter, laser altimeter, gravity measurements) are similar in magnitude. Most climate models suggest that climate warming would result in increased melting from coastal regions in Greenland and an overall increase in snowfall. However, they do not predict the substantial acceleration of some outlet glaciers that we are observing. This results from a fundamental weakness in the existing models, which are incapable of realistically simulating the outlet glaciers that discharge ice into the ocean.

Observations show that Greenland is thickening at high elevations, because of the increase in snowfall, which was predicted, but that this gain is more than offset by an accelerating mass loss, with a large component from rapidly thinning and accelerating outlet glaciers. Although there is no evidence for increasing snowfall over Antarctica, observations show that some higher elevation regions are also thickening, likely as a result of high interannual variability in snowfall. There is little surface melting in Antarctica, and the substantial ice losses from West Antarctica and the Antarctic Peninsula are very likely caused by increased ice discharge as velocities of some glaciers increase. This is of particular concern in West Antarctica, where bedrock beneath the ice sheet is deep below sea level, and outlet glaciers are to some extent "contained" by the ice shelves into which they flow. Some of these ice shelves are thinning, and some have totally broken up, and these are the regions where the glaciers are accelerating and thinning most rapidly.

1.3. Small Glaciers

Within the uncertainty of the measurements, the following generalizations are justifiable. Since the mid-19th century, small glaciers have been losing mass at an average rate equivalent to 0.3-0.4 mm a^{-1} of sea level rise. The rate has varied. There was a period of reduced loss between the 1940s and 1970s, with the average rate approaching zero in about 1970. We know with very high confidence that it has been accelerating. The best estimate of the current (2007) mass balance is near to -380 to -400 Gt a^{-1}, or nearly 1.1 mm SLE a^{-1}; this may be an underestimate if, as suspected, the inadequately measured rate of loss by calving outweighs the inadequately measured rate of gain by "internal"[2] accumulation. Our physical understanding allows us to conclude that if the net gain of radiative energy at the Earth's surface continues to increase, then so will the acceleration of mass transfer from small glaciers to the ocean. Rates of loss observed so far are small in comparison with rates inferred for episodes of abrupt change during the last few hundred thousand years. In a warmer world the main eventual constraint on mass balance will be exhaustion of the supply of ice from glaciers, which may take place in as little as 50-100 years.

1.4. Causes of Change

Potential causes of the observed behavior of ice bodies include changes in snowfall and/or surface melting, long-term response to past changes in climate, and changes in ice dynamics. Smaller glaciers appear to be most sensitive to radiatively induced changes in melting rate, but this may be because of inadequate attention to the dynamics of tidewater glaciers (see Box 2.1 for definitions). Recent observations of the ice sheets have shown that changes in dynamics can occur far more rapidly than previously suspected. There has been a significant increase in meltwater production on the Greenland Ice Sheet for the 1998-2003 time period compared to the previous three decades, but this loss was partly compensated by increased precipitation. Total melt area is continuing to increase during summer and fall and has already reached up to 50% of the Greenland Ice Sheet; further increase in arctic temperatures will continue this process and will add additional runoff. Recent rapid changes in marginal regions of both ice sheets show mainly acceleration and thinning, with some glacier velocities increasing more than twofold. Most of these glacier accelerations closely followed reduction or loss of ice shelves. Total breakup of Jakobshavn Isbræ ice tongue in Greenland was preceded by its very rapid thinning. Thinning of more than 1 meter per year (m a^{-1}), and locally more than 5 m a^{-1}, was observed during the past decade for many small ice shelves in the Amundsen Sea and along the Antarctic Peninsula. Significant changes in ice shelf thickness are most readily caused by changes in basal melting. Recent data show a high correlation between periods of heavy surface melting and increase in glacier velocity. A possible cause is rapid meltwater drainage to the glacier bed, where the water enhances lubrication of basal sliding. Although no seasonal changes in the speeds were found for the rapid glaciers that discharge most ice from Greenland, meltwater remains an essential control on glacier flow, and an increase in meltwater production in a warmer climate could likely have major consequences of increased flow rates and ice mass loss.

1.5. Ocean Influence

The interaction of warm waters with the periphery of the large ice sheets represents one of the most significant possibilities for abrupt change in the climate system. Mass loss through oceanic melting and iceberg calving accounts for more than 95% of the ablation from Antarctica and 40-50% of the ablation from Greenland. Future changes in ocean circulation and ocean temperatures will produce changes in basal melting, but the magnitude of these changes is currently not well modeled or predicted. The susceptibility of ice shelves to high melt rates and to collapse is a function of the presence of warm waters entering the cavities beneath ice shelves. Ocean circulation is driven by density contrasts of water masses and by surface wind forcing. For abrupt climate change scenarios, attention should be focused on the latter. A change in wind patterns could produce large and fast changes in the temperatures of ocean waters. A thinning ice shelf results in glacier ungrounding, which is the main cause of the glacier acceleration because it has a large effect on the force balance near the ice front. Calving, which can originate in fractures far back from the ice front, is very poorly understood. Antarctic ice- shelf area declined by more than 13,500 square kilometers (km^2) in the last 3 decades of the 20th century, punctuated by the collapse of the Larsen A and Larsen B ice shelves. Ice shelf viability is compromised if mean annual air temperature exceeds

−5°C. Observations from the last decade have radically altered the thinking on how rapidly an ice sheet can respond to perturbations at the marine margin. Several-fold increases in discharge followed the collapse of ice shelves on the Antarctic Peninsula; this is something models did not predict *a priori*. No ice sheet model is currently capable of capturing the glacier speedups in Antarctica or Greenland that have been observed over the last decade.

1.6. Sea Level Feedback

The primary factor that raises concerns about the potential of abrupt changes in sea level is that large areas of modern ice sheets are currently grounded below sea level. An important aspect of these marine-based ice sheets which has long been of interest is that the beds of ice sheets grounded below sea level tend to deepen inland, either due to overdeepening from glacial erosion or isostatic adjustment. Marine ice sheets are inherently unstable, whereby small changes in climate could trigger irreversible retreat of the grounding line (locations along the coast where the ice is no longer ground supported and begins to float). For a tidewater glacier, rapid retreat occurs because calving rates increase with water depth. In Greenland, few outlet glaciers remain below sea level very far inland, indicating that glacier retreat by this process will eventually slow down or halt. A notable exception may be Greenland's fastest moving glacier, Jakobshavn Isbræ, Petermann and Humboldt Glacier in the northwest, and 79N Glacier in the northeast, which appears to tap into the central core of Greenland that is below sea level. Given that a grounding line represents the point at which ice becomes buoyant, then a rise in sea level will cause grounding line retreat. This situation thus leads to the potential for a positive feedback to develop between ice retreat and sea level rise. In considering various stabilizing factors, however, we conclude that, provided there is no rapid loss of ice shelves and attendant sea level rise, sea level forcing and feedback is unlikely to be a significant determinant in causing rapid ice-sheet changes in the coming century.

2. What is the Record of Past Changes in Ice Sheets and Global Sea Level?

2.1. Reconstructing Past Changes in Ice Sheets

There are several methods available to reconstruct past changes in ice-sheet area and mass, each with their own strengths and shortcomings. Terrestrial records provide information of former ice-sheet extent, whereby temporary stabilization of an ice margin may be recorded by an accumulation of sediment (moraine) that may be dated by isotopic methods (e.g., ^{10}Be, ^{14}C, etc.). These records are important in identifying the last maximum extent and retreat history of an ice sheet (e.g., *Dyke, 2004*), but most terrestrial records of glaciation prior to the Last Glacial Maximum (LGM) ~21,000 years ago have been removed by erosion, limiting the application of these records to times since the LGM. Moreover, in most cases they only provide information on extent but not thickness, so that potential large changes in volume are not necessarily captured by these records.

Application of this strategy to the retreat of the West Antarctic Ice Sheet (WAIS) from its LGM position provides important context for understanding current ice dynamics. *Conway et al. (1999)* dated recession of the WAIS grounding line in the Ross Sea embayment and found that modern grounding-line retreat is part of an ongoing recession that has been underway for the last ~9,000 years. *Stone et al. (2003)* took a slightly different approach to evaluating WAIS deglaciation whereby they determined the rate of lowering of the ice-sheet surface by dating recessional features preserved on a mountain slope that projected upwards through the ice sheet. Their results complemented those of *Conway et al. (1999)* in showing ice-sheet thinning for the last ~10,000 years that may still be underway. These results are important not only in providing constraints on longterm changes against which to evaluate short-term controls on ice-sheet change but also in providing important benchmarks for modeling ice-sheet evolution. Nevertheless, the spatial coverage of these data from Antarctica remains limited, and additional such constraints are needed.

Another strategy for constraining past ice-sheet history is based on the fact that the weight of ice sheets results in isostatic compensation of the underlying solid Earth, generally referred to as glacial isostatic adjustment (GIA). Changes in ice-sheet mass cause vertical motions that may be recorded along a formerly glaciated coastline where the global sea level serves as a datum. Since changes in ice mass will also cause changes in local (due to gravity) and global (due to volume) sea level, the changes in sea level at a particular coastline record the difference between vertical motions of the land and sea, commonly referred to as near-field relative sea level (RSL) changes. Models that incorporate the physical properties of the solid Earth invert the RSL records to determine the ice-loading history required to produce the isostatic adjustment preserved by these records (e.g., *Peltier, 2004*). Because of the scarcity of such near-field RSL sites from the Antarctic continent, *Ivins and James (2005)* constructed a history of Antarctic ice-mass changes from geologic evidence of ice-margin and ice-thickness changes, such as described above (*Conway et al., 1999; Stone et al., 2003*). This ice-load history was then used to derive a model of present-day GIA.

Regardless how it is derived, the GIA process must be accounted for when using satellite altimetry and gravity data to infer changes in ice mass (e.g., *Velicogna and Wahr, 2006b*) (see Sec. 3). Given the poor constraints from near-field RSL records and geologic records (and their dating) of ice limits and thicknesses for Antarctica, as well as uncertainties in properties of the solid Earth used in these models, uncertainties in this GIA correction is large (*Velicogna and Wahr, 2006b; Barletta et al., 2008*). Accordingly, improvements in understanding present-day GIA are required to improve ice-mass estimates from altimetry and gravity data.

2.2. Reconstructing Past Sea Level

Sea level changes that occur locally, due to regional uplift or subsidence, relative to global sea level are referred to as relative sea level (RSL) changes, whereas changes that occur globally are referred to as eustatic changes. On time scales greater than 100,000 years, eustatic changes occur primarily from changes in ocean-basin volume induced by variations in the rate of sea-floor spreading. On shorter time scales, eustatic changes occur primarily from changes in ice volume, with secondary contributions (order of 1 m) associated with changes in ocean temperature or salinity (steric changes). Changes in global ice volume also cause global changes in RSL in response to the redistribution of mass between land to sea and attendant isostatic compensation and gravitational reequilibration. This GIA process must be accounted for in determining eustatic changes from geomorphic records of former sea level. Because the effects of the GIA process diminish with distance from areas of former glaciation, RSL records from far-field sites provide a close approximation of eustatic changes.

An additional means to constrain past sea level change is based on the change in the ratio of ^{18}O to ^{16}O of seawater (expressed in reference to a standard as $\delta^{18}O$) that occurs as the lighter isotope is preferentially removed and stored in growing ice sheets (and vice versa). These $\delta^{18}O$ changes are recorded in the carbonate fossils of microscopic marine organisms (foraminifera) and provide a near-continuous time series of changes in ice volume and corresponding eustatic sea level. However, because changes in temperature also affect the $\delta^{18}O$ of foraminifera through temperature-dependent fractionation during calcite precipitation, the $\delta^{18}O$ signal in marine records reflects some combination of ice volume and temperature. Figure 2.1 shows one attempt to isolate the ice-volume component in the marine $\delta^{18}O$ record (*Waelbroeck et al., 2002*). Although to a first order this record agrees well with independent estimates of eustatic sea level, this approach fails to capture some of the abrupt changes in sea level that are documented by paleoshoreline evidence (*Clark and Mix, 2002*), suggesting that large changes in ocean temperature may not be accurately captured at these times.

2.3. Sea Level Changes during the Past

The record of past changes in ice volume provides important insight to the response of large ice sheets to climate change. Our best constraints come from the last glacial cycle (120,000 years ago to the present), when the combination of paleoshorelines and the global $\delta^{18}O$ record provides reasonably well-constrained evidence of changes in eustatic sea level (Figure 2.1). Changes in ice volume over this interval were paced by changes in the Earth's orbit around the sun (orbital time scales, 10^4-10^5 a), but amplification from changes in atmospheric CO2 is required to explain the synchronous and extensive glaciation in both polar hemispheres. Although the phasing relationship between sea level and atmospheric CO2 remains unclear (*Shackleton, 2000; Kawamura et al., 2007*), their records are coherent and there is a strong positive relation between the two (Figure 2.2).

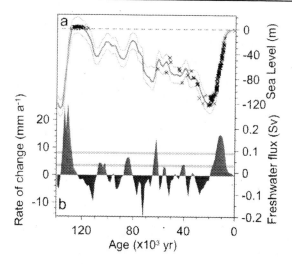

Figure 2.1. (a) Record of sea-level change over the last 130,000 years. Thick blue line is reconstruction from ^{18}O records of marine sediment cores through regression analyses (*Waelbroeck et al., 2002*), with ±13 m error shown by thin gray lines. The × symbols represent individually dated shorelines from Australia (*Stirling et al., 1995, 1998*), New Guinea (*Edwards et al., 1993; Chappell, 2002; Cutler et al., 2003*), Sunda Shelf (*Hanebuth et al., 2000*), Bonaparte Gulf (*Yokoyama et al., 2000*), Tahiti (*Bard et al., 1996*), and Barbados (*Peltier and Fairbanks, 2006*). (b) Rate of sea level change (mm a^{-1}) and equivalent freshwater flux (Sv, where 1 Sv = 106 $_{m}$3 s-1 = 31,500 Gt a^{-1}) derived from sea-level record in (a). Horizontal gray bars represent average rates of sea level change during the 20th century (lower bar) and projected for the end of the 21st century (upper bar) (*Rahmstorf, 2007*).

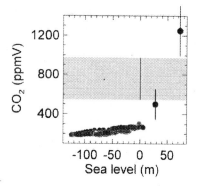

Figure 2.2. Relation between estimated atmospheric CO2 and the ice contribution to eustatic sea level indicated by geological archives and referenced to modern (preindustrial era) conditions [CO2 =280 parts per million by volume (ppmV), eustatic sea level = 0 m]. Horizontal gray box represents range of atmospheric CO2 concentrations projected for the end of the 21st century based on IPCC emission scenarios (lower end is B1 scenario, upper end is A1F1 scenario) (*Nakicenovic et al., 2000*). The vertical red bar represents the IPCC Fourth Assessment Report (AR4) estimate of sea level rise by the end of the 21st century (*Meehl et al., 2007*). The difference between the IPCC AR4 estimate and the high paleo-sea levels under comparable atmospheric CO2 levels of the past (blue dots with vertical bar given as uncertainties) largely reflects the long response time of ice sheets. A central question raised by the dynamic changes in ice sheets described in this chapter (and that are not included in the IPCC AR4 estimates) is how much they will reduce the ice-sheet response time to climate change

A similar correlation holds for earlier times in Earth history when atmospheric CO_2 concentrations were in the range of projections for the end of the 21st century (Figure 2.2). The most recent time when no permanent ice existed on the planet (sea level = +73 m) occurred >35 million years ago when atmospheric CO_2 was 1,250±250 ppmV (*Pagani et al., 2005*). In the early Oligocene (~32 million years ago), atmospheric CO_2 decreased to 500±50 ppmV (*Pagani et al., 2005*), which was accompanied by the first growth of permanent ice on the Antarctic continent, with an attendant eustatic sea level lowering of 45±5 m (*DeConto and Pollard, 2003*). The fact that sea level projections for the end of the 21st century (*Meehl et al., 2007; Rahmstorf, 2007; Horton et al., 2008*) are far below those suggested by this relation (Figure 2.2) reflects the long response time of ice sheets to climate change. With sufficient time at elevated atmospheric CO_2 levels, sea level will continue to rise as ice sheets continue to lose mass (*Ridley et al., 2005*).

During the last interglaciation period (LIG), from ~130,000 years ago to at least 116,000 years ago, CO_2 levels were similar to pre-industrial levels (*Petit et al., 1999; Kawamura et al., 2007*), but large positive anomalies in early summer solar radiation driven by orbital changes caused arctic summer temperatures to be warmer than they are today (*Otto-Bleisner et al., 2006*). Corals on tectonically stable coasts indicate that sea level during the LIG was 4 to 6 m above present (Figure 2.1) (*Stirling et al., 1995, 1998; Muhs et al., 2002*), and ice-core records (*Koerner, 1989; Raynaud et al., 1997*) and modeling (*Cuffey and Marshall, 2000; Otto-Bleisner et al., 2006*) indicate that much of this rise originated from a reduction in the size of the Greenland Ice Sheet, although some contribution from the Antarctic Ice Sheet may be required as well.

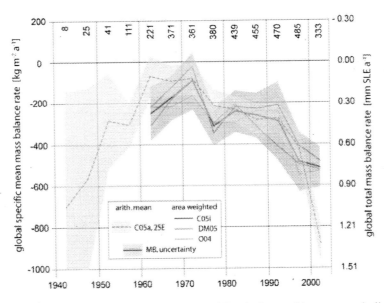

Figure 2.3. Pentadal average mass-balance rates of the world's glaciers and ice caps, excluding Greenland and Antarctica, for the last half century. Specific mass balance (left axis) is converted to total balance and to sea level equivalent (right axis) as described in Table 2.2. C05a: an arithmetic mean over all annual measurements within each pentad, with confidence envelope shaded gray and number of measurements given at top of graph. C05i, DM05, O04: independently obtained spatially corrected series. MB: arithmetic mean of C05i, DM05 and O04, with confidence envelope shaded red. See *Kaser et al. (2006)* for sources and uncertainties; the latter are "2-sigma-like". Estimates are incomplete for the most recent pentad. Copyright American Geophysical Union, 2006; reprinted with permission

At the last glacial maximum, about 21,000 years ago, ice volume and area were about 2.5 times modern, with most of the increase occurring in the Northern Hemisphere (*Clark and Mix, 2002*). Deglaciation was forced by warming from changes in the Earth's orbital parameters, increasing greenhouse gas concentrations, and attendant feedbacks. The record of deglacial sea level rise is particularly well constrained from paleoshoreline evidence (Figure 2.3). Deglacial sea-level rise averaged 10-20 mm a^{-1}, or at least 5 times faster than the average rate of the last 100 years (Figure 2.1), but with variations including two extraordinary episodes at 19,000 years before present (19 ka BP) and 14.5 ka BP, when peak rates potentially exceeded 50 mm a^{-1} *(Fairbanks, 1989; Yokoyama et al., 2000; Clark et al., 2004)* (Figure 2.3), or five times faster than projections for the end of this century (*Rahmstorf, 2007*). Each of these "meltwater pulses" added the equivalent of 1.5 to 3 Greenland ice sheets (~7 m) to the oceans over a one- to five-century period, clearly demonstrating the potential for ice sheets to cause rapid and large sea level changes. A third meltwater pulse may have occurred ~11,700 years ago (*Fairbanks, 1989*), but the evidence for this event is less clear (*Bard et al., 1996; Bassett et al., 2005*).

Recent analyses indicate that the earlier 19-ka event originated from Northern Hemisphere ice (*Clark et al., 2004*). The ~20-m sea level rise ~14,500 years ago (*Fairbanks, 1989; Hanebuth et al., 2000*), commonly referred to as meltwater pulse (MWP) 1A, indicates an extraordinary episode of ice-sheet collapse, with an associated freshwater flux to the ocean of ~0.5 sverdrup (Sv) over several hundred years. The timing, source, and climatic effect of MWP-1A, however, remain widely debated. In one scenario, the event was triggered by an abrupt warming (start of the Bølling warm interval) in the North Atlantic region, causing widespread melting of Northern Hemisphere ice sheets (*Fairbanks et al., 1992; Peltier, 2005*). In another scenario, MWP1A largely originated from the Antarctic Ice Sheet (*Clark et al., 1996, 2002; Bassett et al., 2005*), possibly in response to the ~3,500-year warming in the Southern Hemisphere that preceded the event (*Blunier and Brook, 2001; Clark et al., 2004*). Although the cause of these events has yet to be established, their occurrences following hemispheric warming may implicate short-term dynamic processes activated by that warming, similar to those now being identified around Greenland and Antarctica.

Direct evidence from terrestrial geologic records of one scenario versus the other, however, thus far remains inconclusive. Well-dated terrestrial records of deglaciation of Northern Hemisphere ice sheets, which largely constrain changes in area only, show no acceleration of ice-margin retreat at this time (e.g., *Dyke, 2004; Rinterknecht et al., 2006*), leading some to conclude that the event occurred largely by ice-sheet deflation with little response of the margin (*Simms et al., 2007*). The record of deglaciation of the Antarctic Ice Sheet is less well constrained, and available evidence presents conflicting results, from no contribution (*Ackert et al., 2007; Mackintosh et al., 2007*), to a small contribution (Heroy and Anderson, 2007; Price et al., 2007), to a dominant contribution (Bassett et al., 2007).

The large freshwater fluxes that these events represent also underscore the significance of rapid losses of ice to the climate system through their effects on ocean circulation. An important component of the ocean's overturning circulation involves formation of deep water at sites in the North Atlantic Ocean and around the Antarctic continent, particularly the Weddell and Ross Seas. The rate at which this density-driven thermohaline circulation occurs is sensitive to surface fluxes of heat and freshwater. Eustatic rises associated with the two deglacial meltwater pulses correspond to freshwater fluxes $\geq$ 0.25 Sv, which according to climate models would induce a large change in the thermohaline circulation (*Stouffer et al., 2006; Weaver et al., 2003*).

BOX 2.2. MASS BALANCE, ENERGY BALANCE, AND FORCE BALANCE

The glaciological analyses which we summarize here can all be understood in terms of simple arithmetic.

To determine the mass balance, we add up all the gains of mass, collectively known as accumulation and dominated by snowfall, and all the losses, collectively known as ablation and dominated by melting and calving. The difference between accumulation and ablation is called, by long-established custom, the total mass balance, although the reader will note that we really mean "mass imbalance." That is, there is no reason why the difference should be zero; the same is true of the energy balance and force balance.

The mass balance is closely connected to the energy balance. The temperature of the glacier surface is determined by this balance, which is the sum of gains by the absorption of radiative energy, transfer of heat from the overlying air, and heat released by condensation, and losses by radiative emission, upward transfer of heat when the air is colder than the glacier surface, and heat consumed by evaporation. A negative energy balance means that the ice temperature will drop. A positive energy balance means either that the ice temperature will rise or that the ice will melt.

Ice deformation or dynamics is the result of a balance of forces, which we determine by arithmetic operations comparable to those involved in the mass and energy balances. Shear forces, proportional to the product of ice thickness and surface slope, determine how fast the glacier moves over its bed by shear deformation where the ice is frozen to the bed, or by basal sliding where the bed is wet. Spreading forces, determined by ice thickness, are resisted by drag forces at the glacier bed and its margins, and by forces transmitted upstream from its floating tongue or ice shelf as this pushes seaward past its margins and over locally shoaling seabed. The sum of these forces determines the speed at which the ice moves, together with its direction. However, we must also allow for ice stiffness, which is strongly affected by its temperature, with cold ice much stiffer (more sluggish) than ice near its melting point.

The temperature becomes still more important when we consider basal drag, which is high for a dry-based glacier (one frozen to its bed), but can be very small for wet-based glaciers where their beds have been raised to the melting point by heat conducted from the Earth's interior and frictional heat generated on the spot. Once the bed is at the melting point, any further gain of heat yields meltwater. One of glaciology's bigger surprises is that large parts of the ice sheets, whose surfaces are among the coldest places on Earth, are wet based.

The varying pressure of basal meltwater on the moving ice can alter the force balance markedly. Its general impact is to promote basal sliding, by which mechanism the glacier may flow much more rapidly than it would by shear deformation alone. Basal sliding, in conjunction with the presence of a porous reservoir for meltwater where the bed consists of soft sediment rather than rock, plays a major role in the behavior of ice streams.

There are subtle links between the mass balance and the force balance. The ice flows from where there is net accumulation to where there is net ablation, and the changing size and shape of the glacier depend on the interplay of dynamics and climate, the latter including the climate of the ocean.

3. The Current State of Glaciers, Ice Caps, and Ice Sheets

Rapid changes in ice-sheet mass have surely contributed to abrupt climate change in the past, and any abrupt change in climate is sure to affect the mass balance (see Box 2.2) of at least some of the ice on Earth.

3.1. Mass-Balance Techniques

Traditional estimates of the surface mass balance are from repeated measurements of the exposed length of stakes planted in the snow or ice surface. Temporal change in this length, multiplied by the density of the mass gained or lost, is the surface mass balance at the location of the stake. (In principle the density of mass gained can be measured in shallow cores or snow pits; but in practice there can be considerable uncertainty about density; see, e.g., Sec. 3.1.2.2.) Various means have been devised to apply corrections for sinking of the stake bottom into the snow, densification of the snow between the surface and the stake bottom, and the refreezing of surface meltwater at depths below the stake bottom. Such measurements are time consuming and expensive, and they need to be supplemented at least on the ice sheets by model estimates of precipitation, internal accumulation, sublimation, and melting. Regional atmospheric climate models, calibrated by independent *in situ* measurements of temperature and pressure (e.g., *Steffen and Box, 2001; Box et al., 2006*) provide estimates of snowfall and sublimation. Estimates of surface melting/evaporation come from energy-balance models and degree-day or temperature-index models (reviewed in, e.g., *Hock, 2003*), which are also validated using independent *in situ* measurements. Within each category there is a hierarchy of models in terms of spatial and temporal resolution. Energy-balance models are physically based, require detailed input data, and are more suitable for high resolution in space and time. Degree-day models are advantageous for the purposes of estimating worldwide glacier melt, since the main inputs of temperature and precipitation are readily available in gridded form from Atmosphere-Ocean General Circulation Models (AOGCMs).

Techniques for measuring total mass balance include:

- the mass-budget approach, comparing gains by surface and internal accumulation with losses by ice discharge, sublimation, and meltwater runoff;
- repeated altimetry, or equivalently leveling or photogrammetry, to measure height changes, from which mass changes are inferred;
- satellite measurements of temporal changes in gravity, to infer mass changes directly.

All three techniques can be applied to the large ice sheets; most studies of ice caps and glaciers are annual (or seasonal) mass-budget measurements, with recent studies also using multi-annual laser and radar altimetry. The third technique is applied only to large, heavily glaciated regions such as Alaska, Patagonia, Greenland, and Antarctica. Here, we summarize what is known about total mass balance, to assess the merits and limitations of different approaches to its measurement and to identify possible improvements that could be made over the next few years.

3.1.1. Mass Balance

Snow accumulation is estimated from stake measurements, annual layering in ice cores, sometimes with interpolation using satellite microwave measurements (*Arthern et al., 2006*), or meteorological information (*Giovinetto and Zwally, 2000*) or shallow radar sounding (*Jacka et al., 2004*), or from regional atmospheric climate modeling (e.g., *van de Berg et al., 2006; Bromwich et al., 2004*). The state of the art in estimating snow accumulation for periods of up to a decade is rapidly becoming the latter, with surface data being used mostly for validation, not to drive the models. This is not surprising given the immensity of large ice sheets and the difficulty of obtaining appropriate spatial and temporal sampling of snow accumulation at the large scale by field parties, especially in Antarctica.

Ice discharge is the product of velocity and thickness, with velocities measured *in situ* or remotely, preferably near the grounding line, where velocity is almost depth independent. Thickness is measured by airborne radar, seismically, or from measured surface elevations assuming hydrostatic equilibrium, for floating ice near grounding lines. Velocities are measured by ground-based survey, photogrammetry, or with satellite sensors; the latter are mostly imaging radars operating interferometrically. Grounding lines are poorly known from *in situ* measurement or visible-band imagery but can be mapped very accurately with satellite interferometric imaging radars.

Meltwater runoff (large on glaciers and ice caps, and near the Greenland coast and parts of the Antarctic Peninsula, but small or zero elsewhere) is traditionally inferred from stake measurements but more and more from regional atmospheric climate models validated with surface observations where available (e.g., *Hanna et al., 2005; Box et al., 2006*). The typically small mass loss by melting beneath grounded ice is also estimated from models.

Mass-budget calculations involve the comparison of two very large numbers, and small errors in either can result in large errors in estimated total mass balance. For example, total accumulation over Antarctica, excluding ice shelves, is about 1,850 Gt a^{-1} (*Vaughan et al., 1999; Arthern et al., 2006; van de Berg et al., 2006*), and 500 Gt a^{-1} over Greenland (*Bales et al., 2001*). Associated errors are difficult to assess because of high temporal and spatial variability, but they are probably about ±5% (20-25 Gt a^{-1}) for Greenland. The errors for Antarctica (*Rignot, 2006*) range from 5% in dry interior basins to 20% in wet coastal basins. The total accumulation for Antarctica is approximately 1900 Gt a^{-1} (ranged from 1,811 to 2,076 Gt a^{-1} between 1999 - 2006; Berg et al. (2006)), with an overall uncertainty of 6% or 114 Gt a^{-1}, derived from 93% dry interior region and 7 % wet coastal region, using uncertainties of 5%, and 20% respectively.

Broad interferometric SAR (InSAR) coverage and progressively improved estimates of grounding-line ice thickness have substantially improved ice-discharge estimates, yet incomplete data coverage and residual errors imply errors on total discharge of 2% (*Rignot, 2008*). Consequently, assuming these errors in both snow accumulation and ice losses, current mass-budget uncertainty is ~ ±92 Gt a^{-1} (*Rignot, 2008*) for Antarctica and ±35 Gt a^{-1} for Greenland. Moreover, additional errors may result from accumulation estimates being based on data from the past few decades; at least in Greenland, we know that snowfall is increasing with time. Similarly, it is becoming clear that glacier velocities can change substantially over quite short time periods (*Rignot and Kanagaratnam, 2006*), and the time period investigated (last decade) showed an increase in ice velocities, so these error estimates might well be lower limits.

3.1.2. Repeated altimetry

Rates of surface-elevation change with time (dS/dt) reveal changes in ice-sheet mass after correction for changes in depth/density profiles and bedrock elevation, or for hydrostatic equilibrium if the ice is floating. Satellite radar altimetry (SRALT) has been widely used (e.g., *Shepherd et al., 2002; Davis et al., 2005; Johannessen et al., 2005; Zwally et al., 2005*), together with laser altimetry from airplanes (*Arendt et al., 2002; Krabill et al., 2000*), and from NASA's ICESat (*Zwally et al., 2002a; Thomas et al., 2006*). Modeled corrections for isostatic changes in bedrock elevation (e.g, *Peltier, 2004*) are small (a few millimeters per year) but with errors comparable to the correction. Those for near-surface snow density changes (*Arthern and Wingham, 1998; Li and Zwally, 2004*) are larger (1 or 2 cm a^{-1}) and also uncertain.

3.1.2.1. Satellite radar altimetry

Available SRALT data are from altimeters with a beam width of 20 km or more, designed and demonstrated to make accurate measurements over the almost flat, horizontal ocean. Data interpretation is more complex over sloping and undulating ice- sheet surfaces with spatially and temporally varying dielectric properties. Errors in

SRALT-derived values of dS/dt are typically determined from the internal consistency of the measurements, often after iterative removal of dS/dt values that exceed some multiple of the local value of their standard deviation. This results in small error estimates (e.g., *Zwally et al., 2005, Wingham et al., 2006*) that are smaller than the differences between different interpretations of essentially the same SRALT data (*Johannessen et al., 2005; Zwally et al., 2005*). In addition to processing errors, uncertainties result from the possibility that SRALT estimates are biased by the effects of local terrain or by surface snow characteristics, such as wetness (*Thomas et al., 2008*). Observations by other techniques reveal extremely rapid thinning along Greenland glaciers that flow along depressions where dS/dt cannot be inferred from SRALT data, and collectively these glaciers are responsible for most of the mass loss from the ice sheet (*Rignot and Kanagaratnam, 2006*), implying that SRALT data underestimate near-coastal thinning rates significantly. Moreover, the zone of summer melting in Greenland progressively increased between the early 1990s and 2005 (*Box et al., 2006*), probably raising the radar reflection horizon within near-surface snow by a meter or more over a significant fraction of the ice-sheet percolation facies (*Jezek et al., 1994*). Comparison between SRALT and laser estimates of dS/dt over Greenland show differences that are equivalent to the total mass balance of the ice sheet (*Thomas et al., 2007*)

3.1.2.2. Aircraft and satellite laser altimetry

Laser altimeters provide data that are easier to validate and interpret: footprints are small (about 1 m for airborne laser, and 60 m for ICESat), and there is negligible laser penetration into the ice. However, clouds limit data acquisition, and accuracy is affected by atmospheric conditions and particularly by laser-pointing errors. The strongest limitation by far is that existing laser data are sparse compared to SRALT data.

Airborne laser surveys over Greenland in 1993-94 and 1998-89 yield elevation estimates accurate to ~10 cm along survey tracks (*Krabill et al., 2002*), but with large gaps between flight lines and an incomplete coverage of the glaciers. ICESat orbit-track separation is also quite large compared to the size of a large glacier, particularly in southern Greenland and the

Antarctic Peninsula where rapid changes are occurring, and elevation errors along individual orbit tracks can be large (many tens of centimeters) over sloping ice.

Progressive improvement in ICESat data processing is reducing these errors and, for both airborne and ICESat surveys, most errors are independent for each flight line or orbit track, so that estimates of dS/dt averaged over large areas containing many survey tracks are affected most by systematic ranging, pointing, or platform-position errors, totaling probably less than 5 cm. In Greenland, such conditions typically apply at elevations above 1,500-2,000 m. dS/dt errors decrease with increasing time interval between surveys. Nearer the coast there are large gaps in both ICE Sat and airborne coverage, requiring dS/dt values to be supplemented by degree-day estimates of anomalous melting (*Krabill et al., 2000, 2004*). This supplementation increases overall errors and almost certainly underestimates total losses because it does not take full account of dynamic thinning of unsurveyed outlet glaciers.

In summary, dS/dt errors cannot be precisely quantified for either SRALT data, because of the broad radar beam, limitations with surface topography at the coast, and time- variable penetration, or laser data, because of sparse coverage. The SRALT limitations discussed above will be difficult to resolve. Laser limitations result primarily from poor coverage and can be partially resolved by increasing spatial resolution.

All altimetry mass-balance estimates include additional uncertainties in:

1. The density (rho) assumed to convert thickness changes to mass changes. If changes are caused by recent changes in snowfall, the appropriate density may be as low as 300 kilograms per cubic meter (kg m^{-3}); for long-term changes, it may be as high as 900 kg m^{-3}. This is of most concern for high-elevation regions with small dS/dt, where the simplest assumption is rho = 600 ± 300 kg m^{-3}. For a 1-cm a^{-1} thickness change over the million square kilometers of Greenland above 2,000 m, uncertainty would be ±3 Gt a^{-1}. Rapid, sustained changes, commonly found near the coast, are almost certainly caused by changes in melt rates or glacier dynamics, and for which rho is ~900 kg m^{-3}.

2. Possible changes in near-surface snow density. Densification rates are sensitive to snow temperature and wetness. Warm conditions favor more rapid densification (*Arthern and Wingham, 1998; Li and Zwally, 2004*), and melting is likely to be followed by refreezing as ice. Consequently, recent Greenland warming probably caused surface lowering simply from this effect. Corrections are inferred from largely unvalidated models and are typically <2 cm a^{-1}, with unknown errors. If overall uncertainty is 5 mm a^{-1}, associated mass-balance errors are approximately ±8 Gt a^{-1} for Greenland and ±60 Gt a^{-1} for Antarctica.

3. The rate of crustal uplift. This is inferred from glacio-isostatic models and has uncertain errors. An overall uncertainty of 1 mm a^{-1} would result in mass-balance errors of about ±2 Gt a^{-1} for Greenland and ±12 Gt a^{-1} for Antarctica.

4. The large interannual to decadal changes evidenced in snowfall and hence accumulation in Antarctica (*Monaghan et al., 2006*); the lack of overall trend in net accumulation over the entire continent. This makes it particularly difficult to estimate the mass balance of interior regions because satellite missions have been collecting data for merely 10-15 years. Such investigation clearly requires several decades of data to provide meaningful results.

3.1.3. Temporal variations in earth's gravity

Since 2002, the GRACE satellite has measured Earth's gravity field and its temporal variability. After removing the effects of tides, atmospheric loading, spatial and temporal changes in ocean mass, etc., high-latitude data contain information on temporal changes in the mass distribution of the ice sheets and underlying rock. Because of its high altitude, GRACE makes coarse-resolution measurements of the gravity field and its changes with time. Consequently, resulting mass-balance estimates are also at coarse resolution – several hundred kilometers. But this has the advantage of covering entire ice sheets, which is extremely difficult using other techniques. Consequently, GRACE estimates include mass changes on the many small ice caps and isolated glaciers that surround the big ice sheets; the former may be quite large, being strongly affected by changes in the coastal climate. Employing a surface mass concentration (mascon) solution technique, *Luthcke et al. (2006)* computed multi-year time series of GRACE-derived surface mass flux for Greenland and Antarctica coastal and interior ice sheet sub-drainage systems as well as the Alaskan glacier systems. These mascon solutions provide important observations of the seasonal and interannual evolution of the Earth's land ice.

Error sources include measurement uncertainty, leakage of gravity signal from regions surrounding the ice sheets, interannual variability in snowfall, melt and ice dynamics, and causes of gravity changes other than ice-sheet changes. Of these, the most serious are the gravity changes associated with vertical bedrock motion. *Velicogna and Wahr (2005)* estimated a mass-balance correction of 5 ± 17 Gt a^{-1} for bedrock motion in Greenland, and a correction of 173 ± 7 1 Gt a^{-1} for Antarctica (*Velicogna and Wahr, 2006a*), which may be underestimated (*Horwath and Dietrich, 2006*) or quite reasonable (*Barletta et al., 2008*). Although other geodetic data (variations in length of day, polar wander, etc.) provide constraints on mass changes at high latitudes, unique solutions are not yet possible from these techniques. One possible way to reduce uncertainties significantly, however, is to combine time series of gravity measurements with time series of elevation changes, records of rock uplift from GPS receivers, and records of snow accumulation from ice cores. Yet, this combination requires years to decades of data to provide a significant reduction in uncertainty (see point 4 above).

3.2. Mass Balance of the Greenland and Antarctic Ice Sheets

Ice locked within the Greenland and Antarctic ice sheets (Table 2.1) has long been considered comparatively immune to change, protected by the extreme cold of the polar regions. Most model results suggested that climate warming would result primarily in increased melting from coastal regions and an overall increase in snowfall, with net 21st-century effects probably a small mass loss from Greenland and a small gain in Antarctica, and little combined impact on sea level (*Church et al., 2001*). Observations generally confirmed this view, although Greenland measurements during the 1990s (*Krabill et al., 2000; Abdalati et al., 2001*) began to suggest that there might also be a component from ice-dynamical responses, with very rapid thinning on several outlet glaciers. Such responses had not been seen in prevailing models of glacier motion, primarily determined by ice temperature and basal and lateral drag, coupled with the enormous thermal inertia of a large glacier.

Increasingly, measurements in both Greenland and Antarctica show rapid changes in the behavior of large outlet glaciers. In some cases, once-rapid glaciers have slowed to a virtual standstill, damming up the still-moving ice from farther inland and causing the ice to thicken (*Joughin et al., 2002; Joughin and Tulaczyk, 2002*). More commonly, however, observations reveal glacier acceleration. This may not imply that glaciers have only recently started to change; it may simply mean that major improvements in both quality and coverage of our measurement techniques are now exposing events that also occurred in the past. But in some cases, changes have been very recent. In particular, velocities of tributary glaciers increased markedly very soon after ice shelves or floating ice tongues broke up (e.g., *Scambos et al., 2004; Rignot et al., 2004a*). Moreover, this is happening along both the west and east coasts of Greenland (*Joughin et al., 2004; Howat et al., 2005; Rignot and Kanagaratnam, 2006*) and in at least two locations in Antarctica (*Rignot et al., 2002; Joughin et al., 2003; Scambos et al., 2004; Rignot et al., 2004a*). Such dynamic responses are not explainable in large-scale ice sheet predictive models, nor is the forcing thought responsible for initiating them included in these ice sheet evolutive models. What remains unclear is the response time of large ice sheets. If the ice-dynamical changes observed over the last few years (see Sec. 3) are sustained under global warming, the response time will be significantly shorter.

3.2.1. Greenland

Above ~2,000 m elevation, near-balance between about 1970 and 1995 (*Thomas et al., 2001*) shifted to slow thickening thereafter (*Thomas et al., 2001, 2006; Johannessen et al., 2005; Zwally et al., 2005*). Nearer the coast, airborne laser altimetry surveys supplemented by modeled summer melting show widespread thinning (*Krabill et al., 2000, 2004*), resulting in net loss from the ice sheet of 27 ± 23 Gt a^{-1}, equivalent to ~0.08 mm a^{-1} sea level equivalent (SLE) between 1993-94 and 1998-89 doubling to 55 ± 23 Gt a-₁ for 1997-2003[3]. However, the airborne surveys did not include some regions where other measurements show rapid thinning, so these estimates represent lower limits of actual mass loss.

Table 2.1. Summary of the recent mass balance of Greenland and Antarctica.
(*) 1 km^3 of ice = ~0.92 Gt; (#) Excluding ice shelves; SLE = sea level equivalent

	Greenland	**Antarctica**
Area (106km2)	1.7	12.3
Volume (106km3)*	2.9 (7.3 m SLE)	24.7 (56.6 m SLE)
Total accumulation (Gt a-1)#	500 (1.4 mm SLE)	1850 (5.1 mm SLE)
Mass balance	Since ~1990: Thickening above 2,000 m, at an accelerating rate; thinning at lower elevations also accelerating to cause a net loss from the ice sheet of perhaps >100 Gt a-1after 2000.	Since early 1990s: slow thickening in central regions and southern Antarctic Peninsula; localized thinning at accelerating rates of glaciers in Antarctic Peninsula and Amundsen Searegion. Probable net loss, but close to balance.

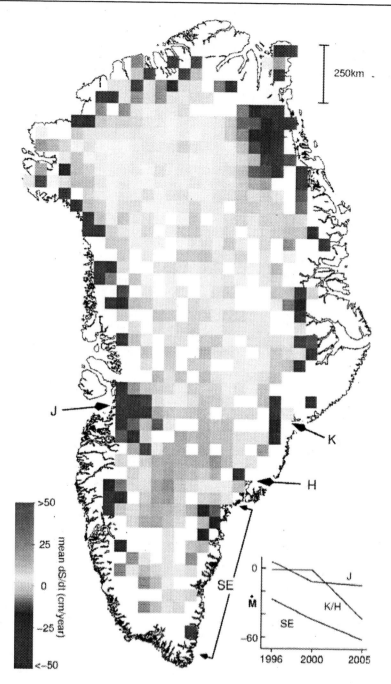

Figure 2.4. Rates of elevation change (dS/dt) for Greenland derived from comparisons at more than 16,000 locations where ICESat data from Oct/Nov and May/June 2004 overlay ATM surveys in 1998/9, averaged over 50-km grid squares. Locations of rapidly thinning outlet glaciers at Jakobshavn (J), Kangerdlugssuaq (K), Helheim (H), and along the southeast coast (SE) are shown, together with plots showing their estimated mass balance (10^6 Gt a^{-1}) versus time (*Rignot and Kanagaratnam, 2006*)

More recently, four independent studies also show accelerating losses from Greenland: (1) Analysis of gravity data from GRACE show total losses of 75±20 Gt a^{-1} between April 2002 and April 2004 rising to 223±3 3 Gt a^{-1} between May 2004 and April 2006 (*Velicogna and Wahr, 2005, 2006a*). (2) Other analyses of GRACE data show losses of 129 ± 15 Gt a^{-1} for July 2002 through March 2005 (*Ramillien et al., 2006*), (3) 219 ± 21 Gt a-ı for April 2002 through November 2005 (*Chen et al., 2006*), and (4) 101 ± 16 Gt a^{-1} for July 2003 to July 2005 (*Luthcke et al., 2006*). Although the large scatter in the estimates for similar time periods suggests that errors are larger than quoted, these results show an increasing trend in mass loss.

Interpretations of SRALT data from ERS-1 and -2 (*Johannessen et al., 2005; Zwally et al., 2005*) show quite rapid thickening at high elevations, with lower elevation thinning at far lower rates than those inferred from other approaches that include detailed observations of these low-elevation regions. The *Johannessen et al. (2005)* study recognized the unreliability of SRALT data at lower elevations because of locally sloping and undulating surface topography. *Zwally et al. (2005)* attempted to overcome this by including dS/dt estimates for about 3% of the ice sheet derived from earlier laser altimetry, to infer a small positive mass balance of 11 ± 3 Gt a^{-1} for the entire ice sheet between April 1992 and October 2002.

Mass-budget calculations for most glacier drainage basins indicate total ice-sheet losses increasing from 83 ± 28 Gt a^{-1} in 1996 to 127 ± 28 Gt a^{-1} in 2000 and 205 ± 3 8 Gt a^{-1} in 2005 (*Rignot and Kanagaratnam, 2006*). Most of the glacier losses are from the southern half of Greenland, especially the southeast sector, east-central, and west-central. In the northwest, losses were already significant in the early 1 990s and did not increase in recent decades. In the southwest, losses are low but slightly increasing. In the north, losses are very low, but also slightly increasing in the northwest and northeast.

Comparison of 2005 ICESat data with 1998-89 airborne laser surveys shows losses during the interim of 80 ± 25 Gt a^{-1} (*Thomas et al., 2006*), and this is probably an underestimate because of sparse coverage of regions where other investigations show large losses.

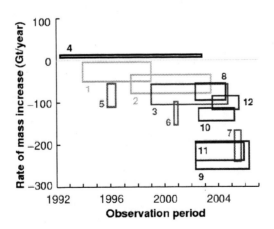

Figure 2.5. Mass-balance estimates for the entire Greenland ice sheet: green—airborne laser altimetry (ATM); purple—ATM/ICESat (summarized in *Thomas et al., 2006*); black—Satellite Radar Altimetry (SRALT) (4: *Zwally et al., 2005*); red—mass budget (5,6,7: *Rignot and Kanagaratnam, 2006*); blue—GRACE (8 and 9: *Velicogna and Wahr, 2005, 2006a*; 10: *Ramillien et al., 2006*; 11: *Chen et al., 2006*; 12: *Luthcke et al., 2006*). The ATM results were supplemented by degree-day estimates of anomalous melting near the coast (*Krabill et al., 2000,; 2004*), and probably underestimate total losses by not taking full account of dynamic thinning of outlet glaciers (*Abdalati et al., 2001*). SRALT results seriously underestimate rapid thinning of comparatively narrow Greenland glaciers, and may also be affected by progressively increased surface melting at higher elevations. Gt, gigatons

The pattern of thickening/thinning over Greenland, derived from laser-altimeter data, is shown in Figure 2.4, with the various mass-balance estimates summarized in Figure 2.5. It is clear that the SRALT-derived estimate differs widely from the others, each of which is based on totally different methods, suggesting that the SRALT interpretations underestimate total ice loss for reasons discussed in Section 3.1.1. Here, we assume this to be the case, and focus on the other results shown in Figure 2.5, which strongly indicate net ice loss from Greenland at rates that increased from at least 27 Gt a^{-1} between 1993- 94 and 1998-99 to about double between 1997 and 2003, to more than 80 Gt a^{-1} between 1998 and 2004, to more than 100 Gt a^{-1} soon after 2000, and to more than 200 Gt a^{-1} after 2005. There are insufficient data for any assessment of total mass balance before 1990, although mass-budget calculations indicated near overall balance at elevations above 2,000 m and significant thinning in the southeast (*Thomas et al., 2001*).

3.2.2. Antarctica

Determination of the mass budget of the Antarctic ice sheet is not as advanced as that for Greenland. Melt is not a significant factor, but uncertainties in snow accumulation are larger because fewer data have been collected, and ice thickness is poorly characterized along outlet glaciers. Instead, ice elevations, which have been improved with ICESat data, are used to calculate ice thickness from hydrostatic equilibrium at the glacier grounding line. The grounding line position and ice velocity are inferred from Radarsat-1 and ERS-1/2 InSAR. For the period 1996-2000, *Rignot and Thomas (2002)* inferred East Antarctic growth at 20 ± 1 Gt a^{-1}, with estimated losses of 44 ± 13 Gt a^{-1} for West Antarctica, and no estimate for the Antarctic Peninsula, but the estimate for East Antarctica was based on only 60% coverage. Using improved data for 1996-2004 that provide estimates for more than 85% of Antarctica (and which were extrapolated on a basin per basin basis to 100% of Antarctica), *Rignot (2008)* found an ice loss of 106 ± 60 Gt a^{-1} for West Antarctica, 28 ± 45 Gt a^{-1} for the peninsula, and a mass gain of 4 ± 61 Gt a-1 for East Antarctica in year 2000. In year 1996, the mass loss for West Antarctica was 83 ± 59 Gt a^{-1}, but the mass loss increased to 132 ± 60 Gt a^{-1} in 2006 due to glacier acceleration. In the peninsula, the mass loss increased to 60 ± 46 Gt a^{-1} in 2006 due to the massive acceleration of glaciers in the northern peninsula following the breakup of the Larsen B ice shelf in the year 2002. Overall, the ice sheet mass loss nearly doubled in 10 years, nearly entirely from West Antarctica and the northern tip of the peninsula, while little change has been found in East Antarctica. Other mass-budget analyses indicate thickening of drainage basins feeding the Filchner-Ronne ice shelf from portions of East and West Antarctica (*Joughin and Bamber, 2005*) and of some ice streams draining ice from West Antarctica into the Ross Ice Shelf (*Joughin and Tulaczyk, 2002*), but mass loss from the northern part of the Antarctic Peninsula (*Rignot et al., 2005*) and parts of West Antarctica flowing into the Amundsen Sea (*Rignot et al., 2004b*). In both of these latter regions, losses are increasing with time.

Although SRALT coverage extends only to within about 900 km of the poles (Figure 2.6), inferred rates of surface elevation change (dS/dt) should be more reliable than in Greenland, because most of Antarctica is too cold for surface melting (reducing effects of changing dielectric properties), and outlet glaciers are generally wider than in Greenland (reducing uncertainties associated with rough surface topography). Results show that interior parts of East Antarctica monitored by ERS- 1 and ERS-2 thickened during the 1990s, equivalent to growth of a few tens of gigatonnes per year, depending on details of the near-surface density structure (*Davis et al., 2005; Wingham et al., 2006; Zwally et al., 2005*), but *Monaghan et al. (2006)* and *van den Broeke et al. (2006)* show no change in accumulation over a longer time period in this region, suggesting that SRALT may be biased by the large decadal variability in snowfall in Antarctica With ~80% SRALT coverage of the ice sheet, and interpolating to the rest, *Zwally et al. (2005)* estimated a West Antarctic loss of 47±4 Gt a^{-1}, East Antarctic gain of 17±11 Gt a^{-1}, and overall loss of 30±12 Gt a^{-1}, excluding the Antarctic Peninsula, a large fraction of the coastal sectors, and with error estimates neglecting potential uncertainties. *Wingham et al. (2006)* interpret the same data to show that mass gain from snowfall, particularly in the Antarctic Peninsula and East Antarctica, exceeds dynamic losses from West Antarctica. More importantly, however, *Monaghan et al. (2006)* and *van den Broeke et al. (2006)* found very strong decadal variability in Antarctic accumulation, which suggests that it will require decades of data to separate decadal variations from long-term trends in accumulation, for instance, associated with climate warming.

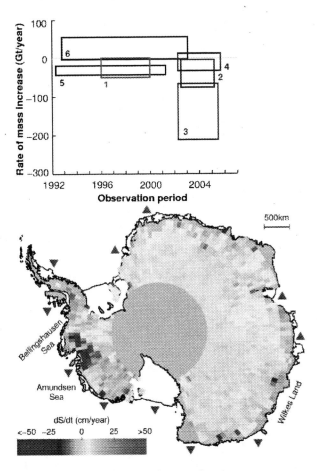

Figure 2.6. Rates of elevation change (dS/dt) derived from ERS radar-altimeter measurements between 1992 and 2003 over the Antarctic Ice Sheet (*Davis et al., 2005*). Locations of ice shelves estimated to be thickening or thinning by more than 30 cm a-1 (*Zwally et al., 2005*) are shown by purple triangles (thinning) and red triangles (thickening). Inset shows mass-balance estimates for the ice sheet: red—mass budget (1: *Rignot and Thomas, 2002*); blue—GRACE (2: *Ramillien et al., 2006*; 3: *Velicogna and Wahr, 2006b*; 4: *Chen et al., 2006*); black—ERS SRALT (5: *Zwally et al., 2005*; 6: *Wingham et al., 2006*)

The present ice mass balance of Antarctica and its deglaciation history from the Last Glacial Maximum are still poorly known. It has been shown recently that the uplift rates derived from the Global Positioning System (GPS) can be employed to discriminate between different ice loading scenarios. There is general agreement that Antarctica was a major participant in the last glacial age within the West Antarctic Ice Sheet (WAIS), perhaps contributing more than 15 m to rising sea level during the last 21,000 years (*Clark et al., 2002*). The main controversy is whether or not the dominant Antarctic melt contribution to sea level rise occurred during the Holocene or earlier, corresponding to the initial deglaciation phase (21–14 ka) of Northern Hemispheric ice sheets (*Peltier, 1998*). Postglacial rebound rates are not well constrained and are an error source for ice mass-balance assessment with GRACE satellite data. Analyses of GRACE measurements for 2002-05 show the ice sheet to be very close to balance with a gain of 3 ± 20 Gt a-1 (*Chen et al., 2006*) or net loss from the sheet ranging from 40 ± 35 Gt a^{-1} (*Ramillien et al., 2006*) to 137 ± 72 Gt a^{-1} (*Velicogna and Wahr, 2006b*), primarily from the West Antarctic Ice Sheet.

Taken together, these various approaches indicate a likely net loss of 80 Gt a^{-1} in the mid 1990s growing to 130 Gt a^{-1} in mid 2000s.

The largest losses are concentrated along the Amundsen and Bellingshausen sectors of West Antarctica, in the northern tip of the Antarctic Peninsula, and to a lesser extent in the Indian Ocean sector of East Antarctica.

A few glaciers in West Antarctic are losing a disproportionate amount of mass. The largest mass loss is from parts of the ice sheet flowing into Pine Island Bay, which represents enough ice to raise sea level by 1.2 m.

In East Antarctica, with the exception of glaciers flowing into the Filchner/Ronne, Amery, and Ross ice shelves, nearly all the major glaciers are thinning, with those draining the Wilkes Land sector losing the most mass. Like much of West Antarctica, this sector is grounded well below sea level.

Observations are insufficient to provide reliable estimates of mass balance before 1990, yet there is evidence for long-term loss of mass from glaciers draining the Antarctic Peninsula (*Pritchard and Vaughan, 2007*) and for speed up of Pine Island Glacier and neighbors since at least the 1970s (*Joughin et al., 2003*) In addition, balancing measured sea level rise since the 1950s against potential causes such as thermal expansion and non-Antarctic ice melting leaves a "missing" source equivalent to many tens of gigatonnes per year.

3.3. Rapid Changes of Small Glaciers

3.3.1. Introduction

Small glaciers are those other than the two ice sheets. Mass balance is a rate of either gain or loss of ice, and so a change in mass balance is an acceleration of the process. Thus we measure mass balance in units such as kg $_{m-2}$ a-1 (mass change per unit surface area of the glacier; 1 kg m^{-2} is equivalent to 1 mm depth of liquid water) or, more conveniently at the global scale, Gt a^{-1} (change of total mass, in gigatonnes per year). A change in mass balance is measured in Gt a^{-2}, gigatonnes per year per year: faster and faster loss or gain.

3.3.2. Mass-balance measurements and uncertainties

Most measurements of the mass balance of small glaciers are obtained in one of two ways. *Direct* measurements are those in which the change in glacier surface elevation is measured directly at a network of pits and stakes. Calving is treated separately. In *geodetic* measurements, the glacier surface elevation is measured at two times with reference to some fixed external datum. Recent advances in remote sensing promise to increase the contribution from geodetic measurements and to improve spatial coverage, but at present the observational database remains dominated by direct measurements. The primary source for these is the World Glacier Monitoring Service (*WGMS; Haeberli et al., 2005*). *Kaser et al. (2006)* (see also *Lemke et al., 2007*; Sec. 4.5) present compilations which build on the WGMS dataset and extend it significantly.

In Figure 2.3 (see also Table 2.2), the three spatially corrected curves agree rather well, which motivated *Kaser et al. (2006)* to construct their consensus estimate of mass balance, denoted MB. The arithmetic-average curve C05a is the only curve extending before 1961 because measurements are too few at those times for area-weighting or spatial interpolation to be practicable. The early measurements suggest weakly that mass balances were negative. After 1961, we can see with greater confidence that mass balance became less negative until the early 1970s, and that thereafter it has been growing more negative.

The uncorrected C05a, a simple arithmetic average of all the measurements, generally tracks the other curves with fair accuracy. Apparently spatial bias, while not negligible, is of only moderate significance. However the C05a estimate for 2001-04 is starkly discordant. The discordance is due in large part to the European heat wave of 2003 and to under-representation of the high arctic latitudes, where measurements are few and 2003 balances were only moderately negative. It illustrates the extent to which spatial bias can compromise global estimates. The other curves, C05i, DM05 and O04, each attempt to correct carefully for spatial bias.

Mass-balance measurements at the glacier surface are relatively simple, but difficulties arise with contributions from other parts of the glacier. Internal accumulation is one of the most serious problems. It happens in the lower percolation zones of cold glaciers (those whose internal temperatures are below freezing) when surface meltwater percolates beneath the current year's accumulation of snow. Internal accumulation is impractical to measure and is difficult to model with confidence. It is a plausible conjecture that there are many more cold glaciers than temperate glaciers (in which meltwater can be expected to run off rather than to refreeze).

Table 2.2. Global small-glacier mass balance for different periods. Consensus estimates (*Kaser et al., 2006*), including small glaciers in Greenland and Antarctica, of global average specific mass balance (*b*); global total mass balance (*B*), equal to *A×b* where *A*=785 ×10⁹ m² is the areal extent of small glaciers; and the sea level equivalent (*SLE*), equal to −*B*/(. *AO*), where .=1,000 kg m⁻³ and ocean area *AO*=362× 10¹² m²

Period	b (kg m⁻² a⁻¹)	B (Gt a⁻¹)	SLE (mm a-1)
1961-2004	-231±101	-182±78	0.50±0.22
1961-1990	-173±89	-136±70	0.37±0.19
1991-2004	-356±121	-280±95	0.77±0.26
2001-2004	-451±89	-354±70	0.98±0.19

The calving of icebergs is a significant source of uncertainty. Over a sufficiently long averaging period, adjacent calving and noncalving glaciers ought not to have very different balances, but the time scale of calving is quite different from the annual scale of surface mass balance, and it is difficult to match the two. Tidewater glaciers tend to evolve by slow growth (over centuries) alternating with brief (decades-long) episodes of rapid retreat. Many tidewater glaciers are undergoing such retreat at present, but in general they are under-represented in the list of measured glaciers. The resulting bias, which is known to be opposite to the internal-accumulation bias, must be substantial.

We can draw on geodetic and gravimetric measurements of multidecadal mass balance to reinforce our understanding of calving rates. To illustrate, *Larsen et al. (2007)* estimated the mass balance in southeastern Alaska and adjacent British Columbia as -16.7±4.4 Gt a1. Earlier, *Arendt et al. (2002)* measured glaciers across Alaska by laser altimetry and estimated an acceleration in mass loss for the entire state from 52 ± 15 Gt a^{-1} (mid-1950s to mid-1990s) to 96 ± 35 Gt a^{-1} (mid-1990s to 2001). These are significantly greater losses than the equivalent direct estimates, and much of the discrepancy must be due to under- representation of calving in the latter. This under-representation is compounded by a lack of basic information. The extent, and even the total terminus length, of glacier ice involved in calving is not known, although a substantial amount of information is available in scattered sources.

Global mass-balance estimates suffer from uncertainty in total glacierized area, and the rate of shrinkage of that area is not known accurately enough to be accounted for. A further problem is delineating the ice sheets so as to avoid double-counting or omitting peripheral ice bodies.

Measured glaciers are a shifting population. Their total number fluctuates, and the list of measured glaciers changes continually. The commonest record length is 1 year; only about 50 are longer than 20 years. These difficulties can be addressed by assuming that each single annual measurement is a random sample. However, the temporal variance of such a short sample is difficult to estimate satisfactorily, especially in the presence of a trend.

On any one glacier, a small number of point measurements must represent the entire glacier. It is usually reasonable to assume that the mass balance depends only on the surface elevation, increasing from net loss at the bottom to net gain above the equilibrium line altitude. A typical uncertainty for elevation-band averages of mass balance is 200 kilograms per square meter per year (kg m^{-2} a^{-1}), but measurements at different elevations are highly correlated, meaning that whole-glacier measurements have intrinsic uncertainty comparable with that of elevation-band averages.

At the global scale, the number of measured glaciers is small by comparison with the total number of glaciers. However the mass balance of any one glacier is a good guide to the balance of nearby glaciers. At this scale, the distance to which single-glacier measurements yield useful information is of the order of 600 km. Glacierized regions with few or no measured glaciers within this distance obviously pose a problem. If there are no nearby measurements at all, we can do no better in a statistical sense than to set the regional average equal to the global average, attaching to it a suitably large uncertainty.

3.3.3. Historical and recent balance rates

To extend the short time series of measured mass balance, *Oerlemans et al. (2007)* have tried to calibrate records of terminus fluctuations (i.e., of glacier length) against the direct measurements by a scaling procedure. This allowed them to interpret the terminus fluctuations back to the mid-19th century in mass-balance units. Figure 2.7 shows modeled mass loss since the middle of the 19th century, at which time mass balance was near to zero for perhaps a few decades. Before then, mass balance had been positive for probably a few centuries. This is the signature of the Little Ice Age, for which there is abundant evidence in other forms. The balance implied by the *Oerlemans et al. (2007)* reconstruction is a net loss of about 110 to 150 Gt a^{-1} on average over the past 150 years. This has led to a cumulative rise of sea level by 50-60 mm.

It is not possible to detect mass-balance acceleration with confidence over this time span, but we do see such an acceleration over the shorter period of direct measurements (Figure 2.7). This signature matches well with the signature seen in records of global average surface-air temperature (*Trenberth et al., 2007*). Temperature remained constant or decreased slightly from the 1 940s to the 1 970s and has been increasing since. In fact, mass balance also responds to forcing on even shorter time scales. For example, there is a detectable small-glacier response to large volcanic eruptions. In short, small glaciers have been evolving as we would expect them to when subjected to a small but growing increase in radiative forcing.

At this point, however, we must recall the complication of calving, recently highlighted by *Meier et al. (2007)*. Small glaciers interact not only with the atmosphere but also with the solid earth beneath them and with the ocean. They are thus subject to additional forcings which are only indirectly climatic. *Meier et al. (2007)* made some allowance for calving when they estimated the global total balance for 2006 as -402±95 Gt a^{-1}, although they cautioned that the true magnitude of loss was probably greater.

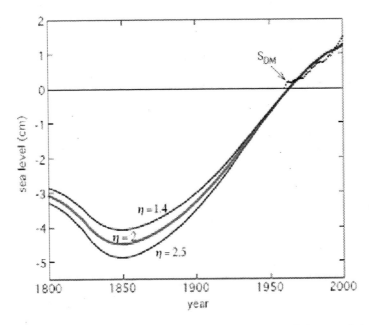

Figure 2.7. Reconstruction of the cumulative glacier contribution to sea level change relative to an arbitrary zero in 1961 (*Oerlemans et al., 2007*). The three smooth curves represent different choices for ., a parameter which regulates the conversion of normalized glacier length to volume. *SDM* (dots) is the cumulative contribution estimated directly from measurements. Copyright of the author; reprinted with permission

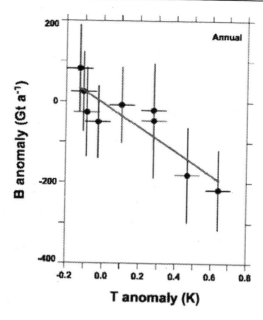

Figure 2.8. Correlation of the anomaly (relative to the 1961-1990 average) in pentadal mean annual mass balance *B* (*Kaser et al., 2006*) with the corresponding anomaly in *T*, surface air temperature over land (CRUTEM3; *Trenberth et al., 2007*). The fitted line suggests a proportionality *dB/dT* of -297±133 Gt $_a$-1 K-1 for the era of direct balance measurements (1961-2004)

"Rapid" is a relative term when applied to the mass balance of small glaciers. For planning purposes we might choose to think that the 1850-2000 average rate of *Oerlemans et al. (2007)* is "not very rapid". After all, human society has grown accustomed to this rate, although it is true that the costs entailed by a consistently nonzero rate have only come to be appreciated quite recently. But a loss of 110 to 150 Gt a-1 can be taken as a useful benchmark. It is greater in magnitude than the net loss of 54±82 Gt a^{-1} estimated by *Kaser et al. (2006)* for 1971-75 and significantly less than the *Kaser et al. (2006)* net loss of 354±70 Gt a^{-1} for 2001-04. So in the last three decades the world's small glaciers have moved from losing mass at half the benchmark rate to rates two or three times faster than the benchmark rate. As far as the measurements are able to tell us, this acceleration has been steady.

Figure 2.8 shows accordance between balance and temperature. Each degree of warming yields about another -300 Gt a^{-1} of mass loss beyond the 1961-90 average, -136 Gt a^{-1}. This suggestion is roughly consistent with the current warming rate, about 0.025 K a^{-1}, and balance acceleration, about -10 Gt a^{-2} (Figure 2.8). To compare with rates inferred for the more distant past, it may be permissible to extrapolate (with caution, because we are neglecting the sensitivity of mass balance to change in precipitation and also the sensitivity of dB/dT, the change in mass balance per degree of warming, to change in the extent and climatic distribution of the glaciers). For example, at the end of the Younger Dryas, about 11,600 years ago, small glaciers could have contributed at least 1,200 Gt a-1 [4 K (300) Gt a^{-1} K^{-1}] of meltwater if we adopt the total summer warming (*Denton et al., 2005*). Such large rates, if reached, could readily be sustained for at least a few decades during the 21st century. At some point the total shrinkage must begin to impact the rate of loss (we begin to run out of small-glacier ice). Against that certain development must be set the probability that peripheral ice caps would also begin to detach from the ice sheets, thus "replenishing" the inventory of small glaciers. *Meier et al. (2007)*, by extrapolating the current acceleration, estimated a total contribution to sea level of 240±128 mm by 2100, implying a negative balance of 1,500 Gt a^{-1} in that year. These figures assume that the current acceleration of loss continues. Alternatively, if loss continues at the current rate of 400 Gt a^{-1}, the total contribution is 104±25 mm. In contrast *Raper and Braithwaite (2006)*, who allowed for glacier shrinkage, estimated only 97 mm by 2100. Part of the difference is due to their exclusion of small glaciers in Greenland and Antarctica. If included, and if they were assumed to contribute at the same rate as the other glaciers, these would raise the *Raper and Braithwaite (2006)* estimate to 137 mm.

3.4. Causes of Changes

Potential causes of the observed behavior of the ice sheets include changes in snowfall and/or surface melting, long-term responses to past changes in climate, and changes in the dynamics, particularly of outlet glaciers, that affect total ice discharge rates. Recent observations have shown that changes in dynamics can occur far more rapidly than previously suspected, and we discuss causes for these in more detail in Section 4.

3.4.1. Changes in snowfall and surface melting

Recent studies find no continentwide significant trends in Antarctic accumulation over the interval 1980-2004 (*van den Broeke et al., 2006; Monaghan et al., 2006*), and surface melting has little effect on Antarctic mass balance. Modeling results indicate probable increases in both snowfall and surface melting over Greenland as temperatures increase (*Hanna et al., 2005; Box et al., 2006*). Model results predict increasing snowfall in a warming climate in Antactica and Greenland, but only the latter could be verified by independent measurements (*Johannessen et al., 2005.*) An update of estimated Greenland Ice Sheet runoff and surface mass balance (i.e., snow accumulation minus runoff) results presented in *Hanna et al. (2005)* shows significantly increased runoff losses for 1998- 2003 compared with the 1961-90 climatologically "normal" period. But this was partly compensated by increased precipitation over the past few decades, so that the decline in surface mass balance between the two periods was not statistically significant. Data from more recent years, extending to 2007, however, suggest a strong increase in the net loss from the surface mass balance. However, because there is summer melting over ~50% of Greenland already (*Steffen et al., 2004b*), the ice sheet is particularly susceptible to continued warming. Small changes in temperature substantially increasing the zone of summer melting, and, a temperature increase by more than 3°C would probably result in irreversible loss of the ice sheet (*Gregory et al., 2004*). Moreover, this estimate is based on imbalance between snowfall and melting and would be accelerated by changing glacier dynamics of the type we are already observing.

In addition to the effects of long-term trends in accumulation/ablation rates, mass-balance estimates are also affected by interannual variability. This increases uncertainties associated with measuring surface accumulation/ablation rates used for mass-budget calculations, and it results in a lowering/raising of surface elevations measured by altimetry (e.g., *van der Veen, 1993*). *Remy et al. (2002)* estimate the resulting variance in surface elevation to be around 3 m over a 30-year time scale in parts of Antarctica. This clearly has implications for the interpretation of altimeter data.

3.4.2. Ongoing dynamic ice sheet response to past forcing

The vast interior parts of an ice sheet respond only slowly to climate changes, with time scales up to 10,000 years in central East Antarctica. Consequently, current ice-sheet response does includes a component from ongoing adjustment to past climate changes. Model results (e.g., *Huybrechts, 2002; Huybrechts et al., 2004*) show only a small longterm change in Greenland ice-sheet volume, but Antarctic shrinkage of about 90 Gt a^{-1}, concomitant with the tail end of Holocene grounding-line retreat since the Last Glacial Maximum. This places a lower bound on present-day ice sheet losses.

3.4.3. Dynamic response to ice-shelf breakup

Recent rapid changes in marginal regions of both ice sheets include regions of glacier thickening and slowdown but mainly acceleration and thinning, with some glacier velocities increasing more than twofold. Most of these glacier accelerations closely followed reduction or loss of ice shelves. Such behavior was predicted almost 30 years ago by *Mercer (1978)*, but was discounted, as recently as the IPCC Third Assessment Report (*Church et al., 2001*) by most of the glaciological community, based largely on results from prevailing model simulations. Considerable effort is now underway to improve the models, but it is far from complete, leaving us unable to make reliable predictions of ice-sheet responses to a warming climate if such glacier accelerations were to increase in size and frequency. It should be noted that there is also a large uncertainty in current model predictions of the atmosphere and ocean temperature changes which drive the ice-sheet changes, and this uncertainty could be as large as that on the marginal flow response.

Total breakup of Jakobshavn Isbræ ice tongue in Greenland was preceded by its very rapid thinning, probably caused by a massive increase in basal melting rates (*Thomas et al., 2003*). Despite an increased ice supply from accelerating glaciers, thinning of more than 1 m a^{-1}, and locally more than 5 m a^{-1}, was observed between 1992 and 2001 for many small ice shelves in the Amundsen Sea and along the Antarctic Peninsula (*Shepherd et al., 2003; Zwally et al., 2005*). Thinning of ~1 m a^{-1} (*Shepherd et al., 2003*) preceded the fragmentation of almost all (3,300 km^2) of the Larsen B ice shelf along the Antarctic Peninsula in fewer than 5 weeks in early 2002 (*Scambos et al., 2003*), and the correlation between long melt seasons and ice shelf breakup was highlighted by *Fahnestock et al. (2002)*. A southward-progressing loss of ice shelves along the Antarctic Peninsula is consistent with a thermal limit to ice-shelf viability (*Mercer*, 1978; *Morris and Vaughan, 1994*). *Cook et al. (2005)* found that no ice shelves exist on the warmer side of the −5°C mean annual isotherm, whereas no ice shelves on the colder side of the − 9°C isotherm have broken up. Before the 2002 breakup of Larsen B ice shelf, local air temperatures increased by more than 1.5°C over the previous 50 years (*Vaughan et al., 2003*), increasing summer melting and formation of large melt ponds on the ice shelf. These may have contributed to breakup by draining into and wedging open surface crevasses that linked to bottom crevasses filled with seawater (*Scambos et al., 2000*).

Most ice shelves are in Antarctica, where they cover an area of ~1.5x10^6 km^2 with nearly all ice streams and outlet glaciers flowing into them. The largest ones in the Weddell and Ross Sea Embayments also occupy the most poleward positions and are currently still far from the viability criteria cited above. By contrast, Greenland ice shelves occupy only a few thousand square kilometers, and many are little more than floating glacier tongues. Ice shelves are nourished by ice flowing from inland and by local snow accumulation, and mass loss is primarily by iceberg calving and basal melting. Melting of up to tens of meters per year has been estimated beneath deeper ice near grounding lines (*Rignot and Jacobs, 2002*). Significant changes in ice-shelf thickness are most readily caused by changes in basal melting or iceberg calving.

Ice-shelf basal melting depends on temperature and ocean circulation within the cavity beneath (*Jenkins and Doake, 1991*). Isolation from direct wind forcing means that the main drivers of below-ice-shelf circulation are tidal and density (thermohaline) forces, but lack of knowledge of bathymetry below the ice has hampered the use of three- dimensional models to simulate circulation beneath the thinning ice shelves as well as a lack of basic data on changes in ocean thermal forcing.

If glacier acceleration caused by thinning ice shelves can be sustained over many centuries, sea level will rise more rapidly than currently estimated. A good example are tidewater glaciers as discussed in Section 3.3.2. But such dynamic responses are poorly understood and, in a warmer climate, the Greenland Ice Sheet margin would quickly retreat from the coast, limiting direct contact between outlet glaciers and the ocean. This would remove a likely trigger for the recently detected marginal acceleration. Nevertheless, although the role of outlet-glacier acceleration in the longer term (multidecade) evolution of the ice sheet is hard to assess from current observations, it remains a distinct possibility that parts of the Greenland Ice Sheet may already be very close to their threshold of viability.

3.4.4. Increased basal lubrication

Observations on some glaciers show seasonal variations in ice velocity, with marked increases soon after periods of heavy surface melting (e.g., *O 'Neel et al., 2001*). Similar results have also been found on parts of the Greenland ice sheet, where ice is moving at ~100 m a^{-1} (*Zwally et al., 2002b*). A possible cause is rapid meltwater drainage to the glacier bed, where it enhances lubrication of basal sliding. If so, there is a potential for increased melting in a warmer climate to cause an almost simultaneous increase in ice- discharge rates. However, there is little evidence for seasonal changes in the speeds of the rapid glaciers that discharge most Greenland ice. In northwest, northeast, southeast, and west-central Greenland, *Rignot and Kanagaratnam (2006)* found an 8-10% increase in monthly velocity over the summer months compared to the winter months, so that abundance of meltwater in the summer is not providing a significant variation in ice discharge compared to the yearly average. However, this does not mean that a doubling of the meltwater production could only drive a 16-20% increase in speed. Meltwater remains an essential control on glacier flow as many studies of mountain glaciers have shown for many decades, so it is quite likely that an increase in meltwater production from a warmer climate could likely have major consequences on the flow rates of glaciers.

4. Potential Mechanism of Rapid Ice Response

4.1. Ocean-Ice Interactions

The interaction of warm ocean waters with the periphery of the large ice sheets represents one of the most significant possibilities for abrupt change in the climate system. Ocean waters provide a source of energy that can drive high melt rates beneath ice shelves and at tidewater glaciers. Calving of icebergs at glacier termini is an additional mechanism of ice loss and has the capacity to destabilize an ice front. Mass loss through oceanic melting and iceberg calving accounts for more than 95% of the ablation from Antarctica and 40-50% of the ablation from Greenland. As described in the previous section, we have seen evidence over the last decade or so, largely gleaned from satellite and airborne sensors, that the most evident changes in the ice sheets have been occurring at their periphery. Some of the changes, for example in the area of the Pine Island Glacier, Antarctica, have been attributed to the effect of warming ocean waters at the margin of the ice sheet (*Payne et al., 2004*). There does not yet exist, however, an adequate observational database against which to definitively correlate ice-shelf thinning or collapse with warming of the surrounding ocean waters.

4.1.1. Ocean circulation

To understand how changes in ocean temperature can impact ice shelves and tidewater glaciers, it is necessary first to understand properties of the global ocean circulation. The polar oceans receive warm salty water originating in the nonpolar oceans. In the North Atlantic Ocean, the northward flowing extension of the Gulf Stream ultimately arrives in the vicinity of the Greenland Ice Sheet, at depth. In the Southern Ocean, the southward extension of the North Atlantic Deep Waters ultimately arrive in the vicinity of the Antarctic Ice Sheet, again at depth. The polar oceans themselves produce cold, fresh water, and salty waters are denser than the cold, fresh waters. The result is that the warm, salty waters are found at depths of several hundred meters in the polar oceans, having subducted beneath the cold, fresh surface polar waters.

Despite the potential of the warm, deep waters to impact the basal melting of ice shelves, little observational progress has been made in studying these waters, nor is there any information on the pre-instrumental (geologic) record of these waters. The main obstacle to progress has been that no sustained observation program can provide a regional and temporal view of the behavior of these deep waters. Instead, for the most part, we have only scattered ship-based observations, poorly sampled in time and space of the locations and temperatures of the deep waters. Limited observations have established that warm, deep waters are present near some Antarctic ice shelves (e.g., Pine Island Glacier, *Jacobs et al., 1996*) and not near others (e.g., Ross Ice Shelf, *Jacobs and Giulivi, 1998*). Greenland's ice shelves follow similarly with some having warm, deep waters present (e.g., Jakobshavn Isbræ, *Holland et al., 2007a*) and others much less so (e.g., Petermann Gletscher, *Steffen et al., 2004a*).

The nature of the circulation of ocean waters beneath an ice shelf can be broadly classified into two regimes. In one regime, only cold ocean waters (i.e., near the freezing point) are found in front of and beneath an ice shelf. These waters produce little melting of the ice shelf base, as for instance, the base of the Ross Ice Shelf, which is estimated to melt at about 0.2 m a^{-1} (*Holland et al., 2003*). In a second regime, warm waters (i.e., a few degrees above the freezing point) are found in front of and beneath the ice shelf. Here, the melt rate can be one-hundredfold stronger, up to 20 m a^{-1}, as for example at the base of the Pine Island Glacier (*Jacobs et al., 1996*). This nonlinear sensitivity of basal mass balance to ocean temperature has recently been highlighted (*Holland et al., 2007b*), as well as the sensitivity of melt rate to the geometry of the environment. The presence of warm water in the vicinity of an ice shelf is a necessary condition for high melting, but it is not sufficient by itself. Additional factors such as the details of the bathymetry can be equally important, as for example, a submarine sill can block access of warm waters while a submarine canyon can facilitate the exchange of warm, deep waters into a cavity beneath an ice shelf. Recent years have seen an increase in the collection of bathymetric data around the Greenland and Antarctic continental shelves, and in some instances even beneath the ice shelves.

4.1.2. Ice-pump circulation

The manner in which ocean waters circulate beneath an ice shelf has loosely become known as the 'ice-pump' circulation (*Lewis and Perkins, 1986*). The circulation can be visualized as dense, salty water (either cold or warm), entering an ice shelf cavity and flowing toward the back of the cavity, to the grounding line where the ice shelf first goes afloat on the ocean. Here at the grounding line, the ice shelf is at its greatest thickness. Because the freezing point of seawater decreases as ocean depth increases, the invading ocean waters have an ever increasing thermal head with respect to the ice as the depth of the ice increases. The thermal head determines the amount of melting at the grounding line. An end result of melting is a cooled and freshened ocean water mass at the grounding line. An empirical consequence of the equation of state for seawater is that this water mass will always be less dense than the source waters that originally fed into the ice-shelf cavity. These light waters subsequently flow upward along the ice-shelf base as a kind of upside-down gravity current, a flow feature termed a plume. As the waters rise, the depth-dependent freezing point also rises, and at some point the rising waters can actually become supercooled with respect to the local freezing point. In this instance some of the meltwaters refreeze to the base of the ice shelf, forming so-called marine ice, in contrast to the meteoric ice (also called snow/ice) that feeds the ice shelf from the inland ice sheet. It is the manner in which ocean waters can melt the deep ice and refreeze ice at shallow depths that has given rise to the term 'ice pump.' In the case of warm waters in the cavity beneath the ice shelf, the term ice pump is a misnomer, as there may be no refreezing of ice whatsoever, just melting. These under-ice circulation processes are clearly important to the stability of ice shelves or ice tongues, but it is difficult to yet predict their impact on Antarctica and Greenland in the coming decades. Future changes in ocean circulation and ocean temperatures will produce changes in basal melting, but the magnitude of these changes is currently not modeled or predicted.

4.2. Ice-Shelf Processes

4.2.1. Ice-Shelf basal melting

A nonlinear response of ice-shelf melting to increasing ocean temperatures is a central tenet in the scenario for abrupt climate change arising from ocean–ice-shelf interaction. The nonlinear response is a theoretical and computational result; observations are yet inadequate to verify this conclusion. Nonetheless, the basis of this result is that the melt rate at the base of an ice shelf is the product of the thermal head and the velocity of the ocean waters at the base. The greater the thermal head or the velocity, then the greater the melt rate. A key insight from the theoretical and modeling research is that as the ocean water temperature is increased, the buoyancy of the plume beneath the ice shelf is increased because greater melting is initiated by the warmer waters. A more buoyant plume rises faster, causes greater melting, and becomes more buoyant. This positive feedback is a key nonlinear response mechanism of an ice-shelf base to warming ocean waters.

The susceptibility of ice shelves to high melt rates and to collapse is a function of the presence of warm waters entering the ice-shelf cavities. But the appearance of such warm waters does not actually imply that the global ocean needs to warm. It is true that observational evidence (*Levitus et al., 2000*) does indicate that the ocean has warmed over the past decades, and that the warming has been modest (approximately 0.5 C globally). While this is one mechanism for creating warmer waters to enter a cavity beneath the ice shelf, a more efficient mechanism for melting is not to warm the global ocean waters but to redirect existing warm water from the global ocean toward ice shelf cavities; however, ocean temeprture measuremens close to the ice margin are lacking. Ocean circulation is driven by density contrasts of water masses and by surface wind forcing. Subtle changes in surface wind forcing (*Toggweiler and Samuels, 1995*) may have important consequence for the redistribution of warm water currents in polar oceans. A change in wind patterns (i.e., a relatively fast process) could produce large and fast changes in the temperatures of ocean waters appearing at the doorstep of the ice shelves.

4.2.2. Ice-Shelf thinning

Changes in the geometry of ice shelves or floating ice tongues can cause a dynamic response that penetrates hundreds of kilometers inland. This can be triggered through high rates of basal melt or through a calving episode, providing the perturbation impacts the ice-sheet grounding zone (*Thomas et al., 2005; Payne et al., 2004; Pattyn et al., 2006*). Grounding-zone thinning can induce rapid and widespread inland ice response if fast-flowing ice streams are present. This has been observed in the Pine Island and Thwaites Glacier systems (*Rignot et al., 2002; Shepherd et al., 2002*). Glacier discharge also increased on the Antarctic Peninsula following the 2002 collapse of the Larsen B ice shelf *(Rott et al., 2002; DeAngelis and Skvarca, 2003; Rignot et al., 2004a).*

Whether or not a glacier will stabilize following a perturbation depends to a large degree on whether it is grounded or floating. Flow rates of more than 300 tidewater glaciers on the Antarctic Peninsula increased by an average of 12% from 1992 to 2005 (*Pritchard and Vaughan, 2007*). Pritchard and Vaughan interpret this as a dynamic response to thinning at the ice terminus. Glaciers in contact with the ocean are likely to see an ongoing response to ice-shelf removal.

A thinning ice shelf results in glacier ungrounding, which is the main cause of the glacier acceleration because it has a large effect on the force balance near the ice front (*Thomas, 2004*). This effect also explains the retreat of Pine Island Glacier (*Thomas et al., 2005*) and the recent acceleration and retreat of outlet glaciers in east Greenland.

4.2.3. Iceberg calving

Calving is the separation of ice blocks from a glacier at a marginal cliff. This happens mostly at ice margins in large water bodies (lakes or the ocean), and the calved blocks become icebergs. The mechanism responsible for iceberg production is the initiation and propagation of fractures through the ice thickness. Calving can originate in fractures far back from the ice front (*Fricker et al., 2005*). This process is incompletely understood, partly because of the difficulty and danger of making observations.

While it is not clear that calving is a deterministic process (because the outcome cannot be predicted exactly from knowledge of initial condition), some internal (ice dynamical) and external influences on calving rates have been qualitatively elucidated. Internal dynamic controls are related to the stiffness and thickness of ice, longitudinal strain rates, and the propensity for fractures to form and propagate. High rates of ice flow promote longitudinal stretching and tensile failure. External influences on calving rates include ocean bathymetry and sea level, water temperature, tidal amplitude, air temperature, sea ice, and storm swell.

These variables may have a role in a general "calving law" that can be used to predict calving rates. Such a law does not yet exist but is important because calving has the capacity to destabilize an ice front. Acceleration of Jakobshavn Isbræ beginning in 2000 has been interpreted as a response to increased calving at the ice front and collapse of the floating tongue following very rapid thinning (*Thomas, 2004; Joughin et al., 2004*).

The external variables that trigger such an event are not well understood. Increased surface melting due to climatic warming can destabilize the ice front and lead to rapid disintegration of an entire ice shelf (*Scambos et al., 2004*). Penetration of surface meltwater into crevasses deepens the fissures and creates areas of weakness that can fail under longitudinal extension.

A number of small ice shelves on the Antarctic Peninsula collapsed in the last three decades of the 20th century. Ice-shelf area declined by more than 13,500 km^2 in this period, punctuated by the collapse of the Larsen A and Larsen B ice shelves in 1995 and 2002 (*Scambos et al., 2004*). This was possibly related to atmospheric warming in the region, estimated to be about 3°C over the second half of the 20th century. *Vaughan and Doake (1996)* suggest that ice-shelf viability is compromised if mean annual air temperature exceeds

−5°C. Above this temperature, meltwater production weakens surface crevasses and rifts and may allow them to propagate through the ice thickness. It is also likely that thinning of an ice shelf, caused by increased basal melting, preconditions it for breakup. Consequently, warming of ocean waters may also be important. The Weddell Sea warmed in the last part of the 20th century, and the role that this ocean warming played in the ice shelf collapses on the Antarctic Peninsula is unknown. Warmer ocean temperatures cause an increase in basal melt rates and ice-shelf thinning. If this triggers enhanced extensional flow, it might cause increased crevassing, fracture propagation, and calving.

Similarly, the impacts of sea-ice and iceberg-clogged fjords are not well understood. These could damp tidal forcing and flexure of floating ice tongues, suppressing calving. *Reeh et al. (1999)* discuss the transition from tidewater outlets with high calving rates in southern Greenland to extended, floating tongues of ice in north Greenland, with limited calving flux and basal melting representing the dominant ablation mechanism. Permanent sea ice in northeast Greenland may be one of the factors enabling the survival of floating ice tongues in the north (*Higgins, 1991*). This is difficult to separate from the effects of colder air and ocean temperatures.

4.3. Ice Stream and Glacier Processes

Ice masses that are warm based (at the melting point at the bed) can move via basal sliding or through deformation of subglacial sediments. Sliding at the bed involves decoupling of the ice and the underlying till or bedrock, generally as a result of high basal water pressures (*Bindschadler, 1983*). Glacier movement via sediment deformation involves viscous flow or plastic failure of a thin layer of sediments underlying the ice (*Kamb, 1991; Tulaczyk et al., 2001*). Pervasive sediment deformation requires large supplies of basal meltwater to dilate and weaken sediments. Sliding and sediment deformation are therefore subject to similar controls; both require warm-based conditions and high basal water pressures, and both processes are promoted by the low basal friction associated with subglacial sediments. In the absence of direct measurements of the prevailing flow mechanism at the bed, basal sliding and subglacial sediment deformation can be broadly combined and referred to as *basal flow*.

4.3.1. Basal flow

Basal flow can transport ice at velocities exceeding rates of internal deformation: 100s to more than 10,000 meters per year, and glacier surges, tidewater glacier flow, and ice stream motion are governed by basal flow dynamics (*Clarke, 1987*). Ice streams are responsible for drainage of as much as 90% of West Antarctica (*Paterson, 1994*), leading to a low surface profile and a mobile, active ice mass that is poorly represented by ice- sheet models that cannot portray these features.

Glaciers and ice sheets that are susceptible to basal flow can move quickly and erratically, making them intrinsically less predictable than those governed by internal deformation. They are more sensitive to climate change because of their high rates of ice turnover, which gives them a shorter response time to climate (or ice-marginal) perturbations. In addition, they may be directly responsive to increased amounts of surface meltwater production associated with climate warming.

This latter process is crucial to predicting dynamic feedbacks to the expanding ablation area, longer melt season, and higher rates of surface meltwater production that are predicted for most ice masses.

Although basal meltwater has traditionally been thought to be the primary source of subglacial water, models have shown that supraglacial streams with discharges of over 0.15 m^3 s^{-1} can penetrate down through 300 m of ice to reach bedrock, via self- propagation of water-filled crevasses (*Arnold and Sharp, 2002*). There are several possible subglacial hydrological configurations: ice-walled conduits, bedrock conduits, water film, linked cavities, soft-sediment channels, porous sediment sheets, and ordinary aquifers (*Mair et al., 2001; Flowers and Clarke, 2002*).

Modern interest in water flow through glaciers can be dated from a pair of theoretical papers published in 1972. In one of these, *Shreve (1972)* discussed the influence of ice pressure on the direction of water flow through and under glaciers, and in the other *Röthlisberger (1972)* presented a theoretical model for calculating water pressures in subglacial conduits. Through a combination of these theoretical considerations and field observations, it is concluded that the englacial drainage system probably consists of an arborescent network of passages. The millimeter-sized finger-tip tributaries of this network join downward into ever larger conduits. Locally, moulins provide large direct connections between the glacier surface and the bed. Beneath a valley glacier the subglacial drainage is likely to be in a tortuous system of linked cavities transected by a few relatively large and comparatively straight conduits. The average flow direction in the combined system is controlled by a combination of ice-overburden pressure and bed topography, and in general is not normal to contours of equal elevation on the bed. Although theoretical studies usually assume that subglacial conduits are semicircular in cross section, there are reasons for believing that this ideal is rarely realized in nature. Much of the progress in subglacial hydrology has been theoretical, as experimental techniques for studying the englacial hydraulic system are few, and as yet not fully exploited, and observational evidence is difficult to obtain.

How directly and permanently do these effects influence ice dynamics? It is not clear at this time. This process is well known in valley glaciers, where surface meltwater that reaches the bed in the summer melt season induces seasonal or episodic speedups (*Iken and Bindschadler, 1986*). Speedups have also been observed in response to large rainfall events (e.g., *O'Neel et al., 2005*).

4.3.2. Flow acceleration and meltwater

Summer acceleration has also been observed in the ablation area of polar icefields (*Copland et al., 2003*), where meltwater ponds drain through moulins and reach the bed through up to 200 m of cold ice (*Boon and Sharp, 2003*). The influx of surface meltwater triggers a fourfold speedup in flow in the lower ablation area each year. There is a clear link between the surface hydrology, seasonal development of englacial drainage connections to the bed, and basal flow, at least at this site.

It is uncertain whether surface meltwater can reach the bed through thick columns of cold ice. Cold ice is impermeable on the intergranular scale (*Paterson, 1994*). However, water flowing into moulins may carry enough kinetic and potential energy to penetrate to the bed and spread out over an area large enough to affect the basal velocity. *Zwally et al. (2002a)* record summertime speedup events near the western margin of the Greenland Ice Sheet, associated with the drainage of large supraglacial lakes in a region where the ice sheet is several hundred meters thick. It is unknown whether the meltwater penetrated all the way to the bed, but this is interpreted to be the cause of the summer speedups and is consistent with observations on valley glaciers.

These observations are unequivocal but the speedups are modest (10%) and localized. Alternative interpretations of the *Zwally et al. (2002a)* data have also been proposed. The region may be influenced by seasonal acceleration at the downstream ice margin or through accelerated summer flow in nearby Jakobshavn Isbræ, rather than local supraglacial lake drainage. Recent summer speedups in Jakobshavn Isbræ are believed to be a response to marine conditions (summer calving, seasonal sea ice, and basal melting on the floating ice tongue).

More studies like that of *Zwally et al. (2002a)* are needed to determine the extent to which supraglacial water actually reaches the bed and influences basal motion. At this time it is still unclear how influential surface meltwater is on polar icefield dynamics, but it may prove to be an extremely important feedback in icefield response to climate change, as it provides a direct link between surface climate and ice dynamics. A modeling study by *Parizek and Alley (2004)* that assumes surface-meltwater-induced speedups similar to those observed by *Zwally et al. (2002a)* found this effect to increase the sensitivity of the Greenland Ice Sheet to specified warmings by 10-15%. This is speculative, as the actual physics of meltwater penetration to the bed and its influence on basal flow are not explicitly modeled or fully understood.

4.4. Modeling

4.4.1. Ice-Ocean modeling

There has been substantial progress in the numerical modeling of the ice-shelf–ocean interaction over the last decade. A variety of ocean models have now been adapted so that they can simulate the interaction of the ocean with an overlying ice shelf (see *ISOMIP Group, 2007,* for summary of modeling activities). The present state of the art in these simulations is termed as static-geometry simulations, as the actual shape of the ice-shelf cavity does not change during these simulations. Such static geometry simulations are a reasonable first step in advancing understanding of such a complex system. Steps are now being taken to co-evolve the ocean and ice shelf (*Grosfeld and Sandhager, 2004; Walker and Holland, 2007*) in what can be termed as dynamic-geometry simulations. It is only the latter type of simulations that can ultimately provide any predictive capability on abrupt change in global sea level as resulting from changing ocean temperatures in cavities beneath the ice shelf. The scientific community presently does not possess an adequate observational or theoretical understating of this problem. Progress is being made, but given the relatively few researchers and resources tackling the problem, the rate of progress is slow. It is conceivable that changes are presently occurring or will occur in the near term (i.e., the present century) in the ice-shelf–ocean interaction that we are not able to observe or model.

4.4.2. Ice Modeling

The extent of impact of ice-marginal perturbations depends on the nature of ice flow in the inland ice. Ice dynamics in the transition zone between inland and floating ice – the grounding zone – are complex, and few whole-ice-sheet models have rigorously addressed the mechanics of ice flow in this zone. *MacAyeal (1989)* introduced a model of ice shelf-ice stream flow that provides a reasonable representation of this transition zone, although the model has only been applied on regional scales. This model, which has had good success in simulating Antarctic ice-stream dynamics, assumes that ice flux is dominated by flow at the bed and longitudinal stretching, with negligible vertical shear deformation in the ice.

The West Antarctic ice sheetcontains enough ice to raise sea levels by about 6 m. It also rests on bedrock below sea level, which leaves it vulnerable to irreversible shrinkage if the rate of ice flow from the grounded ice sheet into the surrounding ice shelves were to increase, causing partial flotation and hence retreat of the grounded ice sheet. A hotly debated hypothesis in glaciology asserts that a marine ice sheet is susceptible to such irreversible shrinkage if its grounding line rests on an upward-sloping bed, because a small retreat in grounding line position should lead to increased discharge, which leads to further retreat and so on. The key to this hypothetical positive feedback is that discharge through the grounding line - where grounded ice lifts off the bed to become an ice shelf - must increase with water depth there. The assertion that this is the case has been around for over 30 years but has not previously been proven. *Schoof (2007)* has been able to use the boundary layer theory to show that the positive feedback does indeed exist.

Recent efforts have explored higher order simulations of ice sheet dynamics, including a full-stress solution that allows modeling of mixed flow regimes *(Pattyn, 2002; Payne et al., 2004)*. The study by *Payne et al. (2004)* examines the inland propagation of grounding-line perturbations in the Pine Island Glacier. The dynamic response has two different time scales:

an instantaneous mechanical response through longitudinal stress coupling, felt up to 100 km inland, followed by an advective-diffusive thinning wave propagating upstream on a decadal time scale, with a new equilibrium reached after about 150 years. These modeling results are consistent with observations of recent ice thinning in this region.

Full-stress solutions have yet to be deployed on continental scales (or applied to the sea-level question), but this is becoming computationally tractable. Improvements may also be possible through nested modeling, with high-resolution grids and high-order physics in regions of interest. Moving-grid techniques for explicit modeling of the ice sheet - ice shelf grounding zone are also needed *(Vieli and Payne, 2005)*. The current suite of models does not handle this well. Most regional-scale models that focus on ice-shelf dynamics use fixed grounding lines, while continental-scale ice sheet models distinguish between grounded and floating ice, but the grounding zone falls into the horizontal grid cell where this transition occurs. At model resolutions of 1 0s of kilometers, this does not capture the details of grounding line migration. *Vieli and Payne (2005)* show that this has a large effect on modeled ground-line stability to external forcing.

Observations from the last decade have radically altered the thinking on how rapidly an ice sheet can respond to perturbations at the marine margin. Severalfold increases in discharge followed the collapse of ice shelves on the Antarctic Peninsula, with accelerations of up to 800% following collapse of the Larsen B ice shelf *(Scambos et al., 2004; Rignot et al., 2004a)*. The effects on inland ice flow are rapid, large, and propagate immediately over very large distances. This is something models did not predict *a priori*, and the modeling community is now scrambling to catch up with the observations. No whole-ice-sheet model is presently capable of capturing the glacier speedups in Antarctica or Greenland that have been observed over the last decade. This means that we have no real idea of how quickly or widely the ice sheets will react if they are pushed out of equilibrium.

4.5. Sea-Level Feedback

Perhaps the primary factor that raises concerns about the potential of abrupt changes in sea level is that large areas of modern ice sheets are currently grounded below sea level (i.e., the base of the ice sheet occurs below sea level) (Figure 2.9). Where it exists, it is this condition that lends itself to many of the processes described in previous sections that can lead to rapid ice-sheet changes, especially with regard to atmosphere-ocean-ice interactions that may affect ice shelves and calving fronts of tidewater glaciers.

An equally important aspect of these marine-based ice sheets which has long been of interest is that the beds of ice sheets grounded below sea level tend to deepen inland, either due to overdeepening from glacial erosion or isostatic adjustment. The grounding line is the critical juncture that separates ice that is thick enough to remain grounded from either an ice shelf or a calving front. In the absence of stabilizing factors, this configuration indicated that marine ice sheets are inherently unstable, whereby small changes in climate could trigger irreversible retreat of the grounding line *(Hughes, 1973; Weertman, 1974; Thomas and Bentley, 1978)*. For a tidewater glacier, rapid retreat occurs because calving rates increase with water depth *(Brown et al., 1983)*. Where the grounding line is fronted by an unconfined ice shelf, rapid retreat occurs because the extensional thinning rate of an ice shelf increases with thickness, such as would accompany grounding-line retreat (*Weertman, 1974*).

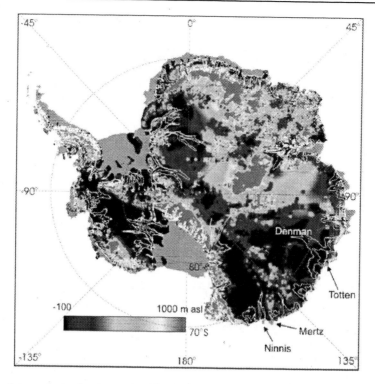

Figure 2.9. Bedrock topography for Antarctica highlighting areas below sea level (in black), fringing ice shelves (in dark gray), and areas above sea level (in rainbow colors). Areas of enhanced flow are identified by contours (in white) of estimated steady-state velocities, known as balance velocities. From *Bamber et al. (2007)*.

The amount of retreat clearly depends on how far inland glaciers remain below sea level. Of greatest concern is West Antarctica, where all the large ice streams are grounded well below sea level, with deeper trenches lying well inland of their grounding lines (Figure 2.9). A similar situation applies to the entire Wilkes Land sector of East Antarctica. In Greenland, few outlet glaciers remain below sea level very far inland, indicating that glacier retreat by this process will eventually slow down or halt. A notable exception may be Greenland's fastest glacier, Jakobshavn Isbræ, which appears to tap into the central core of Greenland that is below sea level (Figure 2.10). Other regions in the northern part of the ice sheet are the Humboldt glacier, the Petermann glacier and the Nioghalvfjerdsfjorden glacier (Figure 2.10).

Several factors determine the position of the grounding line, and thus the stability of marine ice sheets. On time scales that may lead to rapid changes, the two most important of these are the backstress provided by ice-shelf buttressing and sea level (*Thomas and Bentley, 1978*). Given that a grounding line represents the point at which ice becomes buoyant, then a rise in sea level will cause grounding line retreat (and vice versa). Following some initial perturbation, this situation thus leads to the potential for a positive feedback to develop between ice retreat and sea level rise. Recent studies from West Antarctica, however, suggest that for some geological situations, the sensitivity of grounding line retreat to sea level rise may be less important than previously considered. *Anandakrishnan et al. (2007)* documented formation of a wedge of subglacial sediment at the grounding line of the Whillans Ice Stream, resulting in ice to be substantially thicker there than floating ice in hydrostatic equilibrium. *Alley et al. (2007)* showed with numerical ice-flow models that a grounding line sitting on a sedimentary wedge is immune to sea-level changes of up to 10 m. Because the wedges develop by accumulation of debris delivered to the grounding line from a subglacial deforming sediment layer, this stabilizing mechanism only applies to those places where such a process is operating. Today, this likely applies to the Siple Coast ice streams and perhaps those flowing into the Ronne Ice Shelf. It is not clear, however, that it applies to ice streams flowing into other Antarctic ice shelves or to the outlet glaciers draining Greenland.

Of these two factors, the buttressing force of the ice shelf is likely more important than sea level in affecting grounding-line dynamics. If this force is greater than that just caused by seawater pressure, then the grounding line is vulnerable to ice-shelf changes. For thick grounding lines, such as characterize most outlet glaciers and ice streams draining Greenland and Antarctica today, this vulnerability far exceeds that associated with feasible sea-level changes expected by the end of this century (0.5-1.0 m) (*Rahmstorf, 2007*), particularly in the context of the likelihood of substantial climate change that would affect the ice shelves in the same timeframe. In considering the wedge-stability factor as well, we thus conclude that, in the absence of rapid loss of ice shelves and attendant sea level rise, sea level forcing and feedback are unlikely to be significant determinants in causing rapid ice-sheet changes in the coming century.

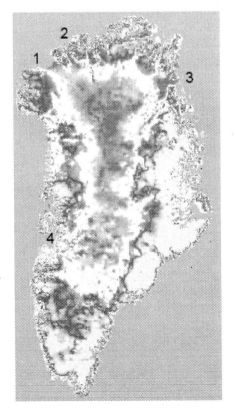

Figure 2.10. Bedrock topography for Greenland; areas below sea level are shown in blue. Note the three channels in the north (1: Humboldt Glacier; 2: Petermann Glacier; 3: 79- North Glacier or Nioghalvfjerdsfjorden Glacier) and at the west coast (4: Jakobshavn Isbrae) connecting the region below sea level with the ocean (Russell Huff and Konrad Steffen, CIRES, University of Colorado at Boulder.)

3. HYDROLOGICAL VARIABILITY AND CHANGE

Key Findings

- Protracted droughts, and their impacts on agricultural production and water supplies, are among the greatest natural hazards facing the United States and the globe today and in the foreseeable future.

- Floods predominantly reflect both antecedent conditions and meteorological events and are often more localized relative to drought in both time and space. On subcontinental-to-continental scales, droughts occur more frequently than floods and can persist for decades and even centuries.

- On interannual to decadal time scales, droughts can develop faster than the time scale needed for human societies to adapt to the change. Thus, a severe drought lasting several years can be regarded as an abrupt change, although it may not reflect a permanent change of state of the climate system.

- Droughts and episodes of regional-scale flooding can both be linked to the large-scale atmospheric circulation patterns over North America, and often occur simultaneously in different parts of the country, compounding their impact on human activities.

- Empirical studies and climate model experiments conclusively show that droughts over North America have been significantly influenced by the state of tropical sea surface temperatures (SSTs). Of particular relevance to North America, cool La Niña-like SSTs in the eastern equatorial Pacific frequently cause development of droughts over the southwestern United States and northern Mexico. Warm subtropical North Atlantic SSTs play a secondary role in forcing drought in southwestern North America.

- Historic droughts over North America have been severe, the "Dust Bowl" drought of the 1930s being the canonical example, but those droughts were not nearly as prolonged as a series of "megadroughts" reconstructed from tree rings since Medieval times (ca. 1,000 years ago) up to about A.D. 1600. Modeling experiments indicate that these megadroughts were likely partly forced by cool SSTs in the eastern equatorial Pacific as well. However, their exceptional duration has not been adequately explained nor has any involvement in forcing from SST changes in other oceans.

- These megadroughts are significant because they occurred in a climate system that was not being perturbed in a major way by human activity (i.e., the ongoing anthropogenic changes in greenhouse gas concentrations, atmospheric dust loadings, and land-cover changes).

- Even larger and more persistent changes in hydroclimatic variability worldwide are indicated throughout the Holocene (the past 11,500 years) by a diverse set of paleoclimatic indicators including some with annual-to-decadal resolution (e.g., speleothems, varved-lake records, high-resolution lake-sediment records). The global-scale controls associated with those changes were quite different from those of the past millennium and today, but they show the additional range of natural variability and abrupt hydroclimatic change that can be expressed by the climate system, including widespread and protracted (multi-century) droughts.

- There is no clear evidence to date of human-induced global climate change on North American precipitation amounts. However, since the IPCC AR4 report, further analysis of climate model scenarios of future hydroclimatic change over North America and the global subtropics indicate that subtropical aridity is likely to intensify and persist due to future greenhouse warming. This projected drying extends poleward into the United States Southwest, potentially increasing the likelihood of severe and persistent drought there in the future. If the model results are correct then this drying may have already begun, but currently cannot be definitively identified amidst the considerable natural variability of hydroclimate in Southwestern North America.

Recommendations

- Research is needed to improve existing capabilities to forecast short- and longterm drought conditions and to make this information more useful and timely for decision making. In the future, drought forecasts should be based on an objective multimodel ensemble prediction system to enhance their reliability and the types of information should be expanded to include soil moisture, runoff, and hydrological variables (See also the *Western Governors' Association (2004)* National Integrated Drought Information System Report).

- The trend toward increasing subtropical aridity indicated by climate model projections needs to be investigated further to determine the degree to which it is likely to happen. If the model projections are correct, strategies for response to this pending aridity, on both regional and global scales, are urgently needed.

- Improved understanding of the dynamical causes of long-term changes in oceanic conditions, the atmospheric responses to these ocean conditions, and the role of soil moisture feedbacks are needed to advance drought prediction capabilities. Ensemble drought prediction is needed to maximize forecast skill and downscaling is needed to bring coarse-resolution drought forecasts from General Circulation Models down to the resolution of a watershed. (See also the *National Integrated Drought Information System Implementation Team, 2007.)*

- High-resolution paleoclimatic reconstructions of past drought have been fundamental to the evaluation of causes over North America in historic times and over the past millennium. This research should be expanded geographically to encompass as much of the global land masses as possible for the development and testing of predictive models.

- The record of past drought from tree rings and other proxies has revealed a succession of megadroughts prior to A.D. 1600 that easily eclipsed the duration of any droughts known to have occurred over North America since that time. Understanding the causes of these extraordinary megadroughts is vitally important.

- An understanding of the seasonality of drought and the relationships between winter and growing season droughts during periods of megadroughts documented in paleoclimatic records is needed. In particular, knowledge about the North American monsoon and how its variability is linked to SSTs and winter precipitation variability over decadal and longer time scales in the Southwestern United States and northern Mexico is critical.

- On longer time scales, significant land-cover changes have occurred in response to persistent droughts, and the role of land-cover changes in amplifying or damping drought conditions should be evaluated.

- Improved understanding of the links among gradual changes in climate (e.g., Meridional Overturning Circulation, or MOC), the role of critical environmental thresholds, and abrupt hydrologic changes is needed to enhance society's ability to plan and manage risks.

- The relationship between climate changes and abrupt changes in water quality and biogeochemical responses is not well understood and needs to be a priority area of study for modern process and paleoclimate research.

- The integration of high-resolution paleoclimate records with climate model experiments requires active collaboration between paleoclimatologists and modelers. This collaboration should be encouraged in future research on drought and climatic change in general.

- In order to reduce uncertainties in the response of floods to abrupt climate change, improvements in large-scale hydrological modeling, enhanced data sets for documenting past hydrological changes, and better understanding of the physical processes that generate flooding are all required.

1. Introduction—Statement of the Problem

A reliable and adequate supply of clean fresh water is essential to the survival of each human being on Earth and the maintenance of terrestrial biotic systems worldwide. However, rapidly growing human populations worldwide are increasing the stresses on currently available water supplies even before we factor in anticipated effects of a changing climate on the availiability of a clean and reliable fresh water supply. Changes in the frequency, intensity, and duration of droughts would have a significant impact on water supplies both for human societies and for terrestrial and inshore marine or estruarine ecosystems. Droughts are defined by the international meteorological community: the "prolonged absence or marked deficiency of precipitation", a "deficiency of precipitation that results in water shortage for some activity or for some group", or a "period of abnormally dry weather sufficiently prolonged for the lack of precipitation to cause a serious hydrological imbalance" (*Heim, 2002*; see also *Peterson et al., 2008* (CCSP SAP 3.3, Box 1.3)). Flooding is another important class of hydrologic variability that tends to affect smaller geographic regions and to last for shorter periods of time compared to drought. Consequently, floods generally have smaller impacts on human activities compared to droughts in North America. See the section on floods in the latter part of this chapter for more details.

Much of the research on climatic change, and most of the public's understanding of that work, has concerned temperature and the term "global warming." Global warming describes ongoing warming in this century by a few degrees Celsius, in some areas a bit more and in some a bit less. In contrast, changes in water flux between the surface of the Earth and the atmosphere are not expected to be spatially uniform but to vary much like the current daily mean values of precipitation and evaporation (*IPCC, 2007*). Although projected spatial patterns of hydroclimatic change are complex, many already wet areas are likely to get wetter and already dry areas are likely to get drier, while some intermediate regions on the poleward flanks of the current subtropical dry zones are likely to become increasingly arid. These anticipated changes will increase problems at both extremes of the water cycle, stressing water supplies in many arid and semi-arid regions while worsening flood hazards and erosion in many wet areas. Changes in precipitation intensity – the proportion of the total precipitation falling in events of different magnitude – have the potential to further challenge the management of water in the future. Moreover, the instrumental, historical, and prehistorical record of hydrological variations indicates that transitions between extremes can occur rapidly relative to the time span under consideration. Within time spans of decades, for example, transitions between wet conditions and dry conditions may occur within a year and can persist for several years.

Hydroclimatic changes are likely to affect all regions in the United States. Semi-arid regions of the Southwest are projected to dry further, and model results suggest that the transition may already be underway (*Hoerling and Kumar, 2003; Seager et al., 2007d*). Intensity of precipitation is also expected to increase across most of the country, continuing its recent trend (*Kunkel et al., 2008,* CCSP SAP 3.3, Sec. 2.2.2.2). The drying in the Southwest is a matter of great concern because water resources in this region are already stretched, new development of resources will be extremely difficult, and the population (and thus demand for water) continues to grow rapidly (see Figure 3.1). This situation raises the politically charged issue of whether the allocation of around 90% of the region's water to agriculture is sustainable and consistent with the course of regional development. Mexico is also expected to dry in the near future, turning this feature of hydroclimatic change into an international and cross-border issue with potential impacts on migration and social stability. The U.S. Great Plains, where deep aquifers are being rapidly depleted, could also experience changes in water supply that affect agricultural practices, grain exports, and biofuel production. Other normally well-watered regions of the United States may also face water shortages caused by short-term droughts when demand outstrips supply and access to new water supplies is severely limited (e.g., Atlanta, GA). Other regions of the United States, while perhaps not having to face a climatic change-induced water shortage, may also have to make changes to infrastructure to deal with the erosion and flooding implications of increases in precipitation intensity.

Increases in the frequency of droughts in response to climate change can in turn produce further climate changes. For example, increased drought frequency may reduce forest growth, decreasing the sequestration of carbon in standing biomass, and increasing its release from the soil (*King et al., 2007* (CCSP SAP 2.2)). Similarly, increasing temperatures and drought will likely promote increased disturbance by fire and insect pathogens, with a consequent impact on ecosystems and their carbon balances (*Backlund et al., 2008* (CCSP SAP 4.3))

In addition, the United States could be affected by hydroclimatic changes in other regions of the world if global climate change becomes a global security issue. Security, conflict, and migration are most directly related to economic, political, social, and demographic factors. However environmental factors, including climate variability and climate change, can also play a role, even if secondary (*Lobell et al., 2008; Nordas and Gleditsch, 2007*). Two recent examples of a quantitative approach to determine the links between conflict and climate are *Raleigh and Urdal (2007)* and *Hendrix and and Glaser (2007)*. Raleigh and Urdal, basing their arguments on statistical relations between late 20th century conflict data and environmental data, find that the influence of water scarcity is at best weak. Hendrix and Glaser focused on sub-Saharan Africa and found that climate variability (e.g. a transition into a dry period) could foster conflict when other conditions (political, economic, demographic, etc.) favored conflict anyway. Hendrix and Glaser also examined a climate projection for sub-Saharan Africa from a single model and found that this led to no significant increase in conflict risk because the year-to-year climate variability did not change. Such quantitative methods need to be applied to other regions where changes in the mean state and variability of climate are occurring now and also to regions where climate change is robustly projected by models. Across different regions of the world, projected increases in flooding risk, potential crop damage and declines in water quality, combined with rising sea level, have the potential to force migration and cause social, economic, and political instability. However, currently there are no comprehensive assessments of the security risk posed by climate change that take account of all the available climate-change projection information and also take into account the multiple causes of conflict and migration. Consequently, no conclusions can yet be drawn on the climate-change impact on global or national security.

The paleoclimatic record reveals dramatic changes in North American hydroclimate over the last millennium that had nothing to do with human-induced changes in greenhouse gases and global warming. In particular, tree ring reconstructions of the Palmer Drought Severity Index (PDSI, see *Kunkel et al., 2008,* CCSP SAP 3.3, Box 2.1) show vast areas of the Southwest and the Great Plains were severely affected by a succession of megadroughts between about A.D. 800 and 1600 that lasted decades at a time and contributed to the development of a more arid climate during the Medieval Period (A.D. 800 to 1300) than in the last century. These megadroughts have been linked to La Niñalike changes in tropical Pacific SSTs, changes in solar irradiance, and explosive volcanic activity. They are dynamically distinct from projected future drying, which is associated with a quite spatially uniform surface warming, based on model projections. However, the paleoclimatic records differ enough from climate model results to suggest that the models may not respond correctly to radiative forcing. The climate system dynamics associated with these prehistoric megadroughts need to be better understood, modeled, and related to the processes involved in future climate change.

Over longer time spans, the paleoclimatic record indicates that even larger hydrological changes have taken place, in response to past changes in the controls of climate, that rival in magnitude those expected during the next several decades and centuries. For example, the mid-continent of North America experienced conditions that were widespread and persistently dry enough to activate sand dunes, lower lake levels, and change the vegetation from forest to grassland for several millennia during the mid-Holocene (roughly 8,000 to 4,000 years ago). These changes were driven primarily by variations in the Earth's orbit that altered the seasonal and latitudinal distribution of incoming solar radiation. Superimposed on these Holocene variations were variations on centennial and shorter time scales that also were recorded by aeolian activity, and by geochemical and paleolimnological indicators.

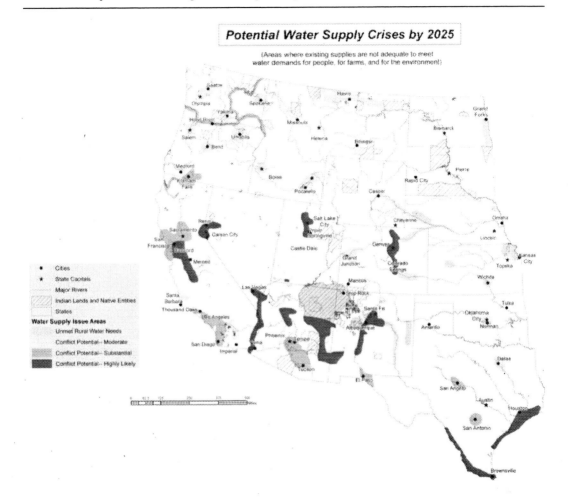

Figure 3.1. Interior Department analysis of regions in the West where water supply conflicts are likely occur by 2025 based on a combination of technical and other factors, including population trends and potential endangered species' needs for water. The red zones indicate areas where the conflicts are most likely to happen. See DOI Water 2025 Status Report (U.S. Department of Interior , *Bureau of Reclamation, 2005*; http://www.usbr.gov/water2025/report.html) for details. Note: There is an underlying assumption of a statistically stationary climate.

The serious hydrological changes and impacts known to have occurred in both historic and prehistoric times over North America reflect large-scale changes in the climate system that can develop in a matter of years and, in the case of the more severe past megadroughts, persist for decades. Such hydrological changes fit the definition of abrupt change because they occur faster than the time scales needed for human and natural systems to adapt, leading to substantial disruptions in those systems. In the Southwest, for example, the models project a permanent drying by the mid-21st century that reaches the level of aridity seen in historical droughts, and a quarter of the projections may reach this level of aridity much earlier. It is not unreasonable to think that, given the complexities involved, the strategies to deal with declining water resources in the region will take many years to develop and implement. If hardships are to be minimized, it is time to begin planning to deal with the potential hydroclimatic changes described here.

2. Causes and Impacts of Hydrological Variability over North America in the Historical Record

After the 1997-98 El Niño, the Western United States entered a drought that has persisted until the time of writing (July 2007). The driest years occurred during the extended La Niña of 1998-2002. Although winter 2004-05 was wet, dry conditions returned afterwards and even continued through the modest 2006-07 El Niño. In spring 2007 the two massive reservoirs on the Colorado River, Lakes Powell and Mead, were only half full. Droughts of this severity and longevity have occurred in the West before and Lake Mead (held back by Hoover Dam which was completed in 1935) was just as low for a few years during the severe 1950s drought in the Southwest. Studies of the instrumental record make clear that western North America is a region of strong meteorological and hydrological variability in which, amidst dramatic year-to-year variability, there are extended droughts and pluvials (wet periods) running from a few years to a decade. These dramatic swings of hydroclimatic variability have tremendous impacts on water resources, agriculture, urban water supply, and terrestrial and aquatic ecosystems. Drought and its severity can be numerically defined using indices that integrate temperature, precipitation, and other variables that affect evapotranspiration and soil moisture. See *Heim (2002)* for details.

2.1. What is our current understanding of the historical record?

Instrumental precipitation and temperature data over North America only become extensive toward the end of the 19th century. Records of sea surface temperatures (SSTs) are sufficient to reconstruct tropical and subtropical ocean conditions starting around A.D. 1856. The large spatial scales of SST variations (in contrast to those of precipitation) allow statistical methods to be used to "fill in" spatial and temporal gaps and provide near-global coverage from this time on (*Kaplan et al., 1998; Rayner et al., 2003*). A mix of station data and tree ring analyses has been used to identify six serious multiyear droughts in western North America during this historical period (*Fye et al., 2003; Herweijer et al., 2006*). Of these, the most famous is the "Dust Bowl" drought that included most of the 1930s decade. The other two in the 20th century are the severe drought in the Southwest from that late 1940s to the late 1950s and the drought that began in 1998 and is ongoing. Three droughts in the mid to late 19th century occurred (with approximate dates) from 1856 to 1865, from 1870 to 1876, and from 1890 to 1896.

In all of these droughts, dry conditions prevailed over most of western North America from northern Mexico to southern Canada and from the Pacific Coast to the Mississippi River and sometimes farther east, with wet conditions farther north and farther south. The pattern of the Dust Bowl drought seemed unique in that the driest conditions were in the central and northern Great Plains and that dry conditions extended into the Pacific Northwest, while anomalies in the Southwest were modest.

BOX 3.1. IMPACTS OF HYDROLOGIC CHANGE: AN EXAMPLE FROM THE COLORADO RIVER

An example of the potential impacts of a rapid change to more drought-prone conditions can be illustrated by the recent drought and its effect on the Colorado River system. The Colorado River basin, as well as much of the WesternUnited States, experienced extreme drought conditions from 1999 to 2004, with inflows into Lake Powell between 25% and 62% of average. In spring 2005, the basin area average reservoir storage was at about 50%, down from over 90% in 1999 (*Fulp, 2005*). Although this most recent drought has caused serious water resource problems, paleoclimatic records indicate droughts as, or more, severe occurred as recently as the mid-19th century (*Woodhouse et al., 2005*). Impacts of the most recent drought were exacerbated by greater demand due to a rapid increase in the populations of the seven Colorado River basin States of 25% over the past decade (*Griles, 2004*). Underlying drought and increases in demand is the fact that the Colorado River resources have been over-allocated since the 1922 Colorado River Compact, which divided water supplies between upper and lower basin States based on a period of flow that has not been matched or exceeded in at least 400 years (*Stockton and Jacoby, 1976; Woodhouse et al., 2006*).

During the relatively short (in a paleoclimatic context) but severe 1999-2004 drought, vulnerabilities of the Colorado River system to drought became evident. Direct impacts included a reduction in hydropower and losses in recreation opportunities and revenues. At Hoover Dam, hydroelectric generation was reduced by 20%, while reservoir levels were at just 71 feet above the minimum power pool at Glen Canyon Dam in 2005 (*Fulp, 2005*). Hydroelectric power generated from Glen Canyon Dam is the source of power for about 200 municipalities (*Ostler, 2005*). Low reservoir levels at Lakes Powell and Mead resulted in the closing of three boat ramps and $10 million in costs to keep others in operation, as well as an additional $5 million for relocation of ferry services (*Fulp, 2005*). Blue ribbon trout fishing and whitewater rafting industries in the upper Colorado River basin (Upper Basin) also suffered due to this drought. In the agricultural sector, depletion of storage in reservoirs designed to buffer impacts of short-term drought in the Upper Basin resulted in total curtailment of 600,000 to 900,000 acre feet a year during the drought (*Ostler, 2005*). As a result of this drought, in combination with current demand, reservoir levels in Lake Mead, under average runoff and normal reservoir operations, are modeled to rise to only 1,120 feet over the next two decades (*Maguire, 2005*). Since the reservoir spills at 1221.4 feet (*Fulp, 2005*), this means the reservoir will not completely fill during this time period.

The Colorado River water system was impacted by the 5-year drought, but water supplies were adequate to meet most needs, with some conservation measures enacted (*Fulp, 2005*). How much longer the system could have handled drought conditions is uncertain, and at some point, a longer drought is certain to have much greater impacts. Under the Colorado River Compact and subsequent legal agreements, the Upper Basin provides 8.25 million acre feet to the Lower Basin each year (although there are some unresolved issues concerning the exact amount). If that amount is not available in storage, a call is placed on the river, and Upper Basin junior water rights holders must forgo their water to fulfill downstream and senior water rights. In the Upper Basin, the junior water rights are held by major water providers and municipalities in the Front Range, including Denver Water, the largest urban water provider in Colorado. Currently, guidelines that deal with the management of the Colorado River system under drought condition are being developed, because supplies are no longer ample to meet all demands during multi-year droughts (*USBR, 2007*). However, uncertainties related to future climate projections make planning difficult.

**TYPICAL JANUARY-MARCH WEATHER ANOMALIES
AND ATMOSPHERIC CIRCULATION
DURING MODERATE TO STRONG
EL NIÑO & LA NIÑA**

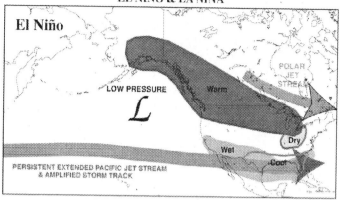

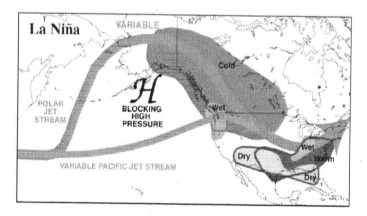

Climate Prediction Center/NCEP/NWS

Figure 3.2. Schematic maps showing the influence of El Niño/Southern Oscillation (ENSO) variability on regional climate over North America. Warm episodes (El Niños) result in cooler-wetter conditions from the Southwest to the Southeast during the winter season. Warm-dry conditions prevail over the same region during cold episodes (La Niñas), also during winter. (From http://www.cpc.noaa.gov/products/analysis_monitoring/ensocycle/nawinter.shtml)

Early efforts used observations to link these droughts to mid-latitude ocean variability. Since the realization of the powerful impacts of El Niño on global climate, studies have increasingly linked persistent, multiyear North American droughts with tropical Pacific SSTs and persistent La Niña events (*Cole and Cook, 1998; Cole et al., 2002; Fye et al., 2004*). This can be appreciated through the schematic maps shown in Figure 3.2 that show the teleconnection patterns of temperature and precipitation over North America commonly associated with the warm and cold phases of the ENSO cycle over the tropical Pacific. Warm ENSO episodes (El Niños) result in cool-wet conditions from the Southwestern over to the Southeastern United States during the winter season. In contrast, cold ENSO episodes (La Niñas) result in the development of warm-dry (i.e., drought) conditions over the same U.S. region, again primarily for the winter season. In contrast, the importance of ENSO on summer season climate is much stronger elsewhere in the world, like over Southeast Asia and Australasia.

BOX 3.2. WAVES IN THE WESTERLIES, WEATHER, AND CLIMATE ANOMALIES

Maps of winds in the upper atmosphere (e.g., at 30,000 feet), shown daily as the jet-stream in newspaper, television, and web-based accounts of current weather, typically show a meandering pattern of air flow with three to five "waves" in the westerly winds that circle the globe in the mid-latitudes. These "Rossby waves" in the westerlies consist of sets of *ridges*, where (in the Northern Hemisphere) the flow pattern in the upper atmosphere broadly curves in the clockwise direction, and *troughs*, where the curvature is in a counter-clockwise direction. Rossby waves are ultimately generated by the temperature and pressure gradients that develop between the tropics and high-latitude regions, and in turn help to redistribute the energy surplus of the tropics through the movement of heat and moisture from the tropics toward the middle and high latitudes. Over North America, an upper-level ridge is typically found over the western third of the continent, with a trough located over the region east of the RockyMountains.

Distinct surface-weather conditions can be associated with the ridges and troughs. In the vicinity of the ridges, air sinks on a large scale, becoming warmer as it does so, while high pressure and diverging winds develop at the surface, all acting to create fair weather and to suppress precipitation. In the vicinity of troughs, air tends to converge around a surface low-pressure system (often bringing moisture from a source region like the subtropical North Pacific or Atlantic Oceans or the Gulf of Mexico) and to rise over a large area, encouraging precipitation.

From one day to the next, the ridges and troughs in the upper-level circulation may change very little, leading to their description as *stationary waves*. Meanwhile, smaller amplitude, shorter wavelength waves, or *eddies*, move along the larger scale stationary waves, again bringing the typical meteorological conditions associated with clockwise turning (fair weather) or counter-clockwise turning (precipitation) air streams, which may amplify or damp the effects of the larger scale waves on surface weather. Standard weather-map features like cold and warm fronts develop in response to the large-scale horizontal and vertical motions. Although uplift (and hence cooling, condensation, and precipitation) may be enhanced along the frontal boundaries between different airmasses, fronts and surface low- and high-pressure centers should be thought of as the symptoms of the large-scale circulation as opposed to being the primary generators of weather. The persistence or frequent recurrence of a particular wave pattern over weeks, months, or seasons then imparts the typical weather associated with the ridges and troughs, creating monthly and seasonal climate anomalies.

The particular upper-level wave pattern reflects the influence of fixed features in the climate system, like the configuration of continents and oceans and the location of major mountain belts like the Cordillera of western North America and the Tibetan Plateau (which tend to anchor ridges in those locations), and variable features like sea-surface temperature patterns, and snow-cover and soil-moisture anomalies over the continents.

However, new research suggests a teleconnection between Pacific SSTs and the North American monsoon as well (e.g., *Castro et al., 2007b*). The North American monsoon (June through September) is a critical source of precipitation for much of Mexico (up to 70% of the annual total) and the Southwestern United States (30%-50%). The means whereby tropical SST anomalies impact climate worldwide are reasonably well understood. The SST anomalies lead to anomalies in the patterns and magnitude of convective heating over the tropical oceans which drive atmospheric circulation anomalies that are transmitted around the world via stationary Rossby waves. The stationary waves then also subsequently impact the propagation of transient eddies thereby altering the patterns of storm tracks, which feeds back onto the mean flow. For a review see Trenberth et al. (1998).

On longer time scales during the Holocene (roughly the past 11,000 years), climatic variations in general, and hydrologic changes in particular, exceeded in both magnitude and duration those of the instrumental period or of the last millennium. In the mid- continent of North America, for example, between about 8,000 and 4,000 years ago, forests were replaced by steppe as the prairie expanded eastward, and sand dunes became activated across the Great Plains. These Holocene paleoclimatic variations occurred in response to the large changes in the controls of global and regional climates that accompanied deglaciation, including changes in ice-sheet size (area and elevation), the latitudinal and seasonal distribution of insolation, and atmospheric composition, including greenhouse gases and dust and mineral aerosols (*Wright et al., 1993*). Superimposed on these orbital-time-scale variations were interannual to millennial time- scale variations, many abrupt in nature (*Mayewski et al., 2004; Viau et al., 2006*), arising from variations in solar output, volcanic aerosols, and internally generated covariations among the different components of the climate system. On longer, or "orbital" time scales, the ice sheets, biogeochemically determined greenhouse gas concentrations, and dust and aerosol loading should be regarded as internal components of the climate system, but over the past 11,000 years, they changed slowly enough relative to other components of the climate system, such as the atmosphere and surface ocean, that they are most appropriately considered as external controls of regional-scale climate variations (*Saltzman, 2002*).

2.1.1. Coupled ocean-atmosphere forcing of north american hydrological variability

The standard approach that uses models to demonstrate a link between SSTs and observed climate variability involves forcing an Atmospheric General Circulation Model (AGCM) with observed SSTs as a lower boundary condition (see *Hoerling et al., in prep* (CCSP SAP 1.3, Box 3.2) for further discussion of this approach). Ensembles of simulations are used with different initial conditions such that the internally generated atmospheric weather in the ensemble members is uncorrelated from one member to the next and, after averaging over the ensemble, the part of the model simulation common to all - the part that is SST forced - is isolated. The relative importance of SST anomalies in different ocean basins can be assessed by specifying observed SSTs only in some areas and using climatological SSTs (or SSTs computed with a mixed layer (ML) ocean) elsewhere.

Schubert et al. (2004a,b) performed a climate model simulation from 1930 to 2004, which suggested that both a cold eastern equatorial Pacific and a warm subtropical Atlantic were the underlying forcing for drought over North America in the 193 0s. *Seager et al.* (2005b) and *Herweijer et al.* (2006) performed ensembles that covered the entire period of SST observations since 1856. These studies conclude that cold eastern equatorial Pacific SST anomalies in each of the three 19th century droughts, the Dust Bowl, and the 1950s drought were the prime forcing factors. *Seager (2007)* has made the same case for the 1998-2002 period of the current drought, suggesting a supporting role for warm subtropical Atlantic in forcing drought in the West. During the 1 930s and 1 950s droughts, the Atlantic was warm, whereas, the 19th century droughts seem to be more solely Pacific driven. Results for the Dust Bowl drought are shown in Figure 3.3, and time series of modeled and observed precipitation over the Great Plains are shown in Figure 3.4. *Hoerling and Kumar (2003)* instead emphasize the combination of a La Niñalike state and a warm Indo-west Pacific Ocean in forcing the 1998-2002 period of the most recent drought. On longer time scales, *Huang et al. (2005)* have shown that models forced by tropical Pacific SSTs alone can reproduce the North American wet spell between the 1976-77 and 1997-98 El Niños. The Dust Bowl drought was unusual in that it did not impact the Southwest. Rather, it caused reduced precipitation and high temperatures in the northern Rocky Mountain States and the western Canadian prairies, a spatial pattern that models generally fail to simulate (*Seager et al., 2007c*).

The SST anomalies prescribed in the climate models that result in reductions in precipitation are small, no more than a fraction of a degree Celsius. These changes are an order of magnitude smaller than the SST anomalies associated with interannual El Niño/Southern Oscillation (ENSO) events or Holocene SST variations related to insolation (incoming solar radiation) variations (~0.50°C; *Liu et al., 2003, 2004*). It is the persistence of the SST anomalies and associated moisture deficits that create serious drought conditions. In the Pacific, the SST anomalies presumably arise naturally from ENSO-like dynamics on time scales of a year to a decade (*Newman et al., 2003*). The warm SST anomalies in the Atlantic that occurred in the 1930s and 1950s (and in between), and usually referred to as part of an Atlantic Multidecadal Oscillation (AMO; Kushnir, 1994; Enfield et al., 2001), are of unknown origin. *Kushnir (1994), Sutton and Hodson (2005),* and *Knight et al. (2005)* have linked them to changes in the Meridional Overturning Circulation (see Chapter 4), which implies that a stronger overturning and a warmer North Atlantic Ocean would induce a drying in southwestern North America. However, others have argued that the AMO-related changes in tropical Atlantic SSTs are actually locally forced by changes in radiation associated with aerosols, rising greenhouse gases, and solar irradiance (*Mann and Emanuel, 2005*).

The dynamics that link tropical Pacific SST anomalies to North American hydroclimate are better understood and, on long time scales, appear as analogs of higher frequency phenomena associated with ENSO. The influence is exerted in two ways: first, through propagation of Rossby waves from the tropical Pacific polewards and eastwards to the Americas (*Trenberth et al., 1998*) and, second, through the impact that SST anomalies have on tropospheric temperatures, the subtropical jets, and the eddy-driven mean meridional circulation (*Seager et al., 2003, 2005a,b; Lau et al., 2006*). During La Niñas both mechanisms force air to descend over western North America, which suppresses precipitation. Although models, and analysis of observations (*Enfield et al., 2001; McCabe et al., 2004; Wang et al., 2006*), support the idea that warm subtropical North Atlantic SSTs can cause drying over western North America, the dynamics that underlay this have not been so clearly diagnosed and explained within model experiments.

The influence of Pacific SSTs on the North American monsoon has been documented at interannual and decadal time scales. In particular, a time-evolving teleconnection response in the early part of the summer appears to influence the strength and position of the monsoon ridge. In contrast to winter precipitation/ENSO relationships, La Niña-like conditions in the eastern and central tropical Pacific favor a wet and early monsoon and corresponding dry and hot conditions in the Central United States (Castro et al., 2001; Schubert et al., 2004a; Castro et al., 2007b). In contrast, El Niño-like conditions favor a dry and delayed monsoon and corresponding wet and cool conditions in the Central United States (op. cit.).

1932-1939 Precipitation Anomalies (wrt 1856-1928 climatology)

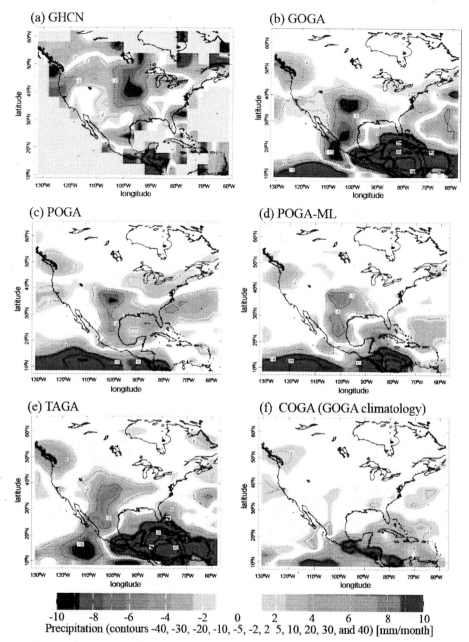

Figure 3.3. The observed (top left) and modeled precipitation anomalies during the Dust Bowl (1932 to 1939) relative to an 1856 to 1928 climatology. Observations are from Global Historical Climatology Network (GHCN). The modeled values are model ensemble means from the ensembles with global sea surface temperature (SST) forcing (GOGA), tropical Pacific forcing (POGA), tropical Pacific forcing and a mixed layer ocean elsewhere (POGA-ML), tropical Atlantic forcing (TAGA), and forcing with land and atmosphere initialized in January 1929 from the GOGA run and integrated forward with the 1856-1928 climatological SST (COGA). The model is the NCAR CCM3. Units are millimeters (mm) per month. From *Seager et al. (2007c)*

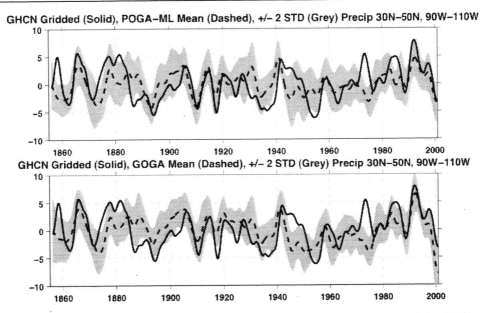

Figure 3.4. (top) The precipitation anomaly (in millimeters per month) over the Great Plains (30°N.-50°N., 90°W.-1 10°W.) for the period 1856 to 2000 from the POGA-ML ensemble mean with only tropical Pacific sea surface temperature (SST) forcing and from gridded station data. (bottom) Same as above but with GOGA ensemble mean with global SST forcing. All data have been 6-year low-pass filtered. The shading encloses the ensemble members within plus or minus of 2 standard deviations of the ensemble spread at any time. From *Seager et al. (2005b)*. GHCN, Global Historical Climatology Network.

2.1.2. Land surface feedbacks on hydroclimate variability

The evidence that multiyear North American droughts appear systematically together with tropical SST anomalies and that atmospheric models forced by these anomalies can reproduce some aspects of these droughts indicates that the ocean is an important driver. In addition to the ocean influence, some modeling and observational studies estimate that soil moisture feedbacks also influence precipitation variability (*Oglesby and Erickson, 1989; Namias, 1991; Oglesby, 1991*). *Koster et al. (2004)* used observations to show that on the time scale of weeks, precipitation in the Great Plains is significantly correlated with antecedent precipitation. *Schubert et al. (2004b)* compared models run with average SSTs, with and without variations in evaporation efficiency, and showed that multiyear North American hydroclimate variability was significantly reduced if evaporation efficiency was not taken into account. Indeed, their model without SST variability was capable of producing multiyear droughts from the interaction of the atmosphere and deep soil moisture. This result needs to be interpreted with caution since *Koster et al. (2004)* also show that the soil moisture feedback in models seems to exceed that deduced from observations. In a detailed analysis of models, observations and reanalyses, *Ruiz-Barradas and Nigam (2005)* and *Nigam and Ruiz-Barradas (2006)* conclude that interannual variability of Great Plains hydroclimate is dominated by transport variability of atmospheric moisture and that the local precipitation recycling, which depends on soil moisture, is overestimated in models and provides a spuriously strong coupling between soil moisture and precipitation.

Past droughts have also caused changes in vegetation. For example, during the Dust Bowl drought there was widespread failure of non-drought-resistant crops that led to exposure of bare soil. Also, during the Medieval megadroughts there is evidence of dune activity in the Great Plains (*Forman et al., 2001*), which implies a reduction in vegetation cover. Conversions of croplands and natural grasses to bare soil could also impact the local hydroclimate through changes in surface energy balance and hydrology. Further, it has been argued on the basis of experiments with an atmosphere model with interactive dust aerosols that the dust storms of the 193 0s worsened the drought, and moved it northward, by altering the radiation balance over the affected area (*Cook et al., 2008*). The widespread devegetation caused by crop failure in the 193 0s could also have impacted the local climate. These aspects of land-surface feedbacks on drought over North America need to be examined further with other models, and efforts need to be made to better quatify the land-surface changes and dust emissions during the Dust Bowl.

2.1.3. Historical droughts over north america and their impacts

According to the National Oceanic and Atmospheric Administration (NOAA; for periodically updated economic information regarding U.S. weather diasters), over the period from 1980 to 2006, droughts and heat waves are among the most expensive natural disaster in the United States along with tropical storms (including the devastating 2005 hurricane season) and wide-spread or regional flooding episodes. The annual cost of drought to the United States is estimated to be in the tens of billions of dollars.

The above describes the regular year-in, year-out costs of drought. In addition, persistent multiyear droughts have had important consequences in national affairs. The icon of drought impacts in North America is the Dust Bowl of the 1930s. In the early 20th century, settlers transferred large areas of the Great Plains from natural prairie grasses, used to some extent for ranching, to wheat farms. After World War I, food demand in Europe encouraged increased conversion of prairie to crops. This was all possible because these decades were unusually wet in the Great Plains. When drought struck in the early 1930s, the non-drought-resistant wheat died, thus exposing bare soil. Faced with a loss of income, farmers responded by planting even more, leaving little land fallow. When crops died again there was little in the way of "shelter belts" or fallow fields to lessen wind erosion. This led to monstrous dust storms that removed vast amounts of top soil and caused hundreds of deaths from dust inhalation (*Worster, 1979; Hansen and Libecap, 2004; Egan, 2006*). As the drought persisted year after year and conditions in farming communities deteriorated, about a third of the Great Plains residents abandoned the land and moved out, most as migrant workers to the Southwest and California, which had not been severely hit by the drought.

The Dust Bowl disaster is a classic case of how a combination of economic and political circumstances interacted with a natural event to create a change of course in national and regional history. It was in the 1930s that the Federal Government first stepped in to provide substantial relief to struggling farm communities heralding policies that remain to this day. The Dust Bowl drought also saw an end to the settlement of the semi-arid lands of the United States based on individual farming families acting independently. In addition, wind erosion was brought under control via collective action, organized within Soil Conservation Districts, while farm abandonment led to buyouts and a large consolidation of land ownership (*Hansen and Libecap, 2004*). Ironically, the population migration to the West likewise provided the manpower needed in the armaments industry after 1941 to support the U.S. World War II effort.

Earlier droughts in the late 19th century have also tested the feasibility of settlement of the West based on provisions within the Homestead Act of 1862. This act provided farmers with plots of land that may have been large enough to support a family in the East but not enough in the arid West, and it also expected them to develop their own water resources. The drought of the early to middle 1 890s led to widespread abandonment in the Great Plains and acceptance, contrary to frontier mythology of "rain follows the plow" (*Libecap and Hansen, 2002*), that if the arid lands were to be successfully settled and developed, the Federal Government was going to have to play an active role. The result was the Reclamation Act of 1902 and the creation of the U.S. Bureau of Reclamation, which in the following decades developed the mammoth water engineering works that sustain agriculture and cities across the West from the Great Plains to the Pacific Coast (*Worster, 1985*).

On a different level, the Great Plains droughts of the 1850s and early 1860s played a role in the combination of factors that led to the near extinction of the American bison (*West, 1995*). Traditionally, bison tried to cope with drought by moving into the better watered valleys and riparian zones along the great rivers that flowed eastward from the Rocky Mountains. However, by the mid-19th century, these areas had become increasingly populated by Native Americans who had recently moved to the Great Plains after being evicted from their villages in more eastern regions by settlers and the U.S. Army, thereby putting increased hunting pressure on the bison herds for food and commercial sale of hides. In addition, the migration of the settlers to California after the discovery of gold there in 1849 led to the virtual destruction of the riparian zones used by the bison for over-wintering and refuge during droughts. The 1 850s and early 1860s droughts also concentrated the bison and their human predators into more restricted areas of the Great Plains still suitable for survival. Drought did not destroy the bison, but it did establish conditions that almost lead to the extinction of one of America's few remaining species of megafauna (*West, 1995; Isenberg, 2000*).

The most recent of the historical droughts, which began in 1998 and persists at the time of writing, has yet to etch itself into the pages of American history, but it has already created a tense situation in the West as to what it portends. Is it like the 1 930s and 1 950s droughts and, therefore, is likely to end relatively soon? Or is it the emergence of the anthropogenic drying that climate models project will impact this region - and the subtropics in general - within the current century and, quite possibly, within the next few years to decades? *Breshears et al. (2005)* noted that the recent Southwest drought was warmer than 1 950s drought and the higher temperatures exacerbated drought impacts in ways that are consistent with expectations for the amplification of drought severity in response to greenhouse forcing. If this drying comes to pass it will impact the future economic, political, and social development of the West as it struggles to deal with declining water resources.

2.1.4. Impacts of change in the atmospheric branch of the hydrological cycle for ground water and river flow

The nature of these impacts ranges from reductions in surface-water supplies affecting reservoir storage and operations, and delivery and treatment of water, to drawdown of aquifers, increased pumping costs, subsidence, and reductions of adjacent or connected surface-water flows. A multitude of water uses, including irrigated and unirrigated agriculture, hydroelectric and thermoelectric power (cooling), municipal and industrial water uses, transportation, and recreation (National Assessment, 2000), can be severely impacted by rapid hydroclimatic changes that promote drought. Reductions in water supplies that affect these uses can have profound impacts on regional economies. For example, drought in the late 1980s and early 1990s in California resulted in a reduction in hydropower and increased reliance on fossil fuels, and an additional $3 billion in energy costs (*Gleick and Nash, 1991*).

Rapid changes in climate that influence the atmospheric part of the hydrological cycle can affect the amount, form, and delivery of precipitation, which in turn influence soil moisture, runoff, ground water, surface flows, and lake levels, as well as atmospheric features such as clouds. Changes can take the form of shifts in state to overall wetter or drier conditions, more persistent drought or flood-causing events, and/or a greater frequency of extreme events. All of these types of rapid changes can have serious societal impacts with far-reaching effects on water availability, quality, and distribution (National Assessment, 2000).

Shifts in the climate background state may modulate, and either constructively or destructively influence, the "typical" hydrologic impacts of seasonal to interannual climate variability. For example, the Southwestern United States, which tends to receive higher than average winter-time precipitation during an El Niño event, and relies on these events to refill water supply reservoirs, could benefit from changes that increase or enhance El Niño events, but suffer from increased droughts if La Niña events, which tend to result in dry winters here, become more frequent (Figure 3.2).

The impacts of these changes can exacerbate scarce water supplies in regions that are already stressed by drought, greater demand, and changing uses. The Department of Interior analysis of Western U.S. water supply issues (USBR, 2005) identifies a number potential water supply crises and conflicts by the year 2025 based on a combination of technical and other factors, including population trends and potential endangered species' needs for water, but under an assumption of a statistically stationary climate (Figure 3.1). Any transient change in climate conditions that leads to an abrupt regime shift to more persistent or more severe drought will only compound these water supply conflicts and impact society.

Abrupt changes in hydroclimate that lead to sustained drought can have enormous impacts on the management of water systems, in particular, the large managed river systems in western areas of the Western United States Many of these managed systems are facing enormous challenges today, even without abrupt changes, due to increased demands, new uses, endangered species requirements, and tribal water-right claims. In addition, many of these systems have been found to be extremely vulnerable to relatively small changes in runoff (e.g., Nemec and Schaake, 1982; Christensen and Lettenmaier, 2006).

2.2. Global Context of North American Drought

When drought strikes North America it is not an isolated event. In "The Perfect Ocean for Drought," *Hoerling and Kumar (2003)* noted that the post-1998 drought that was then impacting North America extended from the western subtropical Pacific across North America and into the Mediterranean region, the Middle East, and central Asia. There was also a band of subtropical drying in the Southern Hemisphere during the same period. It has long been known that tropical SST anomalies give rise to global precipitation anomalies, but the zonal and hemispheric symmetry of ENSO impacts has only recently been emphasized (*Seager et al., 2005a*).

Hemispheric symmetry is expected if the forcing for droughts comes from the tropics. Rossby waves forced by atmospheric heating anomalies in the tropics propagate eastward and poleward from the source region into the middle and high latitudes of both hemispheres (*Trenberth et al., 1998*). The forced wave train will, however, be stronger in the winter hemisphere than the summer hemisphere because the mid-latitude westerlies are both stronger and penetrate farther equatorward, increasing the efficiency of wave propagation from the tropics into higher latitudes. The forcing of tropical tropospheric temperature change by the tropical SST and air-sea heat flux anomalies will also tend to create globally coherent hydroclimate patterns because (1) the temperature change will be zonally uniform and extend into the subtropics (*Schneider, 1977*) and (2) the result will require a balancing change in zonal winds that will potentially interact with transient eddies to create hemispherically and zonally symmetric circulation and hydroclimate changes.

In the tropics the precipitation anomaly pattern associated with North American droughts is very zonally asymmetric with reduced precipitation over the cold waters of the eastern and central equatorial Pacific and increased precipitation over the Indonesian region. The cooler troposphere tends to increase convective instability (*Chiang and Sobel, 2002*), and precipitation increases in most tropical locations outside the Pacific with the exception of coastal East Africa, which dries, possibly as a consequence of cooling of the Indian Ocean (Goddard and Graham, 1999).

North American droughts are therefore a regional realization of persistent near-global atmospheric circulation and hydroclimatic anomalies orchestrated by tropical atmosphere-ocean interactions. During North American droughts, dry conditions are also expected in mid-latitude South America, wet conditions in the tropical Americas and over most tropical regions, and dry conditions again over East Africa. Subtropical to mid- latitude drying should extend across most longitudes and potentially impact the Mediterranean region. However, the signal away from the tropics and the Americas is often obscured by the impact of other climate phenomena such as the North Atlantic Oscillation (NAO) impact on precipitation in the Mediterranean region (*Hurrell, 1995; Fye et al., 2006*). In a similar fashion, the Holocene drought in the mid-continent of
North America (Sec. 4) can be shown to be embedded in global-scale energy balance and atmospheric circulation changes.

2.2.1. The perfect ocean for drought: Gradual climate change resulting in abrupt impacts

The study of the 1998-2002 droughts that spread across the United States, Southern Europe, and Southwest Asia provides an example of a potential abrupt regime shift to one with more persistent and/or more severe drought in response to gradual changes in global or regional climate conditions. Research by *Hoerling and Kumar (2003)* provides compelling evidence that these severe drought conditions were part of a persistent climate state that was strongly influenced by the tropical oceans.

From 1998 through 2002, prolonged below-normal precipitation and above- normal temperatures caused the United States to experience drought in both the Southwest and Western States and along the Eastern Seaboard. These droughts extended across southern Europe and Southwest Asia, with as little as 50% of the average rainfall in some regions (Figure 3.5). The *Hoerling and Kumar (2003)* study used climate model simulations to assess climate response to altered oceanic conditions during the 4-year interval. Three different climate models were run a total of 51 times, and the responses averaged to identify the common, reproducible element of the atmosphere's sensitivity to the ocean. Results showed that the tropical oceans had a substantial effect on the atmosphere (Figure 3.6). The combination of unprecedented warm sea-surface conditions in the western tropical Pacific and 3-plus consecutive years of cold La Niña conditions in the eastern tropical Pacific shifted the tropical rainfall patterns into the far western equatorial Pacific.

Over the 1998 through 2002 period, the cold eastern Pacific tropical sea surface temperatures, though unusual, were not unprecedented. However, the warmth in the tropical Indian Ocean and the west Pacific Ocean was unprecedented during the 20th century, and attribution studies indicate this warming (roughly 1°C since 1950) is beyond that expected of natural variability. The atmospheric modeling results suggest an important role for tropical Indian Ocean and the west Pacific Ocean sea surface conditions in the shifting of westerly jets and storm tracks to higher latitudes with a nearly continuous belt of high pressure and associated drying in the lower mid-latitudes. The tropical ocean forcing of multiyear persistence of atmospheric circulation not only increased the risk for severe and synchronized drying of the mid-latitudes between 1998 and 2002 but may potentially do so in the future, if such ocean conditions occur more frequently.

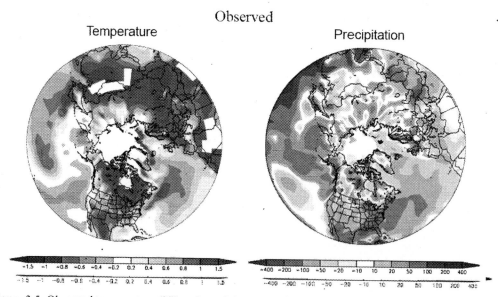

Figure 3.5. Observed temperature (°C) and precipitation (millimeters) anomalies (June 1998-May 2002). Figure from http://www.oar.noaa.gov/spotlite/archive/spot_drought.html.

The *Hoerling and Kumar (2003)* analysis illustrates how changes in regional climate conditions such as slow increases in Indo-Pacific "Warm Pool" SSTs, when exceeding critical environmental thresholds, can lead to abrupt shifts in climate regimes (e.g., the anomalous atmospheric circulation patterns), which in turn alter the hydrologic response to natural variability. The study points out that the overall pattern of warmth in the Indian and west Pacific Oceans was both unprecedented and consistent with greenhouse gas forcing of climate change. Could similar abrupt shifts in climate regimes explain the persistence of droughts in the past? From a paleoclimatic perspective, simulations by *Shin et al. (2006)* using an Atmospheric General Circulation Model (AGCM) with a "slab" ocean, and by *Liu et al. (2003)* and *Harrison et al. (2003)* with a fully coupled Atmosphere-Ocean General

Circulation Model (AOGCM) indicate that a change in the mean state of tropical Pacific SSTs to more La Niña-like conditions can explain North American drought conditions during the mid-Holocene. An analysis of Medieval hydrology by *Seagar et al. (2007b)* suggests the widespread drought in North America occurred in response to cold tropical Pacific SSTs and warm subtropical North Atlantic SSTs externally forced by high irradiance and weak volcanic activity (see *Mann et al., 2005; Emile-Geay et al., 2007*).

GCM

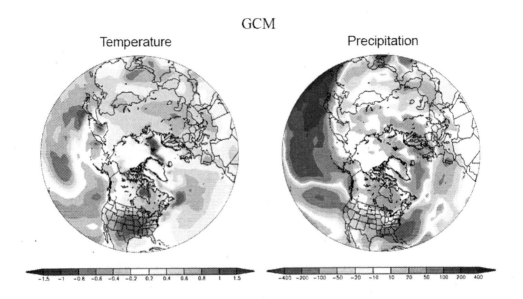

Figure 3.6. Model-simulated temperature (°C) and precipitation (millimeters) anomalies given observed SSTs over the June 1998 – May 2002 period. GCM, General Circulation Model. Figure from http://www.oar.noaa.gov/spotlite/archive/spot_drought

2.3. Is there evidence yet for anthropogenic forcing of drought?

Analyses by *Karoly et al. (2003)* and *Nicholls (2004)* suggest that 2002 drought and associated heat waves in Australia were more extreme than the earlier droughts because the impact of the low rainfall was exacerbated by high potential evaporation. *Zhang et al. (2007)* have suggested that large-scale precipitation trends can be attributed to anthropogenic influences. However there is no clear evidence to date of human-induced global climate change on North American precipitation amounts. The Fourth Assessment Report (AR4) of the IPCC (*IPCC, 2007*) presents maps of the trend in precipitation over the period 1901 to 2005 that shows mostly weak moistening over most of North America and a weak drying in the Southwest. This is not very surprising in that both the first two decades and the last two decades of the 20th entury were anomalously wet over much of North America (*Swetnam and Betancourt, 1998; Fye et al., 2003; Seager et al., 2005b; Woodhouse et al., 2005*). The wettest decades between the 1976/77 and 1997/98 El Niños may have been caused by natural Pacific decadal variability (*Huang et al., 2005*). In contrast to the 20th - century record, the southern parts of North America are projected to dry as a consequence of anthropogenic climate change. After the 1997/98 El Niño, drought has indeed settled into the West, but since it has gone along with a more La Niña-like Pacific Ocean this makes it difficult to determine if some part of the drying is anthropogenic.

Trends based on the shorter period of the post-1950 period show a clear moistening of North America, but this period extends from the 1950s drought to the end of the late-20th century wet period (or pluvial). The 1950s drought has been linked to tropical Pacific and Atlantic SSTs and is presumed to have been a naturally occurring event. Further, the trend from 1950 to the end of the last century is likely to have been caused by the multidecadal change from a more La Niña-like tropical Pacific before 1976 to a more El Niño-like Pacific from 1976 to 1998 (*Zhang et al., 1997*), a transition usually known as the 1976-77 climate or regime shift, which caused wet conditions in the mid-latitude Americas (*Huang et al., 2005*). Again, this change in Pacific SSTs is generally assumed to have been a result of natural Pacific variability, and it has been shown that simple models of the tropical Pacific alone can create multidecadal variations that have this character (*Karspeck et al., 2004*). The warm phase of tropical Pacific decadal variability may have ended with the 1997/98 El Niño, after which La Niña-like conditions prevailed until 2002 followed by weak El Ninos and a return to La Nina in 2007. In these post-1998 years, drought conditions have also prevailed across the West as in previous periods of persistent La Niñas. Consequently, it would be very premature to state that the recent drought heralds a period of anthropogenic drying as opposed to the continuation of natural decadal and multidecadal variations. Detailed analysis of not only precipitation patterns but also patterns of stationary and transient atmospheric circulation, water vapor transports, and SSTs may be able to draw a distinction, but this has not yet been done.

A different view is offered by *Vecchi et al. (2006)*, who used sea level pressure (SLP) data to show a weakening of the along-Equator east-to-west SLP gradient from the late-19th century to the current one. The rapid weakening of this gradient during the 1976-77 climate shift contributes to this trend. *Vecchi et al. (2006)* showed that coupled climate model simulations of the 20th century forced by changes in CO_2, solar irradiance, and other factors also exhibit a weakening of the SLP gradient - a weaker Walker Circulation - which could be taken to mean that the 1976-77 shift, and associated wetting of North America, contained an anthropogenic component. However, as noted in the previous paragraph, it would be very premature to state that the post-2002 period heralds a period of anthropogenic drying as opposed to the continuation of natural decadal and multidecadal variations.

3. North American Drought over the Past Millennia

Historical climate records provide considerable evidence for the past occurrence of exceptional multi-year droughts on the North American continent and their impacts on American history. In addition, modeling experiments have conclusively demonstrated the importance of large-scale tropical SSTs on forcing much of the observed hydroclimatic variability over North America and other global land areas. What is still missing from this narrative is a better understanding of just how bad droughts can become over North America. Is the 1930s Dust Bowl drought the worst that can conceivably occur over North America? Or, is there the potential for far more severe droughts to develop in the future? Determining the potential for future droughts of unprecedented severity can be investigated with climate models (*Seager et al., 2007d*), but the models still contain too much uncertainty in them to serve as a definitive guide. Rather, what we need is an improved understanding of the past occurrence of drought and its natural range of variability. The instrumental and historical data only go back about 130 years with an acceptable degree of spatial completeness over the United States (see the 19th century instrumental data maps in *Herweijer et al., 2006*), which does not provide us with enough time to characterize the full range of hydroclimatic variability that has occurred in the past and could conceivably occur in the future independent of any added effects due to greenhouse warming. To do so, we must look beyond the historical data to longer natural archives of past climate information.

3.1. Tree ring reconstructions of past drought over North America

In the context of how North American drought has varied over the past 2,000 years, an especially useful source of "proxy" climate information is contained in the annual ring-width patterns of long-lived trees (*Fritts, 1976*). A tree can provide information about past climate in its annual ring widths because its growth rate is almost always climate-dependent to some degree. Consequently, the annual ring-width patterns of trees provide proxy expressions of the actual climate affecting tree growth in the past and these expressions can therefore be used to reconstruct past climate. The past 2,000 years is also particularly relevant here because the Earth's climate boundary conditions are not markedly different from those of today, save for the 20th century changes in atmospheric trace gas composition and aerosols that are thought to be responsible for recent observed warming. Consequently, a record of drought variability from tree rings in North America over the past two millennia would provide a far more complete record of extremes for determining how bad conditions could become in the future. Again, this assessment would be independent of any added effects due to greenhouse warming.

An excellent review of drought in the Central and Western United States, based on tree rings and other paleoproxy sources of hydroclimatic variability, can be found in *Woodhouse and Overpeck (1998)*. In that paper, the authors introduced the concept of the "megadrought," a drought that has exceeded the intensity and duration of any droughts observed in the more recent historical records. They noted that there was evidence in the paleoclimate records for several multidecadal megadroughts prior to 1600 that "eclipsed" the worst of the 20th century droughts including the Dust Bowl. The review by *Woodhouse and Overpeck (1998)* was limited geographically and also restricted by the lengths of tree-ring records of past drought available for study. At that time, a gridded set of summer drought reconstructions, based on the Palmer Drought Severity Index (PD SI; *Palmer, 1965*), was available for the conterminous United States, but only back to 1700 (*Cook et al., 1999*). Those data indicated that the Dust Bowl was the worst drought to have hit the U.S. over the past three centuries. However, a subset of the PDSI reconstructions in the western, southeastern, and Great Lakes portions of the United States also extended back to 1500 or earlier. This enabled *Stahle et al. (2000)* to describe in more detail the temporal and spatial properties of the late 16th century megadrought noted earlier by *Woodhouse and Overpeck (1998)* and compare it to droughts in the 20th century. In concurrence with those earlier findings, *Stahle et al. (2000)* showed that even the past 400 years were insufficient to capture the frequency and occurrence of megadroughts that clearly exceeded anything in the historical records in many regions.

3.2. The North American Drought Atlas

Since that time, great progress has been made in expanding the spatial coverage of tree-ring PDSI reconstructions to cover most of North America (*Cook and Krusic, 2004a,b; Cook et al., 2004*). The grid used for that purpose is shown in Figure 3.7. It is a 286-point 2.5° by 2.5° regular grid that includes all of the regions described in *Woodhouse and Overpeck (1998), Cook et al. (1999),* and *Stahle et al. (2000).* In addition, the reconstructions were extended back 1,000 or more years at many locations. This was accomplished by expanding the tree-ring network from the 425 tree-ring chronologies used by *Cook et al. (1999)* to 835 series used by *Cook et al. (2004).* Several of the new series also exceeded 1,000 years in length, which facilitated the creation of new PDSI reconstructions extending back into the megadrought period in the Western United States prior to 1600. Extending the reconstructions back at least 1,000 years was an especially important goal. *Woodhouse and Overpeck (1998)* summarized evidence for at least four widespread multi-decadal megadroughts in the Great Plains and the Western United States during the A.D. 750-1300 interval. These included two megadroughts lasting more than a century each during "Medieval" times in California's Sierra Nevada (*Stine, 1994*). Therefore, being able to characterize the spatial and temporal properties of these megadroughts in the Western United States was extremely important.

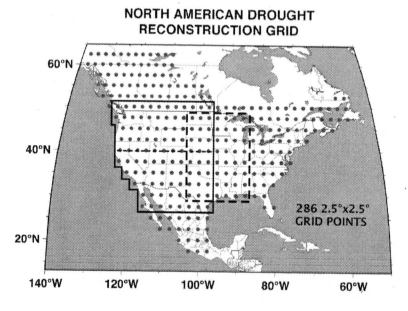

NORTH AMERICAN DROUGHT RECONSTRUCTION GRID

286 2.5°x2.5° GRID POINTS

Figure 3.7. Map showing the distribution of 286 grid points of drought reconstructed for much of North America from long-term tree-ring records. The large, irregular polygon over the West is the area analyzed by *Cook et al. (2004)* in their study of long-term aridity changes. The dashed line at 40°N. divides that area into Northwest and Southwest zones. The dashed-line rectangle defines the Great Plains region that is also examined for long-term changes in aridity here

Using the same basic methods as those in *Cook et al. (1999)* to reconstruct drought over the conterminous United States, new PDSI reconstructions were developed on the 286- point North American grid (Figure 3.7) and incorporated into a North American Drought Atlas (NADA; *Cook and Krusic, 2004a,b; Cook et al., 2007*). The complete contents of NADA can be accessed and downloaded at http://iridl.ldeo.columbia.edu/SOURCES/.LDEO/.TRL/ .NADA2004/.pdsi-atlas.html. In Figure 3.7, the irregular polygon delineates the boundaries of the area we refer to as southwestern North America and includes the southwestern United States and northern Mexico. It encompasses all grid points on and within 27.5°-50°N. latitude and 97.5°- 125°W. longitude and was the area used by *Cook et al. (2004).* The dashed line along the 40th parallel separates the West into northwest and southwest sectors, which will be compared later.

3.3. Medieval Megadroughts in the Western United States

Cook et al. (2004) examined the NADA contents back to A.D. 800 for the West to place the current turn-of-the-century drought there *(Seager, 2007)* in a long-term context. In so doing, a period of elevated aridity was found in the A.D. 900-1300 period that included four particularly widespread and prolonged multi-decadal megadroughts (Figure 3.8). This epoch of large-scale elevated aridity was corroborated by a number of independent, widely scattered, proxy records of past drought in the West *(Cook et al., 2004)*. In addition, the four identified megadroughts agreed almost perfectly in timing with those identified by *Woodhouse and Overpeck (1998)*, which were based on far fewer data. These findings were rather sobering for the West because they (1) verified the occurrence of several past multidecadal megadroughts prior to 1600, (2) revealed an elevated background state of aridity that lasted approximately four centuries, and (3) demonstrated that there are no modern analogs to the A.D. 900-1300 period of elevated aridity and its accompanying megadroughts. This is clearly a cause for concern because the data demonstrate that the West has the capacity to enter into a prolonged state of dryness without the need for greenhouse gas forcing.

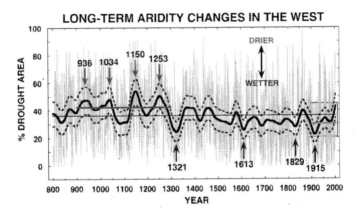

Figure 3.8. Percent area affected by drought (Palmer Drought Severity Index (PDSI) <-1) in the area defined as the West in Figure 3:7 (redrawn from *Cook et al., 2004*). More- positive numbers mean large areas in the West affected by drought. Annual data are in gray and a 60-year low-pass filtered version is indicated by the thick smooth curve. Dashed blue lines are 2-tailed 95% confidence limits based on bootstrap resampling. The modern (mostly 20th century) era is highlighted in yellow for comparison to a remarkable increase in aridity prior to about A.D. 1300

The timing of the A.D. 900-1300 period of elevated aridity is especially worrisome because it occurred during what has historically been referred to as the 'Medieval Warm Period' (MWP; *Lamb, 1965*), a time of persistently above-average warmth over large parts of the Northern Hemisphere *(Esper et al., 2002)*, including the Western United States *(LaMarche, 1974)*. *Stine (1994)* also noted the association of his prolonged Sierra Nevada droughts with the MWP. Given that his particular climate expression was more related to hydroclimatic variability than to pure temperature change, *Stine (1994)* argued that a more appropriate name for this unusual climate period should be the 'Medieval Climate Anomaly'

(MCA) period. We will use MCA from here on out when referring to drought during the Medieval period.

Herweijer et al. (2007) made some detailed examinations of the NADA in order to determine how the megadroughts during the MCA differed from droughts of more modern times. That analysis was restricted to effectively the same spatial domain as that used by *Cook et al. (2004)* for the West, in this case the grid points in the 25°-50°N. latitude, 95°-125°W. longitude box (cf. Figure 3.7). *Herweijer et al. (2007)* also restricted their analyses to a subset of 106 grid points within this domain with reconstructions available since A.D. 1000. This restriction had no appreciable effect on their results (see also *Cook et al., 2004*). *Herweijer et al. (2007)* compared the average PDSI over the 106 grid points for two distinct periods: A.D. 1000-1470 and 1470-2003. Even without any further analyses, it was clear that the earlier period, especially before 1300, was distinctly more drought-prone than the later period. Of particular interest was the fact that the range of annual drought variability during the MCA was not any larger than that seen after 1470. So, the climate conditions responsible for droughts each year during the MCA were apparently no more extreme than those conditions responsible for droughts during more recent times. This can be appreciated by noting that only 1 year of drought during the MCA was marginally more severe than the 1934 Dust Bowl year. This suggests that the 1934 event may be used as a worst-case scenario for how bad a given year of drought can get over the West.

So what differentiates MCA droughts from modern droughts? As shown by *Herweijer et al. (2007)*, the answer is **duration**. Droughts during the MCA lasted much longer, and it is this characteristic that most clearly differentiates megadroughts from ordinary droughts in the Western United States *Herweijer et al. (2007)* identified four megadroughts during the MCA — A.D. 1021-1051, 1130-1170, 1240-1265, and 1360-1382 — that lasted 31, 41, 26, and 23 years, respectively. In contrast, the four worst droughts in the historic period — A.D. 1855-1865, 1889-1896, 1931-1940, and 1950-1957 — lasted only 11, 8, 9, and 8 years, respectively. The difference in duration is striking.

The research conducted by *Cook et al. (2004), Herweijer et al. (2006, 2007),* and *Stahle et al. (2007)* was based on the first version of NADA. Since its creation in 2004, great improvements have been made in the tree-ring network used for drought reconstruction with respect to the total number of chronologies available for use in the original NADA (up from 835 to 1825) and especially the number extending back into the MCA (from 89 to 195 beginning before A.D. 1300). In addition, better geographic coverage during the MCA was also achieved, especially in the Northwest and the Rocky Mountain States of Colorado and New Mexico. Consequently, it is worth revisiting the results of *Cook et al. (2004)* and *Herweijer et al. (2007)*.

Figure 3.9 shows the updated NADA results now divided geographically into Northwest (Figure 3.9A), Southwest (Figure 3.9B), and the Great Plains (Figure 3.9C). See Figure 3.7 for the sub-areas of the overall drought grid that define these three regions. Unlike the drought area index series shown in Figure 3.8, where more positive values indicate large areas affected by drought, the series shown in Figure 3.9 are simple regional averages of reconstructed PDSI. Thus, greater drought is indicated by more negative values in accordance with the original PDSI scale of *Palmer (1965)*. When viewed now in greater geographic detail, the intensity of drought during the MCA is focused more clearly toward the Southwest, with the Northwest much less affected. This geographic shift in emphasis toward the

Southwest during the MCA aridity period is into the region where drought is more directly associated with forcing from the tropical oceans (*Cole et al., 2002; Seager et al., 2005b; Herweijer et al., 2006, 2007*).

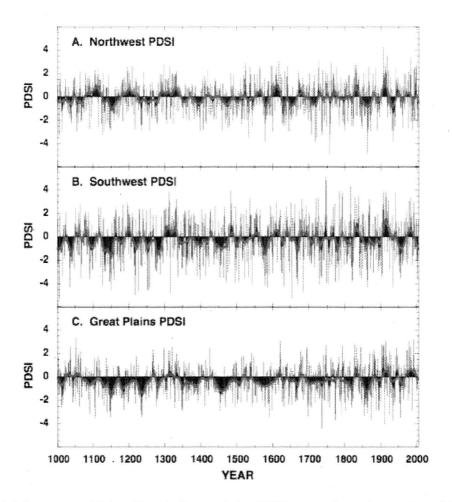

Figure 3.9. Reconstructed Palmer Drought Severity Index (PDSI) averaged over three regions of the Western and Central United States: the Northwest (A), the Southweset (B), and the Great Plains (C). Time is in calendar years A.D. See Figure 3.7 for the geographic locations of these regions. These series are all based on the updated version of North American Drought Atlas (NADA). Unlike the drought area index in Figure 3.8, where more-positive numbers mean larger areas affected by drought, more-negative numbers mean drier conditions here in accordance with the original PDSI scale devised by Palmer (1965)

Aside from the shift of geographic emphasis in the West during the MCA, the updated version of NADA still indicates the occurrence of multidecadal megadroughts that mostly agree with those of *Herweijer et al. (2007)* and the overall period of elevated aridity as described by *Cook et al. (2004)*. From Figure 3.9B, two of those megadroughts stand out especially strong in the Southwest: A.D. 1130-1158 (~29 years) and 1270-1297 (~28 years). The latter is the "Great Drouth" documented by *A.E. Douglass (1929, 1935)* for its association with the abandonment of Anasazi dwellings in the Southwest. Another prolonged drought in A.D. 1434-1481 (~48 years) is also noteworthy. *Herweijer et al. (2007)* did not mention it because it falls after the generally accepted end of the MCA. This megadrought is the same as the "15th century megadrought" described by *Stahle et al. (2007)*.

3.4. Possible Causes of the Medieval Megadroughts

The causes of the Medieval megadroughts are now becoming unraveled and appear to have similar origin to the causes of modern droughts, which is consistent with the similar spatial patterns of Medieval and modern droughts (*Herweijer et al., 2007*). *Cobb et al. (2003)* have used modern and fossil coral records from Palmyra, a small island in the tropical Pacific Ocean, to reconstruct eastern and central equatorial Pacific SSTs for three time segments within the Medieval period. These results indicate that colder—La Niñalike—conditions prevailed, which would be expected to induce drought over western North America. *Graham et al. (2007)* used these records, and additional sediment records in the west Pacific, to create an idealized pattern of Medieval tropical Pacific SST which, when it was used to force an AGCM, did create a drought over the Southwest. Adopting a different approach, *Seager et al. (2007a)* used the Palmyra modern and fossil coral records to reconstruct annual tropical Pacific SSTs for the entire period of 1320 to 1462 A.D. and forced an AGCM with this record. They found that the overall colder tropical Pacific implied by the coral records forced drying over North America with a pattern and amplitude comparable to that inferred from tree ring records, including for two megadroughts (1360-1400 A.D. and 1430-1460 A.D.). Discrepancies between model and observations can be explained through the combined effect of potential errors in the tropical Pacific SST reconstruction role for SST anomalies from other oceans, other unaccounted external forcings, and climate model deficiencies.

The modeling work suggests that the Medieval megadroughts were driven, at least in part, by tropical Pacific SST patterns in a way that is familiar from studies of the modern droughts. Analyses of the global pattern of Medieval hydroclimate also suggest that it was associated with a La Niña-like state in combination with a warm subtropical North Atlantic and a positive North Atlantic Oscillation (*Seager et al., 2007b; Herweijer et al., 2007*). For example, *Haug et al. (2001)* used the sedimentary record from the Cariaco basin in the Caribbean Sea to argue that northern South America experienced several wet centuries during the Medieval period, which is consistent with a La Nina-like Pacific Ocean. As another example, *Sinha et al. (2007)* used a speleothem (a secondary mineral deposit formed in a cave) record from India to show that at the same time the Indian monsoon was generally strong, especially compared to the subsequent Little Ice Age.

It has been suggested that the tropical Pacific adopted a more La Niña-like mean state during the Medieval period, relative to subsequent centuries, as a response to a relatively strong Sun and weaker volcanic activity (*Mann et al., 2005; Emile-Geay et al., 2007*; see also *Adams et al., 2003*). This follows because a positive radiative forcing warms the western equatorial Pacific by more than the east because in the latter region strong upwelling and ocean heat divergence transports a portion of the absorbed heat toward the subtropics. The stronger east-west gradient then strengthens the Walker Circulation, increasing the thermocline tilt and upwelling in the east such that actual cooling can be induced.

Further support for positive radiative forcing over the tropical Pacific Ocean inducing La Niña-like SSTs and drought over the Southwest comes from analyses of the entire Holocene recorded in a New Mexico speleothem, which shows a clear association between increased solar irradiance (as deduced from the atmospheric ^{14}C content recorded in ice cores) and dry conditions (*Asmerom et al., 2007*). However, the theory for the positive radiative forcing-La Niña link rests on experiments with intermediate complexity models (*Clement et al., 1996, 2000; Cane et al., 1997*). In contrast, the coupled GCMs used in the IPCC process do not, however, respond in this way to rising greenhouse gases and may actually slow the Walker Circulation (*Vecchi et al., 2006*). This apparent discrepancy could arise because the tropical response to changes in solar irradiance is different from the response to rising greenhouse gases or it could be that the coupled GCMs respond incorrectly due to the many errors in simulations of the tropical Pacific mean climate, not the least of which is the notorious double-intertropical convergence zone (ITCZ) problem.

3.5. Megadroughts in the Great Plains and U.S. "Breadbasket"

The emphasis up to now has been on the semi-arid to arid Western United States because that is where the late-20th century drought began and has largely persisted up to the present time. The present drought has therefore largely missed the important crop producing States in the Midwest and Great Plains. Yet, previous studies (*Laird et al., 1996; Woodhouse and Overpeck, 1998; Stahle et al., 2000, 2007*) indicate that megadroughts have also occurred in those regions as well. To illustrate this, we have used the updated NADA to produce an average PDSI series for the Great Plains rectangle indicated in Figure 3.7. That series is shown in Figure 3.9C and it is far more provocative than even the Southwest series. The MCA period shows even more persistent drought, now on the centennial time scale, and the 15th century megadrought stands out more strongly as well. The duration of the MCA megadrought in our record is highly consistent with the salinity record from Moon Lake in North Dakota that likewise shows centennial time scale drought around that time. More ominously, in comparison, the 20th century has been a period of relatively low hydroclimatic variability, with the 1930s Dust Bowl and 1950s southern Great Plains droughts being rather unexceptional when viewed from a paleoclimate perspective. The closest historical analog to the extreme past megadroughts is the Civil War drought (*Herweijer et al., 2006*) from 1855 to 1865 (11 years), followed closely by a multiyear drought in the 1 870s. Clearly, there is a great need to understand the causes of long-term drought variability in the Great Plains and the U.S. "Breadbasket" to see how the remarkable past megadroughts indicated in Figure 3.9C developed and persisted. That these causes may be more complicated than those identified with the tropical oceans is suggested by the work of *Fye et al. (2006)*, who found that drought variability in the Mississippi River valley is significantly coupled to variations in the NAO (see also Sec. 2.2).

3.6. Drought in the Eastern United States

Up to this point, the emphasis on drought over the past 2,000 years has been restricted to the Western United States and Great Plains. This choice was intentional because of the current multiyear drought affecting the West, historic droughts of remarkable severity that have struck there (e.g., the 1930s "Dust Bowl" drought), and that region's susceptibility to multi-decadal megadroughts based on the tree-ring evidence (*Herweijer et al., 2007*). Even so, the normally well-watered Eastern United States is also vulnerable to severe droughts, both historically (*Hoyt, 1936; Namias, 1966; Karl and Young, 1987; Manuel, 2008*) and in tree-ring records (*Cook and Jacoby, 1977; Cook et al., 1988; Stahle et al., 1998*), but they have tended to be much shorter in duration compared to those in the West. Does this mean that the Eastern United States has not experienced megadroughts of similar duration as those during the MCA in the West? Evidence from high-resolution sediment core samples from the lower Hudson Valley in New York (*Pederson et al., 2005*) suggest that there was indeed a period of prolonged dryness there centered around the MCA. *Stahle et al. (1988)* also found tree-ring evidence for unusually persistent drought in North Carolina again during the MCA, as did *Seager et al. (2008)* for the greater Southeast based on the updated NADA. So it appears that megadroughts have also occurred in the Eastern United States, especially during the MCA. The cause of these extended-duration droughts in the Eastern United States during the MCA is presently not well understood.

4. Abrupt Hydrologic Changes during the Holocene

Examination of abrupt climate change during the Holocene (i.e., prior to the beginning of the instrumental or dendroclimatological records) can be motivated by the observation that the projected changes in both the radiative forcing and the resulting climate of the 21st century far exceed those registered by the either the instrumental records of the past century or by the proxy records of the past few millennia (*Jansen et al., 2007; Hegerl et al., 2003, 2007; Jones and Mann, 2004*). In other words, all of the variations in climate over the instrumental period and over the past millennium reviewed above have occurred in a climate system whose controls have not differed much from those of the most of the 20th century. In particular, variations in global-averaged radiative forcing as described in the IPCC Fourth Assessment (*IPCC, 2007*) include:

- values of roughly ±0.5 watts per meter squared (W m^{-2}) (relative to a 1500 to 1899 mean) related to variations in volcanic aerosol loadings and inferred changes in solar irradiance, i.e., from natural sources (*Jansen et al., 2007, Figure 6.13*);
- total anthropogenic radiative forcing of about 1.75 W m^{-2} from 1750 to 2005 from long-lived greenhouse gases, land-cover change, and aerosols (*Forster et al., 2007, Figure 2.20b*);
- projected increases in anthropogenic radiative forcing from 2000 to 2100 of around 6 W m^{-2} (*Meehl et al., 2007, Figure 10.2*).

In the early Holocene, annual-average insolation forcing anomalies (at 8 ka relative to present) range from -1.5 W m^{-2} at the equator to over +5 W m^{-2} at high latitudes in both hemispheres, with July insolation anomalies around +20 W m^{-2} in the mid-latitudes of the Northern Hemisphere (*Berger, 1978; Berger and Loutre, 1991*). Top-of-the-atmosphere insolation is not directly comparable with the concept of radiative forcing as used in the IPCC Fourth Assessment (*Committee on Radiative Forcing Effects on Climate, 2005*), owing to feedback from the land surface and atmosphere, but the relative size of the anomalies supports the idea that potential future changes in the controls of climate exceed those observed over the past millennium (*Joos and Sphani, 2008*). Consequently, a longer term focus is required to describe the behavior of the climate system under controls as different from those at present as those of the 21st century will be, and to assess the potential for abrupt climate changes to occur in response to gradual changes in large-scale forcing.

The controls of climate during the 21st century and during the Holocene differ from one another, and from those of the 20th century, in important ways. The major contrast in controls of climate between the early 20th, late 20th, and 21st century are in atmospheric composition (with an additional component of land-cover change), while the major contrast between the controls in the 20th century and those in the early to middle Holocene were in the latitudinal and seasonal distribution of insolation. In the Northern Hemisphere in the early Holocene, summer insolation was around 8% greater than present, and winter about 8% less than present, related to the amplification of the seasonal cycle of insolation due to the occurrence of perihelion in summer then, while in the Southern Hemisphere the amplitude of the seasonal cycle of insolation was reduced (*Webb et al., 1993b*). In both hemispheres in the early Holocene, annual insolation was greater than present poleward of 45°, and less than present between 45°N. and 45°S., related to the greater tilt of Earth's axis than relative to today. The energy balance of the Northern Hemisphere during the early Holocene thus features a large increase in seasonality relative to that of the 20th century. This contrast between the past and future will increase throughout the 21st century owing to the ongoing and projected further reduction in snow and ice cover in the Northern Hemisphere winter.

Consequently, climatic variations during the Holocene should not be thought of either as analogs for future climates or as examples of what might be observable under present-day climate forcing if records were longer, but instead should be thought of as a "natural experiment" (i.e., an experiment not purposefully performed by humans) with the climate system that features large perturbations of the controls of climate, similar in scope (but not in detail) to those expectable in the future. In particular, the climates of both the Holocene and the 21st century illustrate the response of the climate system to significant perturbations of radiative forcing relative to that of the 20th or 21st century.

4.1. Examples of large and rapid hydrologic changes during the Holocene

From the perspective of the present and with a focus on the northern mid-latitudes, the striking spatial feature of Holocene climate variations was the wastage and final disappearance of the middle- to high-latitude North American and Eurasian ice sheets. However, over the much larger area of the tropics and adjacent subtropics, there were equally impressive hydrologic changes, ultimately related to insolation-driven variations in the global monsoon (*COHMAP Members, 1988; Liu et al., 2004*). Two continental- scale hydrologic changes that featured abrupt (on a Holocene time scale) transitions between humid and arid conditions were those in northern Africa and in the mid- continent of North America. In northern Africa, the "African humid period" began after 12 ka with an intensification of the African-Asian monsoon, and ended around 5 ka (*deMenocal et al., 2000; Garcin et al., 2007*), with the marked transition from a "green" (vegetated) Sahara, to the current "brown" (or sparsely vegetated) state. This latter transition provides an example of a climate change that would have significant societal impact if it were to occur today in any region, and provides an example of an abrupt transition to drought driven by gradual changes in large-scale external controls.

In North America, drier conditions than those at present commenced in the mid-continent between 10 and 8 ka (*Thompson et al., 1993; Webb et al., 1993a; Forman et al., 2001*), and ended after 4 ka. This "North American mid-continental Holocene drought" was coeval with dry conditions in the Pacific Northwest, and wet conditions in the south and southwest, in manner consistent (in a dynamic atmospheric circulation sense) with the amplification of the monsoon then (*Harrison et al., 2003*). The mid-Holocene drought in mid-continental North America gave way to wetter conditions after 4 ka, and like the African humid period, provides an example of major, and sometimes abrupt hydrological changes that occurred in response to large and gradual changes in the controls of regional climates.

BOX 3.3. PALEOCLIMATIC DATA/MODEL COMPARISONS

Two general approaches and information sources for studying past climates have been developed. *Paleoclimatic observations* (also known as proxy data) consist of paleoecological, geological, and geochemical data, that when assigned ages by various means, can be interpreted in climatic terms. Paleoclimatic data provide the basic documentation of what has happened in the past, and can be synthesized to reconstruct the patterns history of paleoclimatic variations. *Paleoclimatic simulations* are created by identifying the configuration of large-scale controls of climate (i.e. solar radiation, and its latitudinal and seasonal distribution, or the concentration of greenhouse gases in the atmosphere) at a particular time in the past, and then supplying these to a global or regional climate model to generate sequences of simulated meteorological data, in a fashion similar to the use of a numerical weather forecasting model today. (See CCSP SAP 3.1 for a discussion on climate models.) Both approaches are necessary for understanding past climatic variations—the paleoclimatic observations document past climatic variations but can't explain them without some kind of model, and the models that could provide such explanations must first be tested and shown to be capable of simulating the patterns in the data.

The two approaches are combined in *paleoclimatic data/model comparison* studies, in which syntheses of paleoclimatic data from different sources and suites of climate- model simulations performed with different models are combined in an attempt to replicate a past "natural experiment" with the real climate system, such as those provided by the regular changes in incoming solar radiation related to Earth's orbital variations. Previous generations of data/model comparison studies have focused on key times in the paleoclimatic record, such as the Last Glacial Maximum (21,000 years ago) or mid-Holocene (6000 years ago), but attention is now turning to the study of paleoclimatic variability as recorded in high-resolution time series of paleoclimatic data and generated by long "transient" simulations with models.

Paleoclimatic data/model comparisons contribute to our overall perspective on climate change, and can provide critically needed information on how realistically climate models can simulate climate variability and change, what the role of feedbacks in the climate system are in amplifying or damping changes in the external controls of climate, and the general causes and mechanisms involved in climate change.

These continental-scale hydrologic changes obviously differ in the sign of the change (wet to dry from the middle Holocene to present in Africa and dry to wet from the middle Holocene to present in North America), and in the specific timing and spatial coherence of the hydrologic changes, but they have several features in common, including:

- the initiation of the African humid period and the North American Holocene drought were both related to regional climate changes that occurred in response to general deglaciation and to variations in insolation;
- the end of the African humid period and the North American Holocene drought were both ultimately related to the gradual decrease in Northern Hemisphere summer insolation during the Holocene, and to the response of the global monsoon;
- paleoclimatic simulations suggest that ocean-atmosphere coupling played a role in determining the moisture status of these regions, as it has during the 20th century and the past millennium;
- feedback from local land-surface (vegetation) responses to remote (sea-surface temperature, ocean-atmosphere interaction) and global (insolation, global ice volume, atmospheric composition) forcing may have played a role in the magnitude and rapidity of the hydrological changes.

Our understanding of the scope of the hydrologic changes and their potential explanations for both of these regions have been informed by interactions between paleoclimatic data syntheses and climate-model simulations (e.g., *Wright et al., 1993; Harrison et al., 2003; Liu et al., 2007,* see box 3.3). In this interaction, the data syntheses have driven the elaboration of both models and experimental designs, which in turn have led to better explanations of the patterns observed in the data (see *Bartlein and Hostetler, 2004*).

4.2. The African Humid Period

One of the major environmental variations over the past 10,000 years, measured in terms of the area affected, the magnitude of the overall climatic changes and their rapidity, was the reduction in magnitude around 5,000 years ago of the African-Asian monsoon from its early to middle Holocene maximum, and the consequent reduction in vegetation cover and expansion of deserts, particularly in Africa south of the Saraha. The broad regional extent of enhanced early Holocene monsoons is revealed by the status of lake levels across Africa and Asia (Figure 3.10), and the relative wetness of the interval is further attested to by similarly broad-scale vegetation changes (*Jolly et al., 1998; Kohfeld and Harrison, 2000*). Elsewhere in the region influenced by the African-Asian monsoon, the interval of enhanced monsoonal circulation and precipitation also ended abruptly, in the interval between 5.0 and 4.5 ka across south and east Asia (*Morrill et al., 2003*), demonstrating that the African humid period was embedded in planetary-scale climatic variations during the Holocene.

A general conceptual model has emerged (see *Ruddiman, 2006*) that relates the intensification of the monsoons to the differential heating of the continents and oceans that occurs in response to orbitally induced amplification of the seasonal cycle of insolation (i.e., increased summer and decreased winter insolation in the Northern Hemisphere) (*Kutzbach and Otto-Bliesner, 1982; Kutzbach and Street-Perrott, 1985; Liu et al., 2004*). In addition to the first-order response of the monsoons to insolation forcing, other major controls of regional climates, like the atmospheric circulation variations related to the North American ice sheets, to ocean/atmospheric circulation reorganization over the North Atlantic (*Kutzbach and Ruddiman, 1993; Weldeab et al., 2007*), and to tropical Pacific ocean/atmosphere interactions (*Shin et al., 2006; Zhao et al., 2007*) likely also played a role in determining the timing and details of the response. In many paleoenvironmental records, the African humid period (12 ka to 5 ka) began rather abruptly (relative to the insolation forcing), but with some spatial variability in its expression (*Garcin et al., 2007*), and similarly, it ended abruptly (*deMenocal et al., 2000*; and see the discussion in *Liu et al., 2007*).

The robust expression of the wet conditions (Figure 3.10) together with the amplitude of the "signal" in the paleoenvironmental data has made the African humid period a prime focus for synthesis of paleoenvironmental data, climate-model simulations, and the systematic comparison of the two (*COHMAP Members, 1988*), in particular as a component of the Palaeoclimatic Modeling Intercomparison Project (PMIP and PMIP 2; *Joussaume et al., 1999; Crucifix et al., 2005; Braconnot et al., 2007a,b*). The aim of these paleoclimatic data-model comparisons is twofold: (1) to "validate" the climate models by examining their ability to correctly reproduce an observed environmental change for which the ultimate controls are known and (2) to use the mechanistic aspects of the models and simulations produced with them to explain the patterns and variations recorded by the data. Mismatches between the simulations and observations can arise from one or more sources, including inadequacies of the climate models, misinterpretation of the paleoenvironmental data, and incompleteness of the experimental design (i.e., failure to include one or more controls or processes that influenced the real climate) (*Peteet, 2001; Bartlein and Hostetler, 2004*).

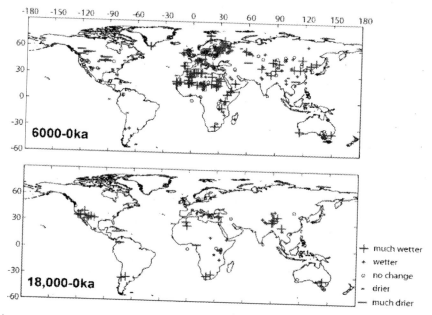

Figure 3.10. Global lake status at 6 ka (6,000 years ago) showing the large region that extends from Africa across Asia, where lake levels were higher than those of the present day as related to the expansion of the African-Asian monsoon. Note also the occurrence of much drier than present conditions over North America. (The most recent version of the Global Lake Surface Database is available on the PMIP 2 website http://pmip2.lsce.ipsl.fr/share/synth/glsdb/lakes

In general, the simulations done as part of PMIP, as well as others, show a clear amplification of the African-Asian monsoon during the early and middle part of the Holocene, but one that is insufficient to completely explain the magnitude of the changes in lake status, and the extent of the observed northward displacement of the vegetation zones into the region now occupied by desert (*Joussaume et al., 1999; Kohfeld and Harrison, 2000*). The initial PMIP simulations were "snapshot" or "time-slice" simulations of the conditions around 6 ka, and as a consequence are able to only indirectly comment on the mechanisms involved in the abrupt beginning and end of the humid period. In addition, the earlier simulations were performed using AGCMs, with present-day land-surface characteristics, which therefore did not adequately represent the full influence of the ocean or terrestrial vegetation on the simulated climate.

As a consequence, climate-simulation exercises that focus on the African monsoon or the African humid period have evolved over the past decade or so toward models and experimental designs that (1) include interactive coupling among the atmosphere, ocean, and terrestrial biosphere and (2) feature transient, or time-evolving simulations that, for example, allow explicit examination of the timing and rate of the transition from a green to a brown Sahara. Two classes of models have been used, including (1) Atmosphere Ocean General Circulation Models with interactive oceans (AOGCMs), Atmosphere terrestrial Vegetation General Circulation Models (AVGCMs), or both (AOVGCMs) that typically have spatial resolutions of a few degrees of latitude and longitude and (2) coarser resolution EMICs, or Earth-system models of intermediate complexity, that include representation of components of the climate system that are not amenable to simulation with the higher resolution GCMs (See *Claussen, 2001*, and *Bartlein and Hostetler, 2004*, for a discussion of the taxonomy of climate models.)

The coupled AOGCM simulations have illuminated the role that sea surface temperatures likely played in the amplification of the monsoon. Driven by both the insolation forcing and by ocean-atmosphere interactions, the picture emerges of a role for the oceans in modulating the amplified seasonal cycle of insolation during the early and mid-Holocene in such a way as to increase the summertime temperature contrast between continent and ocean that drives the monsoon, thereby strengthening it (*Kutzbach and Liu, 1997; Zhao et al., 2005*). In addition, there is an apparent role for teleconnections from the tropical Pacific in determining the strength of the monsoon, in a manner similar to the "atmospheric bridge" teleconnection between the tropical Pacific ocean and climate elsewhere at present (*Shin et al., 2006; Zhao et al., 2007; Liu and Alexander, 2007*).

The observation of the dramatic vegetation change motivated the development of simulations with coupled vegetation components, first by asynchronously coupling equilibrium global vegetation models (EGVMs, *Texier et al., 1997*), and subsequently by using fully coupled AOVGCMs (e.g., *Levis et al., 2004; Wohlfahrt et al., 2004; Gallimore et al., 2005; Braconnot et al., 2007a,b; Liu et al., 2007*). These simulations, which also included investigation of the synergistic effects of an interactive ocean and vegetation on the simulated climate (*Wohlfahrt et al., 2004*), produced results that still underrepresented the magnitude of monsoon enhancement, but to a lesser extent than the earlier AGCM or AOGCM simulations. These simulations also suggest the specific mechanisms through which the vegetation and the related soil-moisture conditions (*Levis et al., 2004; Liu et al., 2007*) influence the simulated monsoon.

The EMIC simulations, run as transient or continuous (as opposed to time-slice) simulations over the Holocene, are able to explicitly reveal the time history of the monsoon intensification or deintensification, including the regional-scale responses of surface climate and vegetation (*Claussen et al., 1999; Hales et al., 2006; Renssen et al., 2006*). These simulations typically show abrupt decreases in vegetation cover, and usually also in precipitation, around the time of the observed vegetation change (5 ka), when insolation was changing only gradually. The initial success of EMICs in simulating an abrupt climate and land-cover change in response to a gradual change in forcing influenced the development of a conceptual model that proposed that strong nonlinear feedbacks between the land surface and atmosphere were responsible for the abruptness of the climate change, and, moreover, suggested the existence of multiple stable states of the coupled climate-vegetation-soil system that are maintained by positive vegetation feedback (*Claussen et al, 1999; Foley et al., 2003*). In such a system, abrupt transitions from one state to another (e.g., from a green Sahara to a brown one), could occur under relatively modest changes in external forcing, with a green vegetation state and wet conditions reinforcing one another, and likewise a brown state reinforcing dry conditions and vice versa. The positive feedback involved in maintaining the green or brown states would also promote the conversion of large areas from one state to the other at the same time.

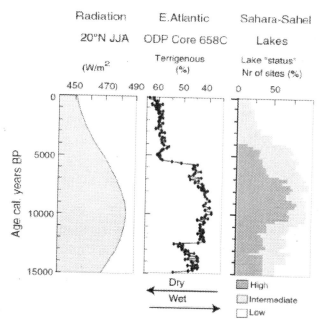

Figure 3.11. African Humid Period records (*Liu et al., 2007*).

A different perspective on the way in which abrupt changes in the land-surface cover of west Africa may occur in response to gradual insolation changes is provided by the simulations by *Liu et al. (2006, 2007)*. They used a coupled AOVGCM (FOAM-LPJ) run in transient mode to produce a continuous simulation from 6.5 ka to present. They combined a statistical analysis of vegetation-climate feedback in the AOVGCM, and an analysis of a simple conceptual model that relates a simple two-state depiction of vegetation to annual precipitation (*Liu et al., 2006*), and argue that the short-term (i.e. year-to-year) feedback between vegetation and climate is negative (see also *Wang et al., 2007; Notaro et al., 2008*), such that a sparsely or unvegetated state (i.e., a brown Sahara) would tend to favor precipitation through the recycling of moisture from bare-ground evapotranspiration. In this view, the negative vegetation feedback would act to maintain the green Sahara against the general drying trend related to the decrease in the intensity of the monsoon and amount of precipitation, until such time that interannual variability results in the crossing of a moisture threshold beyond which the green state could no longer be maintained (see *Cook et al., 2006,* for further discussion of this kind of behavior in response to interannual climate variability (i.e., ENSO). In this conceptual model, the transition between states, while broadly synchronous (owing to the large-scale forcing), might be expected to show a more time-transgressive or diachronous pattern owing to the influence of landscape (soil and vegetation) heterogeneity.

These two conceptual models of the mechanisms that underlie the abrupt vegetation change—strong feedback and interannual variability/threshold crossing—are not that different in terms of their implications, however. Both conceptual models relate the overall decrease in moisture and consequent vegetation change to the response of the monsoon to the gradually weakening amplification of the seasonal cycle of insolation, and both claim a role for vegetation in contributing to the abruptness of the land-cover change, either explicitly or implicitly invoking the nonlinear relationship between vegetation cover and precipitation (Figure 3.11 from *Liu et al., 2007*). The conceptual models differ mainly in their depiction of the precipitation change, with the strong- feedback explanation predicting that abrupt changes in precipitation will accompany the abrupt changes in vegetation, while the interannual variability/threshold crossing explanation does not. It is interesting to note that the *Renssen et al. (2006)* EMIC simulation generates precipitation variations for west Africa that show much less of an abrupt change around 5 ka than did earlier EMIC simulations, which suggests that the strong-feedback perspective may be somewhat model dependent. A recent analysis of a paleolimnological record from the eastern Sahara (*Kröpelin et al., 2008*) shows a more gradual transition from the green to brown state than would be inferred from the marine record of dust flux, which also supports the variability/threshold crossing model.

There is thus some uncertainty in the specific mechanisms that link the vegetation response to climate variations on different time scales, and also considerable temporal- spatial variability in the timing of environmental changes. However, the African humid period and its rapid termination illustrates how abrupt, widespread, and significant environmental changes can occur in response to gradual changes in a large-scale or ultimate control—in this case the amplification of the seasonal cycle of insolation in the Northern Hemisphere and its impact on radiative forcing.

4.3. North American Mid-Continental Holocene Drought

At roughly the same time as the African humid period, large parts of North America experienced drier-than-present conditions that were sufficient in magnitude to be registered in a variety of paleoenvironmental data sources. Although opposite in sign from those in Africa, these moisture anomalies were ultimately related to the same large- scale control - greater-than-present summer insolation in the Northern Hemisphere. In North America, however, the climate changes were also strongly influenced by the shrinking (but still important regionally) Laurentide Ice Sheet. In contrast to the situation in Africa, and likely related to the existence of additional large-scale controls (e.g., the remnant ice sheet, and Pacific ocean-atmosphere interactions), the onset and end of the middle Holocene moisture anomaly was more spatially variable in its expression, but like the African humid period, it included large-scale changes in land cover in addition to effective-moisture variations. Also in contrast to the African situation, the vegetation changes featured changes in the type of vegetation or biomes (e.g., shifts between grassland and forest, *Williams et al., 2004*), as opposed to fluctuations between vegetated and nonvegetated or sparsely vegetated states. There are also indications that, as in Africa and Asia, the North American monsoon was amplified in the early and middle

Holocene (*Thompson et al., 1993; Mock and Brunelle-Daines, 1999; Poore et al., 2005*), although as in the case of the dry conditions, there probably was significant temporal and spatial variation in the strength of the enhanced monsoon (*Barron et al., 2005*). The modern association of dry conditions across central North America and somewhat wetter conditions in North Africa during a La Niña phase (*Palmer and Brankovic, 1989*), led *Forman et al. (2001)* to hypothesize that changes in tropical sea surface variability, in particular the persistence of La Niña-type conditions (generally colder and warmer than those at present in the eastern and western parts of the basin, respectively), might have played an important role in modulating the regional impacts of mid-Holocene climate.

A variety of paleoenvironmental indicators reflect the spatial extent and timing of these moisture variations (Figures 3.12 and 3.13), and in general suggest that the dry conditions increased in their intensity during the interval from 11 ka to 8 ka, and then gave way to increased moisture after 4 ka, and during the middle of this interval (around 6 ka) were widespread. Lake-status indicators at 6 ka indicate lower-than-present levels (and hence drier-than-present conditions) across much of the continent (*Shuman et al., in review*), and quantitative interpretation of the pollen data in *Williams et al. (2004)* shows a similar pattern of overall aridity, but again with some regional and local variability, such as moister-than-present conditions in the Southwestern United States (see also *Thompson et al., 1993*). Although the region of drier-than-present conditions extends into the Northeastern United States and eastern Canada, most of the multiproxy evidence for middle Holocene dryness is focused on the midcontinent, in particular the Great Plains and Midwest, where the evidence for aridity is particularly clear. There, the expression of middle Holocene dry conditions in paleoenvironmental records has long been known, as was the case for the "Prairie Period" evident in fossil-pollen data (see *Webb et al., 1983*), and the recognition of significant aeolian activity (dune formation) on the Great Plains (*Forman et al., 2001; Harrison et al., 2003*) that would be favored by a decrease in vegetation cover.

Temporal variations in the large-scale controls of North American regional climates as well as some of the paleoenvironmental indicators of the moisture changes are shown in Figure 3.13. In addition to insolation forcing (Figure 3.1 3A,B), the size of the Laurentide Ice Sheet was a major control of regional climates, and while diminished in size from its full extent at the Last Glacial Maximum (21 ka), the residual ice sheets at 11 ka and 9 ka (Figure 3.1 3C) still influence atmospheric circulation over eastern and central North America in climate simulations for those times (*Bartlein et al., 1998; Webb et al., 1998*). In addition to depressing temperatures generally around the Northern Hemisphere, the ice sheets also directly influenced adjacent regions. In those simulations, the development of a "glacial anticyclone" over the ice sheet (while not as pronounced as earlier), acted to diminish the flow of moisture from the Gulf of Mexico into the interior, thus keeping the midcontinent cooler and drier than it would have been in the absence of an ice sheet.

Superimposed on these "orbital time scale" variations in controls and regional responses are millennial-scale variations in atmospheric circulation related to changes in the Atlantic Meriodional Overturning Circulation (AMOC) and to other ocean-atmosphere variability (*Shuman et al., 2005, 2007; Viau et al., 2006*). Of these millennial-scale variations, the "8.2 ka event" (Figure 3.1 3D) is of interest, inasmuch as the climate changes associated with the "collapse" of the Laurentide Ice Sheet (*Barber et al., 1999*) have the potential to influence the mid-continent region directly, through regional atmospheric circulation changes (*Dean et al.,*

2002; Shuman et al., 2002), as well as indirectly, through its influence on AMOC, and related hemispheric atmospheric circulation changes.

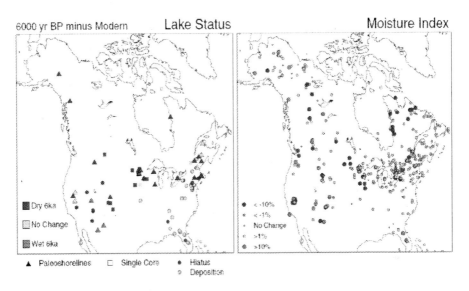

Figure 3.12. North American lake status (left) and moisture-index (AE/PE) anomalies (right) for 6 ka. Lake (level) status can be inferred from a variety of sedimentological and limnological indicators (triangles and squares), and from the absence of deposition (hiatuses, circles) (*Shuman and Finney, 2006*). The inferred moisture-index values are based on modern analog technique applied to a network of fossil-pollen data. Figure adapted from *Shuman et al. (in review)*.

The record of aridity indicators for the midcontinent reveals a more complicated history of moisture variations than does the African case, with some locations remaining dry until the late Holocene, and others reaching maximum aridity during the interval between 8 ka and 4 ka, but in general showing relatively dry conditions between 8 ka and 4 ka. Lake-status records (Figure 3.13E, *Shuman et al., in review*) show the highest frequency of lakes at relatively low levels during the interval between 8 ka and 4 ka, and a higher frequency of lakes at relatively high levels before and after that interval. Records of widespread and persistent aeolian activity and loess deposition (dust transport) increase in frequency from 10 ka to 8 ka, and then gradually fall to lower frequency in the late Holocene, with a noticeable decline between 5 ka and 4 ka. Pollen records of the vegetation changes that reflect dry conditions (Figure 3.1 3G; *Williams, 2002; Williams et al., 2004*) show a somewhat earlier onset of dryness than do the aeolian or lake indicators, reaching maximum frequency around 9 ka. Increased aeolian activity can also be noted during the last 2000 years (Figure 3.13F, *Forman et al., 2001; Miao et al., 2007*), but was less pronounced than during the mid-Holocene.

The pollen record from Steel Lake, MN, expressed in terms of tree-cover percentages (see *Williams, 2002*, for methods) provides an example to illustrate a pattern of moisture-related vegetation change that is typical at many sites in the Midwest, with an abrupt decline in tree cover at this site around 8 ka, and over an interval equal to or less than the sampling resolution of the record (about 200 years, Figure 3.13H). This decrease in tree cover and inferred moisture levels is followed by relatively low but slightly increasing inferred moisture levels for about 4,000 years, with higher inferred moisture levels in the last 4,000 years.

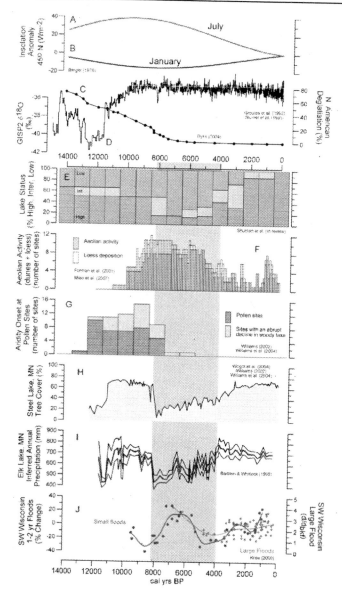

Figure 3.13. Time series of large-scale climate controls (A-D) and paleoenvironmental indicators of North American midcontinental aridity (E-I). A, B, July and January insolation anomalies (differences relative to present) (*Berger, 1978*). C, right-hand scale: Deglaciation of North America, expressed as ice-sheet area relative to that at the Last Glacial Maximum (21 ka) (*Dyke, 2004*). D, left-hand scale: Oxygen-isotope data from the GISP 2 Greenland ice core (*Grootes et al, 1993; Stuiver et al., 1995*). Increasingly negative values indicate colder conditions. The abrupt warming at the end of the Younger Dryas chronozone (GS1/Holocene transition, 11.6 ka) is clearly visible, as is the "8.2 ka event" that marks the collapse of the Laurentide Ice Sheet. E, Lake status in central North America (*Shuman et al., in review*). Colors indicate the relative proportions of lake-status records that show lake levels that are at relatively high, intermediate, or low levels. F, Eolian activity indicators (orange, digitized from Figure 13 in *Forman et al., 2001*) and episodes of loess deposition (yellow, digitized from Figure 3 of *Miao et al., 2007*). G, Pollen indicators of the onset of aridity. Light-green bars indicate the number of sites with abrupt decreases in the abundance of woody taxa (data from *Williams, 2002; Williams et al., 2004*). H, Inferred tree-cover percentage at one of the sites (Steel Lake, MN) summarized in panel G (*Williams, 2002; Williams et al., 2004*; based on pollen data from *Wright et al., 2004*). I, Inferred

annual precipitation values for Elk Lake, MN, a site close to Steel Lake (*Bartlein and Whitlock, 1993*). The inferred annual precipitation values here (as well as inferences made using other paleoenvironmental indicators) suggest that the precipitation anomaly that characterized the middle Holocene aridity is on the order of 350 mm y^{-1}, or about 1 mm d^{-1}. J, Frequency and magnitude of floods across a range of watershed sizes tracks climate variation during the Holocene. The gray shading indicates the interval of maximum aridity.

The magnitude of this moisture anomaly can be statistically inferred from the fossil-pollen data using modern relationships between pollen abundance and climate, as was done for the pollen record at Elk Lake, MN, which is near Steel Lake (Figure 3.13I; *Bartlein and Whitlock, 1993*; see also *Webb et al., 1998*). Expressed in terms of precipitation, the moisture decrease in the midcontinent needed for these vegetation changes is about 350 millimeters per year (mm y^{-1}), or about 1 millimeter per day (mm d^{-1}), or levels between 50 and 80 percent of the present-day values.

As is the case for the African humid period, the effective-moisture variations recorded by paleoenvironmental data from the midcontinent of North America provide a target for simulation by climate models, and also as was the case for Africa, those simulations have evolved over time toward models with increased coupling among systems. The first generation of simulations with AGCMs featured models that were of relatively coarse spatial resolution, had fixed SSTs, and land cover that was specified to match that of the modern day. These simulations, focusing on 6 ka, revealed some likely mechanisms for developing dry conditions in the midcontinent, such as the impact of the insolation forcing on surface energy and water balances and the direct and indirect effects of insolation on atmospheric circulation (*Webb et al., 1993b; Bartlein et al., 1998; Webb et al., 1998*). However, the specific simulations of precipitation or precipitation minus evapotranspiration (P-E) indicated little change in moisture or even increases in some regions. Given the close link between SST variations and drought across North America at present, and the inability of these early simulations to simulate such mechanisms because they had fixed SSTs, this result is not surprising.

What can be regarded as the current-generation simulations for 6 ka include those done with fully coupled AOGCMs (FOAM and CSM 1, *Harrison et al., 2003*; CCSM 3, *Otto-Bliesner et al., 2006*), and an AGCM with a mixed-layer ocean (CCM 3.10, *Shin et al., 2006*). These simulations thereby allow the influence of SST variations to be registered in the simulated climate either implicitly, by calculating them in the ocean component of the models (FOAM, CSM 1, CCSM 3), or explicitly, by imposing them either as present-day long-term averages, or as perturbations of those long-term averages intended to represent extreme states of, for example, ENSO (CCM 3.10). The trade-off between these approaches is that the fully coupled, implicit approach will reflect the impact of the large- scale controls of climate (e.g., insolation) on SST variability (if the model simulates the joint response of the atmosphere and ocean correctly), while the explicitly specified AGCM approach allows the response to a hypothetical state of the ocean to be judged.

These simulations produce generally dry conditions in the interior of North America during the growing season (and an enhancement of the North American monsoon), but as was the case for Africa, the magnitude of the moisture changes is not as large as that recorded by the paleoenvironmental data (with maximum precipitation-rate anomalies on the order of 0.5 mm d^{-1}, roughly half as large as it would need to be to match the paleoenvironmental observations). Despite this, the simulations reveal some specific mechanisms for generating the dry conditions; these include (1) atmospheric circulation responses to the insolation and SST forcing/feedback that favor a "package" of circulation anomalies that include expansion of the subtropical high-pressure systems in summer, (2) the development of an upper-level ridge and large-scale subsidence over central North America (a circulation feature that favors drought at the present), and (3) changes in surface energy and water balances that lead to reinforcement of this circulation configuration. Analyses of the 6 ka simulated and present-day "observed" (i.e., reanalysis data) circulation were used by *Harrison et al. (2003)* to describe the linkage that exists in between the uplift that occurs in the Southwestern United States and northern Mexico as part of the North American monsoon system, and subsidence on the Great Plains and Pacific Northwest (*Higgins et al., 1997*; see also *Vera et al., 2006*).

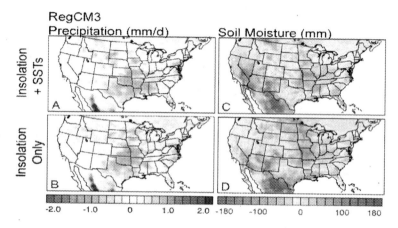

Figure 3.14. Regional climate model (RegCM3) simulations of precipitation rate (A, B) and soil moisture (C, D) for 6,000 years before present (6 ka) (*Diffenbaugh et al., 2006*, land grid points only). RegCM is run using lateral boundary conditions supplied by CAM3, the atmospheric component of CCSM3. In panels A and C, the CAM3 boundary conditions included 6 ka-insolation, and time-varying sea surface temperatures (SSTs) provided by a fully coupled Atmosphere-Ocean General Circulation Model (AOGCM) simulation for 6 ka using CCSM3 (*Otto-Bliesner et al., 2006*). In panels B and D, the CAM3 boundary conditions included 6-ka insolation, and time-varying SSTs provided by a fully coupled CCSM simulation for the present. The differences between simulations reveal the impact of the insolation-forced differences in SST variability between 6 ka and present. mm, millimeters; mm/d, millimeters per day.

The summertime establishment of the upper-level ridge, the related subsidence over the middle of the North American continent, and the onshore flow and uplift in the Southwestern United States and northern Mexico are influenced to a large extent by the topography of western North America, which is greatly oversimplified in GCMs (see Figure 4 in *Bartlein and Hostetler, 2004*). This potential "built-in" source of mismatch between the paleoclimatic simulations and observations can be reduced by simulating climate with regional climate models (RCMs). Summer (June, July, and August) precipitation and soil moisture simulated using RegCM3 (*Diffenbaugh et al., 2006*) is shown in Figure 3.14, which illustrates moisture anomalies that are more comparable in magnitude to those recorded by the paleoenvironmental data than are the GCM simulations. RegCM as applied in these simulations has a spatial resolution of 55 km, which resolves climatically important details of the topography of the Western United States. In these simulations, the "lateral boundary conditions" or inputs to the RCM, were supplied by a simulation using an AGCM (CAM 3), that in turn used the SSTs simulated by the fully coupled AOGCM simulation for 6 ka (and present) by *Otto-Bliesner et al. (2006)*. These SSTs were also supplied directly to RegCM3. The simulations thus reveal the impact of the insolation forcing, as well as the influence of the insolation-related changes on interannual variability in SSTs (over the 30 years of each simulation). The results clearly show the suppression of precipitation over the mid-continent and enhancement over the Southwestern U.S. and northern Mexico, and the contribution of the precipitation anomaly to that of soil moisture (Figure 3.14). In contrast to the GCM simulations, the inclusion of 6 ka SST variability in the RCM simulations reduces slightly the magnitude of the moisture anomalies, but overall these anomalies are close to those inferred from paleoenvironmental observations and reinforce the conceptual model linking the North American mid-continental Holocene drought to increased subsidence (see also *Shinker et al., 2006; Harrison et al., 2003*).

The potential of vegetation feedback to amplify the middle Holocene drought has not been as intensively explored as it has for Africa, but those explorations suggest that it should not be discounted. *Shin et al. (2006)* prescribed some subjectively reconstructed vegetation changes (e.g., *Diffenbaugh and Sloan, 2002*) in their AGCM simulations and noted a reduction in spring and early summer precipitation (that could carry over into reduced soil moisture during the summer), but also noted a variable response in precipitation during the summer to the different vegetation specifications. *Wohlfahrt et al. (2004)* asynchronously coupled an equilibrium global vegetation model, Biome 4 (*Kaplan et al., 2003*), to an AOGCM and observed a larger expansion of grassland in those simulations than in ones without the vegetation change simulated by the EGVM. Finally, *Gallimore et al. (2005)* examined simulations using the fully coupled AOVGCM (FOAM-LPJ), and while the overall precipitation change for summer was weakly negative, the impact of the simulated vegetation change (toward reduced tree cover at 6 ka), produced a small positive precipitation change.

An analysis currently in progress with RegCM3 suggests that the inclusion of the observed middle Holocene vegetation in the boundary conditions for the 6 ka simulation described above (*Diffenbaugh et al., 2006*) further amplifies the negative summer precipitation anomaly in the core region of the Holocene drought, and also alters the nature of the seasonal cycle of the dependence of soil moisture on precipitation. The magnitude of the drought in these simulations is relatively close to that inferred from the paleoenvironmental data.

The North American midcontinental drought during the middle Holocene thus provides an illustration of a significant hydrologic anomaly with relatively abrupt onset and ending that occurred in response to gradual changes in the main driver of Holocene climate change (insolation), reinforced by regional- and continental-scale changes in atmospheric circulation related directly to deglaciation. As was the case for the African humid period, feedback from the vegetation change that accompanied the climate changes could be important in reinforcing or amplifying the climate change, and work is underway to evaluate that hypothesis.

There are other examples of abrupt hydrological responses to gradual or large-scale climatic changes during the Holocene. For example, the development of wetlands in the Northern Hemisphere began relatively early in the course of deglaciation but accelerated during the interval high summer insolation between 12 ka and 8 ka (*Gajewski et al., 2001; MacDonald et al., 2006*). The frequency and magnitude of floods across a range of different watershed sizes also tracks climate variations during the Holocene (Figure 3.13J; *Knox 1993, 2000; Ely, 1997*), albeit in a complicated fashion, owing to dependence of flooding on long-term climate and land-cover conditions as well as on short-term meteorological events (see Sec. 6).

4.4. Century-Scale Hydrologic Variations

Hydrologic variations, many abrupt, occur on time scales intermediate between the variations over millennia that are ultimately related to orbitally governed insolation variations and the interannual- to decadal-scale variations documented by annual- resolution proxy records. A sample of time series that describe hydrologic variations on decadal-to-centennial scales over the past 2,000 years in North America appears in Figure 3.15 and reveals a range of different kinds of variation, including:

- generalized trends across several centuries (Figure 3.15C,F,G);
- step-changes in level or variability (independent of sampling resolution) (Figure 3.15A,B,F);
- distinct peaks in wet (Figure 3.15A) or dry conditions (Figure 3.13F, Figure 3.15B,G);
- a tendency to remain persistently above or below a long term mean (Figure 3.1 5C-F), often referred to as "regime changes"; and
- variations in all components of the hydrologic cycle, including precipitation, evaporation, storage, and runoff, and in water quality (e.g., salinity).

Hydrological records that extend over the length of the Holocene, in particular those from hydrologically sensitive speleothems, demonstrate similar patterns of variability throughout (e.g., *Asmerom et al., 2007*), including long-term trends related to the Holocene history of the global monsoon described above (e.g., *Wang et al., 2005*).

The ultimate controls of these variations include (1) the continued influence of the long-term changes in insolation that appear to be ultimately responsible for the mid-Holocene climate anomalies discussed above, (2) the integration of interannual variations in climate that arise from ocean-atmosphere coupling, and (3) the impact of the variations in volcanism, solar irradiance, long-lived greenhouse gases and aerosols, and land-cover responsible for climatic variations over the past two millennia (*Jansen et al., 2007*, IPCC AR4 WG1, Sec. 6.6) or some combination of these three controls. (See also *Climate Research Committee, National Research Council, 1995*).

No one of these potential controls can account for all of the variations observed in hydrological indicators over the past two millennia. By the late Holocene, the amplitude of the insolation anomalies is quite small (Figure 3.1 3 A-B), and the impact of deglaciation is no longer significant (Figure 3.13C-D). Variations in indices that describe decadal-timescale ocean-atmosphere interactions, often known as "teleconnection" or "climate-mode" indices (e.g., the PDO or "Pacific Decadal Oscillation" or the NAM or "Northern Annular Mode;" see *Trenberth et al., 2007*, IPCC AR4 WG1 Sec. 3.6 for review), are sometimes invoked to explain apparent periodicity or "regime changes" in proxy records (e.g., *Stone and Fritz, 2006; Rasmussen et al., 2006*). However, the observational records that are used to define those indices are not long enough to discriminate among true cyclical or oscillatory behavior, recurrent changes in levels (or regime shifts), and simple red-noise or autocorrelated variations in time series (*Rudnick and Davis, 2003; Overland et al., 2006*), so perceived

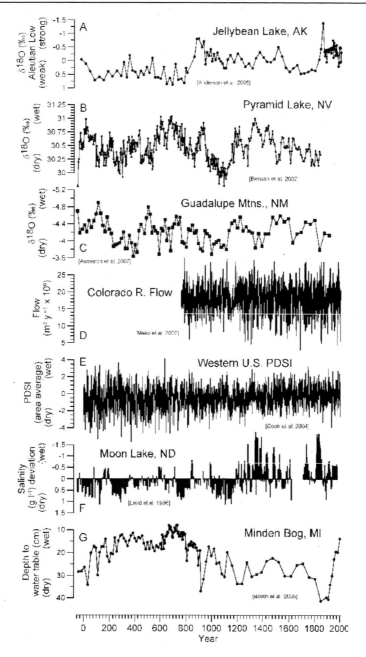

Figure 3.15. Representative hydrological time series for the past 2,000 years. A, oxygen- isotope composition of lake-sediment calcite from Jellybean Lake, AK, an indirect measure of the strength of the Aleutian Low, and hence moisture (*Anderson et al., 2005*). B, oxygen-isotope values from core PLC97-1, Pyramid Lake, NV, which reflect lake- level status (*Benson et al., 2002*); C, oxygen-isotope values from a speleothem from the Guadalupe Mountains., NM, which reflect North American monsoon-related precipitation (*Asmerom et al., 2007*); D, dendroclimatological reconstructions of Colorado River flow (*Meko et al., 2007*); E, area averages for the Western United States of dendroclimatological reconstructions of PDSI (Palmer Drought Severity Index, *Cook et al., 2004*); F, diatom-inferred salinity estimates for Moon Lake, ND, expressed as deviations from a long-term average (*Laird et al., 1996*); G, depth-to-water-table values inferred from testate amoeba samples from a peat core from Minden Bog, MI (*Booth et al., 2006*). Abbreviations: ‰, per mil; m^3 y^{-1}, cubic meters per year; g l^{-1}, grams per liter; cm, centimeter.

periodicities in paleoenvironmental records could arise from sources other than, for example, solar irradiance cycles inferred from ^{14}C-production records. Moreover, there are no physical mechanisms that might account for decadalscale variations over long time spans in, for example, the PDO, apart from those that involve the integration of the shorter time-scale variations (i.e., ENSO; *Newman et al., 2003; Schneider and Cornuelle, 2005*). Finally, although the broad trends global or hemispheric-average temperatures over the past millennium seem reasonably well accounted for by the combinations of factors described in (3) above, there is little short- term agreement among different simulations. Consequently, despite their societal importance (e.g., *Climate Reseach Committee, 1995*), the genesis of centennial-scale climatic and hydrologic variations remains essentially unexplained.

5. Future Subtropical Drying: Dynamics, Paleocontext, and Implications

It is a robust result in climate model projections of the climate of the current century that many already wet areas of the planet get wetter – such as in the oceanic Intertropical Convergence Zone (ITCZ), the Asian monsoonal region, and equatorial Africa - and already dry areas get drier – such as the oceanic subtropical high pressure zones, southwestern North America, the Intra-America Seas, the Mediterranean region, and southern Africa (*Held and Soden, 2006*); see also *Hoerling et al. (2006)*. Drying and wetting as used here refer to the precipitation minus the surface evaporation, or P-E. P-E is the quantity that, in the long-term mean over land, balances surface and subsurface runoff and, in the atmosphere, balances the vertically integrated moisture convergence or divergence. The latter contains components due to the convergence or divergence of water vapor by the mean flow convergence or divergence, the advection of humidity by the mean flow, and the convergence or divergence of humidity by the transient flow. A warmer atmosphere can hold more moisture, so the pattern of moisture convergence or divergence by the mean flow convergence or divergence intensifies. This makes the deep tropical regions of the ITCZ wetter and the dry regions of the subtropics, where there is descending air and mean flow divergence, drier (*Held and Soden, 2006*).

While a warming-induced intensification of hydrological gradients is a good first start for describing hydrological change, there are many exceptions to this simple picture. For example the Amazon is a wet region where models do not robustly predict either a drying or a wetting. Here the models create more El Niño-like tropical Pacific SSTs that tend to make the Amazon drier, highlighting the potential importance of tropical circulation changes in climate change (*Li et al., 2006*). The Sahel region of West Africa dried dramatically in the latter half of the last century (*Nicholson et al., 2000*), which has been attributed to changes in SSTs throughout the tropics (*Giannini et al., 2003*). The models within the IPCC AR4 generally reproduce these changes in SST and Sahel drying as a consequence of anthropogenic climate change during the late-20th century (*Biasutti and Giannini, 2006*). However the same models have widely varying projections for how precipitation will change in the Sahel over the current century with some predicting a return to wetter conditions (*Biasutti and Giannini, 2006; Hoerling et al., 2006*). It is unknown why the modeled response in the Sahel to 20th century radiative forcing is different to the response to current-century forcing. However, it is worth noting that the one climate model that best simulates the 20th century drying continues to dry the Sahel in the current century (*Held et al., 2005*). In this tropical region, as in the Amazon, hydrological change appears to potentially involve nonlocal controls on the atmospheric circulation as well as possible complex land-surface feedbacks.

The greater southwestern regions of North America, which include the American Southwest and northern Mexico, are included within this region of subtropical drying.

Seager et al. (2007d) show that there is an impressive agreement amongst the projections with 19 climate models (and 47 individual runs) (Figure 3.16). These projections collectively indicate that this region progressively dries in the future and that the transition to a more arid climate begins in the late 20th century and early current century (Figure 3.17). The increased aridity becomes equivalent to the 1950s Southwest drought in the early part of the current century in about a quarter of the models and half of the models by mid-century. *Seager et al. (2007d)* also showed that intensification of the existing pattern of atmospheric water-vapor transport was only responsible for about half the Southwest drying and that half was caused by a change in atmospheric circulation. They linked this fact to a poleward expansion of the Hadley Cell and dry subtropical zones and a poleward shift of the mid-latitude westerlies and storm tracks, both also robust features of a warmer atmosphere (*Yin, 2005; Bengtsson et al., 2006; Lu et al., 2007*). The analysis of satellite data by *Seidel et al. (2008)* suggests such a widening of Earth's tropical belt over the past quarter century as the planet has warmed. This analysis is consistent with climate model simulations that suggest future subtropical drying as the jet streams and the associated wind and precipitation patterns move poleward with global warming. Note, however, that GCMs are unable to capture the mesoscale processes that underlie the North American monsoon (*e.g., Castro et al., 2007a*), so there is uncertainty regarding the impact of these changes on monsoon season precipitation in the American Southwest and northern Mexico.

The area encompassing the Mediterranean regions of southern Europe, North Africa, and the Middle East dries in the model projections even more strongly, with even less disagreement amongst models, and also beginning toward the end of the last century. Both here and in southwestern North America, the drying is not abrupt in that it occurs over the same time scale as the climate forcing strengthens. However, the severity is such that the aridity equivalent to historical droughts — but as a new climate rather than a temporary state — is reached within the coming years to a few decades. Assessed on the time scale of water-resource development, demographic trends, regional development, or even political change, this could be described as a "rapid" if not abrupt climate change and, hence, is a cause for immediate concern.

The future subtropical drying occurs in the models for reasons that are distinct from the causes of historical droughts. The latter are related to particular patterns of tropical SST anomalies, while the former arises as a consequence of overall, near-uniform warming of the surface and atmosphere and how that impacts water-vapor transports and atmospheric circulation. Both mechanisms involve a poleward movement of the mid-latitude westerlies and similar changes to the eddy-driven mean meridional circulation. However, a poleward expansion of the Hadley Cell has not been invoked to explain the natural droughts. Further future drying is expected to be accompanied by a maximum of warming in the tropical upper troposphere (a consequence of moist convection in the deep tropics), whereas natural droughts have gone along with cool temperatures in the tropical troposphere. Hence, past droughts are not analogs of future drying, which should make identification of anthropogenic drying easier when it occurs.

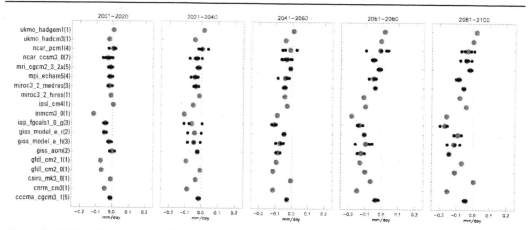

Figure 3.16. The change in annual mean precipitation minus evapotranspiration (P-E) over the American Southwest (125°W. – 95°W., 25°N., – 40°N., land areas only) for 19 models relative to model climatologies for 1950-2000. Results are averaged over 20-year segments of the current century. The number of ensemble members for the projections is listed by the model name at left. Black dots represent ensemble members, where available, and red dots represent the ensemble mean for each model. Units are in millimeters per day

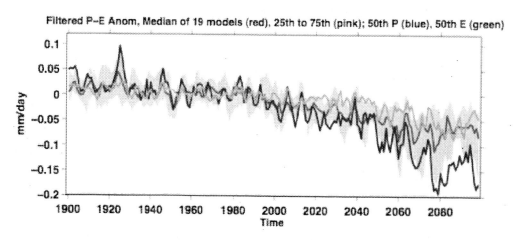

Figure 3.17. Modeled changes in annual mean precipitation minus evaporation (P-E) over southwestern North America (125° – 95° W., 25° – 40°N., land areas only) averaged over ensemble members for 19 models participating in IPCC AR4. The historical period used known and estimated climate forcings and the projections used the SResA1B emissions scenario (*IPCC, 2007*). Shown are the median (red line) and 25th and 75th percentiles (pink shading) of the P-E distribution amongst the 19 models, and the ensemble medians of P (blue line) and E (green line) for the period common to all models (1900-2098). Anomalies for each model are relative to that model's climatology for 1950-2000. Results have been 6-year low-pass filtered to emphasize low frequency variations. Units are millimeters per day (mm/day). The model ensemble mean P-E in this region is around 0.3 mm/day. From *Seager et al. (2007d)*

It is unclear how apt the Medieval megadroughts are as analogs of future drying. As mentioned above, it has been suggested that they were caused by tropical Pacific SSTs being La Niña-like for up to decades at a time during the Medieval period, as well as by the subtropical North Atlantic being warm. The tropical Pacific SST change possibly arose as a response to increased surface solar radiation. If this is so, then future subtropical drying will likely have no past analogs. However, it cannot be ruled out that the climate model projections are wrong in not producing a more La Niña-like state in response to increased radiative forcing. For example, the current generation of models has well known and serious biases in their simulations of tropical Pacific climate, and these may compromise the model projections of climate change. If the models are wrong, then it is possible that the future subtropical drying caused by general warming will be augmented by the impacts of an induced more La Niña-like state in the tropical Pacific. However, the association between positive radiative forcing, a more La Niña-like SST state, and dry conditions in southwestern North America that has been argued for using paleoclimate proxy data is for solar forcing, whereas future climate change will be driven by greenhouse forcing. It is not known if the tropical climate system responses to solar and greenhouse gas forcing are different. These remaining problems with our understanding of, and ability to model, the tropical climate system in response to radiative forcing mean that there remains uncertainty in how strong the projected drying in the Southwest will be, an uncertainty that includes the possibility that it will be more intense than in the model projections.

Future drying in southwestern North America will have significant social impacts in both the U.S. and Mexico. To date there are no published estimates of the impact of reduced P-E on the water-resource systems of the region that take full account of the climate projections. To do so would involve downscaling to the river basin scale from the projections with global models using either statistical methods or regional models, a problem of considerable technical difficulty. However both *Hoerling and Eischeid (2007)* and *Christensen and Lettenmaier (2006)* have used simpler methods to suggest that the global model projections imply that Colorado River flow will drop by between several percent and a quarter. While the exact number cannot, at this point, be known with any certainty at all, our current ability to model hydrology in this region unambiguously projects reduced flow.

Reduced flow in the Colorado and the other major rivers of the Southwest will come at a time when the existing flow is already fully allocated and when the population in the region is increasing. Current allocations of the Colorado River are also based on proportions of a fixed flow that was measured early in the last century at a time of unusual high flow (*Woodhouse et al., 2005*). It is highly likely that it will not be possible to meet those allocations in the projected drier climate of the relatively near future. In this context it needs to be remembered that agriculture uses some 90% of Colorado River water and about the same amount of total water use throughout the region, but even in California with its rich, productive, and extensive farmland, agriculture accounted for no more than 2% of the State economy.

6. Floods: Present, Past, and Future

Like droughts, floods, or episodes of much wetter-than-usual conditions, are embedded in large-scale atmospheric circulation anomalies that lead to a set of meteorological and hydrological conditions that support their occurrence. In contrast to droughts, floods are usually more localized in space and time, inasmuch as they are related to a specific combination of prior hydrologic conditions (e.g., the degree of soil saturation prior to the flood) upon which specific short-term meteorological events (intense rainstorms or rapid snowmelt) are superimposed (*Hirschboeck, 1989; Mosley and McKerchar, 1993; Pilgrim and Codery, 1993*) Floods are also geomorphologically constrained by drainage-basin and floodplain characteristics (*Baker et al., 1988; O'Connor and Costa, 2003*). However, when climatic anomalies are large in scope and persistent, such as those that occurred during 1993 in the Upper Mississippi Valley (*Kunkel et al., 1994; Anderson et al., 2003*) (and again in 2008), regionally extensive episodes of flooding can occur. When climate significantly changes, as it has in the past (*Knox, 2000*), and will likely do in the future, changes in the overall flood regime, including the frequency of different size floods and the areas affected, will also occur (*Kundzewicz et al., 2007*).

6.1. The 1993 Mississippi Valley Floods—Large-Scale Controls and Land-Surface Feedback

The flooding that occurred in the Upper Mississippi Valley of central North America in the late spring and summer of 1993 provides a case study of the control of a major flood event by large-scale atmospheric circulation anomalies. Significant feedback from the unusually wet land surface likely reinforced the wet conditions, which contributed to the persistence of the wet conditions. The 1993 flood ranks among the top five weather disasters in the United States,and was generated by the frequent occurrence of large areas of moderate to heavy precipitation, within which extreme daily total rainfall events were embedded. These meteorological events were superimposed on an above-normal soil- moisture anomaly at the beginning June of that year (*Kunkel et al., 1994*). These events were supported by the occurrence of a large-scale atmospheric circulation anomaly that featured the persistent flow of moisture from the Gulf of Mexico into the interior of the continent (*Bell and Janowiak, 1995; Trenberth and Guillemot, 1996*). The frequency of seasonal (90-day long) excessive (i.e. exceeding a 20-year return period) precipitation anomalies has generally been increasing over time in the United States (*Kunkel et al., 2008* (CCSP SAP 3.3, Sec. 2.2.2.3, Figure 2.9).

The atmospheric circulation features that promoted the 1993 floods in the Mississippi Valley, when contrasted with the widespread dry conditions during the summer of 1988, provide a "natural experiment" that can be used to evaluate the relative importance of remote (e.g., the tropical Pacific) and local (over North America) forcing, and of the importance of feedback from the land surface to reinforce the unusually wet or dry conditions. For example, *Trenberth and Guillemot (1996)* used a combination of observational and "reanalysis" data (*Kalnay et al., 1996*), along with some diagnostic analyses to reveal the role of large-scale moisture transport into the midcontinent, with dryness ocurring in response to less flow and flooding in response to greater-than-normal flow. *Liu et al. (1998)* used a combination of reanalysis data and simple models to examine the interactions among the different controls of the atmospheric circulation anomalies in these 2 years.

Although initial studies using a regional climate model pointed to a small role for feedback from the wet land surface in the summer of 1993 to increase precipitation over the midcontinent (*Giorgi et al., 1996*), subsequent studies exploiting the 1988/1993 natural experiment using both regional climate models and general circulation models point to an important role for the land surface in amplifying the severity and persistence of floods and droughts (*Bonan and Stillwell-Soller, 1998; Bosilovich and Sun, 1999; Hong and Pan, 2000; Pal and Eltahir, 2002*). These analyses add to the general pattern that emerges for large moisture anomalies (both wet and dry) in the midcontinent of North America to have a) local controls (i.e. atmosoheric circulation and moisture flux over North America), b) remote controls (e.g. Pacific SST anomalies) and c) a significant role for feedback that can reinforce the moisture anomalies. The 1993 floods continue to be a focus for climate model intercomparisons (*Anderson et al., 2003*).

6.2. Paleoflood Hydrology

The largest floods observed either in the instrumental or in the paleorecord have a variety of causes (*O'Connor and Costa, 2004*), for the most part related to geological processes. However, some the largest floods are meteorological floods, which are relevant for understanding the nature of abrupt climate changes (*Hirschboeck 1989; House et al. 2002*) and potential changes in the environmental hazards associated with flooding (*Benito et al., 2004; Wohl, 2000*). Although sometimes used in an attempt to extend the instrumental record for operational hydrology purposes (i.e., fitting flood-distribution probability density functions; *Kochel and Baker, 1982; Baker et al., 1988*), paleoflood hydrology also provides information on the response of watersheds to long-term climatic variability or change (*Ely, 1997; Ely et al., 1993; Knox, 2000*), or to joint hydrological- climatological constraints on flood magnitude (*Enzel et al., 1993*).

Knox *(2000,* see also *Knox, 1985, 1993*) reconstructed the relative (to present) magnitude of small floods (i.e., those with frequent return intervals) in southwestern Wisconsin during the Holocene using radiocarbon-dated evidence of the size of former channels in the floodplains of small watersheds, and the magnitude (depth) of larger overbank floods using sedimentological properties of flood deposits. The variations in flood magnitude can be related to the joint effects of runoff (from precipitation and snowmelt) and vegetation cover (Figure 3.13). The largest magnitudes of both sizes of flood occurred during the mid-Holocene drought interval, when tree cover was low, permitting more rapid runoff of flood-generating snowmelt and precipitation (see *Knox, 1972*). As tree cover increased with increasing moisture during the interval from 6 ka to 4 ka, flood magnitudes decreased, then increased again after 3.5 ka as effective moisture increased further in the late Holocene.

The paleoflood record in general suggests a close relationship between climatic variations and the flood response. This relationship may be quite complex, however, inasmuch as the hydrologic response to climate changes is mediated by vegetation cover, which itself is dependent on climate. In general, runoff from forested hillslopes is lower for the same input of snowmelt or precipitation than from less well-vegetated hillslopes (*Pilgrim and Cordery, 1993*). Consequently, a shift from dry to wet conditions in a grassland may see a large response (i.e., an increase) in flood magnitude at first (until the vegetation cover increases), while a shift from wet to dry conditions may see an initial decrease in flood magnitude, followed by an increase as vegetation cover is reduced (*Knox, 1972, 1993*). This kind of relationship makes it difficult to determine the specific link between climate variations and potentially abrupt responses in flood regime without the development of appropriate process models. Such models will require testing under conditions different from the present, as is the case for models of other environmental systems. Paleoflood data are relatively limited relative to other paleoenvironmental indicators, but work is underway to assemble a working database (*Hirschboeck, 2003*).

6.3. Floods and Global Climate Change

One of the main features of climate variations in recent decades is the emergence of a package of changes in meteorological and hydrological variables that are consistent with global warming and its impact on hydrological cycle and the frequency of extreme events (*Trenberth et al., 2007,* IPCC AR4, WG4, Ch. 3). The mechanisms underlying these changes include the increase in atmospheric moisture, the intensity of the hydrologic cycle, and the changes in atmospheric circulation as the atmosphere warms (*Knight et al., 2008*). As described in one of the key findings of *Gutowski et al., (2008*; CCSP SAP 3.3, Ch. 3) "Heavy precipitation events averaged over North America have increased over the past 50 years, consistent with the increased water holding capacity of the atmosphere in a warmer climate and observed increases in water vapor over the ocean." (See also *Easterling et al., 2000, Kunkel, 2003; Kunkel et al., 2003*). In addition, the frequency of season-long episodes of greater-than-average precipitation is increasing (*Kunkel et al., (2008*; CCSP SAP 3.3, Sec. 2.2.2.3), and the timing of snowmelt is changing in many parts of the country (see Sec. 7). All of the meterological controls of flooding (short- and long-duration heavy precipitation, snowmelt) are thus undergoing long-term changes. However, there is considerable uncertainty in the specific hydrologic response and its temporal and spatial pattern, owing to the auxiliary role that atmospheric circulation patterns and antecedent conditions play in generating floods, and these factors experience interannual- and decadal-scale variations themselves (*Kunkel, 2003*).

These changes in the state of the atmosphere in turn lead to the somewhat paradoxical conclusion that both extremely wet events (floods) and dry events (droughts) are likely to increase as the warming proceeds (*Kundzewicz et al., 2007*, IPCC AR4 WG2 Ch. 3). The extreme floods in Europe in 2002, followed by the extreme drought and heatwave in 2003, have been used to illustrate this situation (*Pal et al., 2004*). They compared observed 20th-century trends in atmospheric circulation and precipitation with the patterns of these variables (and of extreme-event characteristics: dry-spell length and maximum 5-day precipitation) projected for the 21st century using a regional climate model, and noted their internal consistency and consistency with the general aspects of anthropogenic global climate changes.

Projections of future hydrological trends thus emphasize the likely increase in hydrological variability in the future that includes less frequent precipitation, more intense precipitation, increased frequency of dry days, and also increased frequency of extremely wet days (CCSP SAP 3.3, Sec. 3.6.6, in prep.). Owing to the central role of water in human-environment interactions, it is also likely that these hydrological changes, and increases in flooding in particular, will have synergistic impacts on such factors as water quality and the incidence of water-borne diseases that could amplify the impact of basic hydrologic changes (*Field et al., 2007*, IPCC AR4, WG2, Ch. 14.4.1, 14.4.9). The great modifications by humans that have taken place in watersheds around the world further complicate the problem of projecting the potential for future abrupt changes in flooding.

6.4. Assessment of Abrupt Change in Flood Hydrology

Assessing the likelihood of abrupt changes in flood regime is a difficult proposition that is compounded by the large range in temporal and spatial scales of the controls of floods, and the consequent need to scale down the large-scale atmospheric and water- and energy-balance controls, and scale up the hillslope- and watershed-scale hydrological responses. Nevertheless, there is work underway to combine the appropriate models and approaches toward this end (e.g. *Jones et al., 2006; Fowler and Kilsby, 2007; Maurer, 2007*). This work could be enhanced by several developments, including:

- Enhanced modeling capabilities. The attempts that have been made thus far to project the impact of global climate change on hydrology, including runoff, streamflow, and floods and low-flows, demonstrate that the range of models and the approaches for coupling them are still in an early developmental stage (relative to, for example, coupled Atmosphere-Ocean General Circulation Models). Sufficient computational capability must be provided (or made available) to facilitate development and use of enhanced models.

- Enhanced data sets. Basic data on the flood response to climatic variations, both presentday and prehistoric, are required to understand the nature of that response across a range of conditions different from those of the present. Although human impacts on watersheds and recent climatic variability have provided a number of natural experiments that illustrate the response of floods to controls, the impact of larger environmental changes than those found in the instrumental record are required to test the models and approaches than could be used.

- Better understanding of physical processes. The complexity of the response of extreme hydrologic events to climatic variations, including as it does the impacts on both the frequency and magnitude of meteorological extremes, and mediation by land cover and watershed characteristics that themselves are changing, suggests that further diagnostic studies of the nature of the response should be encouraged.

7. Other Aspects of Hydroclimate Change

The atmosphere can hold more water vapor as it warms (as described by the Clausius-Clapeyron equation), to the tune of about 7% per degree Celsius of warming. With only small changes projected for relative humidity (*Soden et al., 2002*), the specific humidity content of the atmosphere will also increase with warming at this rate. This is in contrast to the global mean precipitation increase of about 1-2% per degree Celsius of warming. The latter is caused when evaporation increases to balance increased downward longwave radiation associated with the stronger greenhouse trapping. For both of these constraints to be met, more precipitation has to fall in the heaviest of precipitation events as well explained by *Trenberth et al. (2003)*.

The change in precipitation intensity seems to be a hydrological change that is already evident (*Kunkel et al., 2008* (CCSP SAP 3.3, Secs. 2.2.2.2 and 2.2.2.3). *Groisman et al. (2004)* demonstrate that daily precipitation records over the last century in the United States show a striking increase, beginning around 1990, in the proportion of precipitation within very heavy (upper 1% of events) and extreme (upper 0.1%) of events. In the annual mean there is a significant trend to increased intensity in the southern and central plains and in the Midwest, and there is a significant positive trend in the Northeast in winter. In contrast the Rocky Mountain States show an unexplained significant trend to decreasing intensity in winter.

Groisman et al. (2005) show that the observed trend to increasing precipitation intensity is seen across much of the world, and both they and *Wilby and Wigley (2002)* show that climate model projections of the current century show that this trend will continue. *Groisman et al. (2005)* make the point that the trends in intensity are greater than the trends in mean precipitation, that there is good physical reason to believe that they are related to global warming, and that they are likely to be more easily detected than changes in the mean precipitation.

Increases in precipitation intensity can have significant social impacts as they increase the potential for flooding and overloading of sewers and wastewater treatment plants. See *Rosenzweig et al. (2007)* for a case study of New York City's planning efforts to deal with water-related aspects of climate change. Increasing precipitation intensity can also lead to an increase of sediment flux, including potentially harmful pathogens, into water- supply reservoirs, thus necessitating more careful water-quality management, a situation already being faced by New York City.

Another aspect of hydroclimatic change that can be observed in many regions is the general decrease in snowpack and snow cover (*Mote et al., 2005; Déry and Brown, 2007; Dyer and Mote, 2006*; see also *Lettenmaier et al., 2008*; CCSP SAP 4.3, Sec. 4.2.4). Winter snowfall and the resulting accumulated snowpack depend on temperature in complicated ways. Increasing temperatures favor greater moisture availability and total precipitation (in much the same way that precipitation intensity depends on temperature) and hence greater snow accumulation (if winter temperatures are cold enough), but greater snowmelt and hence a reduced snowpack if temperatures increase enough. Regions with abundant winter precipitation, and winter temperatures close to freezing could therefore experience an overall increase in winter precipitation as temperatures increase but also an overall decrease in snow cover as the balance of precipitation shifts from snow to rain, along with an earlier occurrence of spring snowmelt. Such trends seem to be underway in many regions (*Moore et al., 2007*), but particularly in the Western United States (*Mote et al., 2005, 2008*).

As a consequence of reduced snowpack and earlier spring snowmelt, a range of other hydrologic variables can be affected, including the amount and timing of runoff, evapotranspiration, and soil moisture (*Hamlet et al., 2007; Moore et al., 2007*). Although gradual changes in snowcover and snowmelt timing could be the rule, the transition from general winter-long snowcover, to transient snowcover, to occasional snow cover, could appear to be quite abrupt, from the perspective of the hydrology of individual watersheds.

Most studies of past and modern impacts on water resources focus on abrupt changes in the physical system such as the duration of ice cover and timing of snow melt, lake thermal structure, evaporation, or water level, with considerably less attention on abrupt changes in water quality (e.g. *Lettenmaier et al., 2008*; CCSP SAP 4.3, Sec. 4.2.5). Assessing recent climate impacts on water quality has been complicated by human land use. For example, analysis of contemporary data in the northern Great Plains suggests that climate impacts are small relative to land use (*Hall et al., 1999*). A similar conclusion has been reached in Europe based on the paleoclimate literature, where humans have been impacting the environment for thousands of years (*Hausmann et al., 2002*). Some of the best evidence for climate changes resulting in changes in water quality and on aquatic biological communities comes from work in the Experimental Lakes Area in Canada where land use changes have been more limited (*Schindler, 1996a,b*). This work showed how climate changes affect ion concentration, nutrients, and dissolved organic carbon concentrations, often amplifying acidification and other external perturbations. Other evidence suggests that that climate warming might affect water quality (phytoplankton biomass and nutrient concentrations) indirectly by affecting lake thermal structure (*Lebo et al., 1994; Gerten and Adrian, 2000*). The climate changes may lead to abrupt changes in salinity and water quality for drinking, irrigation, and livestock. The recent paleolimnological records of abrupt changes in salinity have been inferred from changes in diatoms in the sediments of Moon Lake, ND (*Laird et al., 1996*), and the Aral Sea (*Austin et al., 2007*); however, determining if the magnitude of these abrupt changes represents a significant degradation of water quality is difficult to discern.

8. Conclusions

Drought is among the greatest of recurring natural hazards facing both the people of the United States and humanity worldwide today and in the foreseeable future. Its causes are complex and not completely understood, but its impact on agriculture, water supply, natural ecosystems, and other human needs for survival can be severe and long lasting in human terms, making it one of the most pressing scientific problems to study in the field of climatic change. Floods, though generally more localized in time and space than droughts, are also a major natural hazard, and share with droughts many of same large- scale controls and the potential for experiencing major changes in these controls in the future.

Droughts can develop faster than the time scale needed for human societies and natural systems to adapt to the increase in aridity. Thus, a severe drought lasting several years may be experienced as an abrupt change to drier conditions even though wetter conditions will eventually return. The 1930s Dust Bowl drought, which resulted in a mass exodus from the parched Great Plains to more favorable areas in the West, is one such example. The drought eventually ended when the rains returned, but the people did not. For them it was a truly abrupt and permanent change in their lives. Thus, it is a major challenge of climate research to find ways to help reduce the impact of future droughts through improved prediction and the more efficient use of the limited available water resources.

For examples of truly abrupt and long-lasting changes in hydroclimatic variability over midcontinental North America and elsewhere in the world, we must go back in time to the middle Holocene, when much larger changes in the climate system occurred. The climate boundary conditions responsible for those changes were quite different from those today, so the magnitude of change that we might conceivably expect in the future under "natural" forcing of the climate system might not to be as great. However, the rising level of greenhouse gas forcing that is occurring now and in the foreseeable future is truly unprecedented, even over the Holocene. Therefore, the abrupt hydrologic changes in the Holocene ought to be viewed as useful examples of the magnitude of change that could conceivably occur in the future, and the mechanisms through which that change occurs.

The need for improved drought prediction on time scales of years to decades is clear now. To accomplish this will require that we develop a much better understanding of the causes of hydroclimatic variability worldwide. It is likely that extended periods of anomalous tropical ocean SSTs, especially in the eastern equatorial Pacific ENSO region, strongly influence the development and duration of drought over substantial land areas of the globe. As the IPCC AR4 concluded, "the palaeoclimatic record suggests that multiyear, decadal and even centennial-scale drier periods are likely to remain a feature of future North American climate, particularly in the area west of the Mississippi River." Multiple proxies indicate the past 2,000 years included periods with more frequent, longer and/or geographically more extensive droughts in North America than during the 20th century. However, the record of past drought from tree rings offers a sobering picture of just how severe droughts can be under natural climate conditions. Prior to A.D. 1600, a succession of megadroughts occurred that easily eclipsed the duration of any droughts known to have occurred over North America since that time. Thus, understanding the causes of these extraordinary megadroughts is of paramount importance. Increased solar forcing over the tropical Pacific has been implicated, as has explosive volcanism, but the uncertainties remain large.

However significant enhanced solar forcing has been in producing past megadroughts, the level of current and future radiative forcing due to greenhouse gases is very likely to be of much greater significance. It is thus disquieting to consider the possibility that drought-inducing La Niña-like conditions may become more frequent and persistent in the future as greenhouse warming increases. We have no firm evidence that this is happening now, even with the serious drought that has gripped the West since about 1998. Yet, a large number of climate models suggest that future subtropical drying is a virtual certainty as the world warms and, if they are correct, indicate that it may have already begun. The degree to which this is true is another pressing scientific question that must be answered if we are to know how to respond and adapt to future changes in hydroclimatic variability.

4. THE POTENTIAL FOR ABRUPT CHANGE IN THE ATLANTIC MERIDIONAL OVERTURNING CIRCULATION

Key Findings

The Atlantic Meridional Overturning Circulation (AMOC) is an important component of the Earth's climate system, characterized by a northward flow of warm, salty water in the upper layers of the Atlantic, and a southward flow of colder water in the deep Atlantic. This ocean circulation system transports a substantial amount of heat from the Tropics and Southern Hemisphere toward the North Atlantic, where the heat is transferred to the atmosphere. Changes in this circulation have a profound impact on the global climate system, as indicated by paleoclimate records. These include, for example, changes in African and Indian monsoon rainfall, atmospheric circulation of relevance to hurricanes, and climate over North America and Western Europe. In this chapter, we have assessed what we know about the AMOC and the likelihood of future changes in the AMOC in response to increasing greenhouse gases, including the possibility of abrupt change. We have five primary findings:

- It is very likely that the strength of the AMOC will decrease over the course of the 21st century in response to increasing greenhouse gases, with a best estimate decrease of 25-30%.
- Even with the projected moderate AMOC weakening, it is still very likely that on multidecadal to century time scales a warming trend will occur over most of the European region downstream of the North Atlantic Current in response to increasing greenhouse gases, as well as over North America.
- No current comprehensive climate model projects that the AMOC will abruptly weaken or collapse in the 21st century. We therefore conclude that such an event is very unlikely. Further, an abrupt collapse of the AMOC would require either a

sensitivity of the AMOC to forcing that is far greater than current models suggest or a forcing that greatly exceeds even the most aggressive of current projections (such as extremely rapid melting of the Greenland ice sheet). However, we cannot completely exclude either possibility.

- We further conclude it is unlikely that the AMOC will collapse beyond the end of the 21st century because of global warming, although the possibility cannot be entirely excluded.

- Although our current understanding suggests it is very unlikely that the AMOC will collapse in the 21st century, the potential consequences of such an event could be severe. These would likely include sea level rise around the North Atlantic of up to 80 centimeters (in addition to what would be expected from broad-scale warming of the global ocean and changes in land-based ice sheets due to rising CO_2), changes in atmospheric circulation conditions that influence hurricane activity, a southward shift of tropical rainfall belts with resulting agricultural impacts, and disruptions to marine ecosystems.

The above conclusions depend upon our understanding of the climate system, and on the ability of current models to simulate the climate system. However, these models are not perfect, and the uncertainties associated with these models form important caveats to our conclusions. These uncertainties argue for a strong research effort to develop the observations, understanding, and models required to predict more confidently the future evolution of the AMOC.

Recommendations

We recommend the following activities to advance both our understanding of the AMOC and our ability to predict its future evolution:

- Improve long-term monitoring of the AMOC. This monitoring would likely include observations of key processes involved in deep water formation in the Labrador and Norwegian Seas, and their communication with the rest of the Atlantic (such as the Nordic Sea inflow, and overflow across the Iceland-Scotland Ridge), along with observing the more complete three-dimensional structure of the AMOC, including sea surface height. Such a system needs to be in place for decades to properly characterize and monitor the AMOC.

- Improve understanding of past AMOC changes through the collection and analysis of those proxy records that most effectively document AMOC changes and their impacts in past climates (hundreds to many thousands of years ago). Among these proxy records are geochemical tracers of water masses such as 13C and dynamic tracers that constrain rates of the overturning circulation such as the protactinium/thorium (Pa/Th) proxy. These records provide important insights on how the AMOC behaved in substantially different climatic conditions and thus greatly facilitate our understanding of the AMOC and how it may change in the future.

- Accelerated development of climate system models incorporating improved physics and resolution, and the ability to satisfactorily represent small-scale processes that are important to the AMOC. This would include the addition of models of land-based ice sheets and their interactions with the global climate system.
- Increased emphasis on improved theoretical understanding of the processes controlling the AMOC, including its inherent variability and stability, especially with respect to climate change. Among these important processes are the role of small-scale eddies, flows over sills, mixing processes, boundary currents, and deep convection. In addition, factors controlling the large-scale water balance are crucial, such as atmospheric water-vapor transport, precipitation, evaporation, river discharge, and freshwater transports in and out of the Atlantic. Progress will likely be accomplished through studies combining models, observational results, and paleoclimate proxy evidence.
- Development of a system to more confidently predict the future behavior of the AMOC and the risk of an abrupt change. Such a prediction system should include advanced computer models, systems to start model predictions from the observed climate state, and projections of future changes in greenhouse gases and other agents that affect the Earth's energy balance. Although our current understanding suggests it is very unlikely that the AMOC will collapse in the 21st century, this assessment still implies up to a 10% chance of such an occurrence. The potentially severe consequences of such an event, even if very unlikely, argue for the rapid development of such a predictive system.

1. Introduction

The oceans play a crucial role in the climate system. Ocean currents move substantial amounts of heat, most prominently from lower latitudes, where heat is absorbed by the upper ocean, to higher latitudes, where heat is released to the atmosphere. This poleward transport of heat is a fundamental driver of the climate system and has crucial impacts on the distribution of climate as we know it today. Variations in the poleward transport of heat by the oceans have the potential to make significant changes in the climate system on a variety of space and time scales. In addition to transporting heat, the oceans have the capacity to store vast amounts of heat. On the seasonal time scale this heat storage and release has an obvious climatic impact, delaying peak seasonal warmth over some continental regions by a month after the summer solstice. On longer time scales, the ocean absorbs and stores most of the extra heating that comes from increasing greenhouse gases (*Levitus et al., 2001*), thereby delaying the full warming of the atmosphere that will occur in response to increasing greenhouse gases.

One of the most prominent ocean circulation systems is the Atlantic Meridional Overturning Circulation (AMOC). As described in subsequent sections, and as illustrated in Figure 4.1, this circulation system is characterized by northward flowing warm, saline water in the upper layers of the Atlantic (red curve in Figure 4.1), a cooling and freshening of the water at higher northern latitudes of the Atlantic in the Nordic and Labrador Seas, and southward flowing colder water at depth (light blue curve). This circulation transports heat from the South Atlantic and tropical North Atlantic to the subpolar and polar North Atlantic, where that heat is released to the atmosphere with substantial impacts on climate over large regions.

The Atlantic branch of this global MOC (see Figure 4.1) consists of two primary overturning cells: (1) an "upper" cell in which warm upper ocean waters flow northward in the upper 1,000 meters (m) to supply the formation of North Atlantic Deep Water (NADW), which returns southward at depths of approximately 1,500-4,500 m and (2) a "deep" cell in which Antarctic Bottom Waters (ABW) flow northward below depths of about 4,500 m and gradually rise into the lower part of the southward-flowing NADW. Of these two cells, the upper cell is by far the stronger and is the most important to the meridional transport of heat in the Atlantic, owing to the large temperature difference (~1 5° C) between the northward-flowing upper ocean waters and the southward-flowing NADW.

In assessing the "state of the AMOC," we must be clear to define what this means and how it relates to other common terminology. The terms Atlantic Meridional Overturning Circulation (AMOC) and Thermohaline Circulation (THC) are often used interchangeably but have distinctly different meanings. The AMOC is defined as the total (basin-wide) circulation in the latitude depth plane, as typically quantified by a meridional transport streamfunction. Thus, at any given latitude, the maximum value of this streamfunction, and the depth at which this occurs, specifies the total amount of water moving meridionally above this depth (and below it, in the reverse direction). The AMOC, by itself, does not include any information on what drives the circulation.

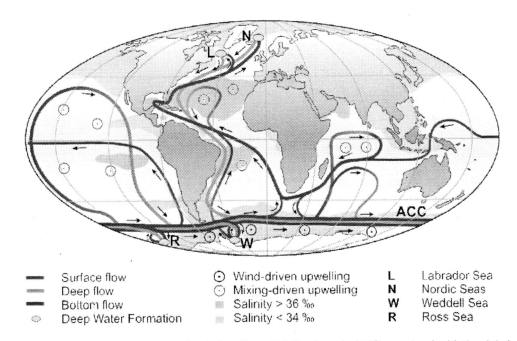

Figure 4.1. Schematic of the ocean circulation (from *Kuhlbrodt et al., 2007*) associated with the global Meridional Overturning Circulation (MOC), with special focus on the Atlantic section of the flow (AMOC). The red curves in the Atlantic indicate the northward flow of water in the upper layers. The filled orange circles in the Nordic and Labrador Seas indicate regions where near-surface water cools and becomes denser, causing the water to sink to deeper layers of the Atlantic. This process is referred to as "water mass transformation," or "deep water formation." In this process heat is released to the atmosphere. The light blue curve denotes the southward flow of cold water at depth. At the southern end of the Atlantic, the AMOC connects with the Antarctic Circumpolar Current (ACC). Deep water

formation sites in the high latitudes of the Southern Ocean are also indicated with filled orange circles. These contribute to the production of Antarctic Bottom Water (AABW), which flows northward near the bottom of the Atlantic (indicated by dark blue lines in the Atlantic). The circles with interior dots indicate regions where water upwells from deeper layers to the upper ocean (see Section 2 for more discussion on where upwelling occurs as part of the global MOC)

In contrast, the term "THC" implies a specific driving mechanism related to creation and destruction of buoyancy. *Rahmstorf (2002)* defines this as "currents driven by fluxes of heat and fresh water across the sea surface and subsequent interior mixing of heat and salt." The total AMOC at any specific location may include contributions from the THC, as well as contributions from wind-driven overturning cells. It is difficult to cleanly separate overturning circulations into a "wind-driven" and "buoyancy-driven" contribution. Therefore, nearly all modern investigations of the overturning circulation have focused on the strictly quantifiable definition of the AMOC as given above. We will follow the same approach in this chapter, while recognizing that changes in the thermohaline forcing of the AMOC, and particularly those taking place in the high latitudes of the North Atlantic, are ultimately most relevant to the issue of abrupt climate change.

There is growing evidence that fluctuations in Atlantic sea surface temperatures (SSTs), hypothesized to be related to fluctuations in the AMOC, have played a prominent role in significant climate fluctuations around the globe on a variety of time scales. Evidence from the instrumental record (based on the last ~130 years) shows pronounced, multidecadal swings in SST averaged over the North Atlantic. These multidecadal fluctuations may be at least partly a consequence of fluctuations in the AMOC. Recent modeling and observational analyses have shown that these multidecadal shifts in Atlantic temperature exert a substantial influence on the climate system ranging from modulating African and Indian monsoonal rainfall to influencing tropical Atlantic atmospheric circulation conditions relevant to hurricanes. Atlantic SSTs also influence summer climate conditions over North America and Western Europe.

Evidence from paleorecords (discussed more completely in subsequent sections) suggests that there have been large, decadal-scale changes in the AMOC, particularly during glacial times. These abrupt change events have had a profound impact on climate, both locally in the Atlantic and in remote locations around the globe. Research suggests that these abrupt events were related to massive discharges of freshwater into the North Atlantic from collapsing land-based ice sheets. Temperature changes of more than $10°$ C on time scales of a decade or two have been attributed to these abrupt change events.

In this chapter, we assess whether such an abrupt change in the AMOC is likely to occur in the future in response to increasing greenhouse gases. Specifically, there has been extensive discussion, both in the scientific and popular literature, about the possibility of a major weakening or even complete shutdown of the AMOC in response to global warming, along with rapid changes in land-based ice sheets (see Chapter 2) and Arctic sea ice (see Box 4.1). As will be discussed more extensively below, global warming tends to weaken the AMOC both by warming the upper ocean in the subpolar North Atlantic and through enhancing the flux of freshwater into the Arctic and North Atlantic. Both processes reduce the density of the upper ocean in the North Atlantic, thereby stabilizing the water column and weakening the AMOC. These processes could cause a weakening or shutdown of the AMOC that could significantly reduce the poleward transport of heat in the Atlantic, thereby possibly leading to regional cooling in the Atlantic and surrounding continental regions, particularly Western Europe.

In this chapter, we examine (1) our present understanding of the mechanisms controlling the AMOC, (2) our ability to monitor the state of the AMOC, (3) the impact of the AMOC on climate from observational and modeling studies, and (4) model-based studies that project the future evolution of the AMOC in response to increasing greenhouse gases and other changes in atmospheric composition. We use these results to assess the likelihood of an abrupt change in the AMOC. In addition, we note the uncertainties in our understanding of the AMOC and in our ability to monitor and predict the AMOC. These uncertainties form important caveats concerning our central conclusions.

BOX 4.1. POSSIBILITY FOR ABRUPT TRANSITIONS IN SEA ICE COVER

Because of certain properties of sea ice, it is quite possible that the ice cover might undergo rapid change in response to modest forcing. Sea ice has a strong inherent threshold in that its existence depends on the freezing temperature of sea water. Additionally, strong positive feedbacks associated with sea ice act to accelerate its change. The most notable of these is the positive surface albedo feedback in which changes in ice cover and surface properties modify the surface reflection of solar radiation. For example, in a warming climate, reductions in ice cover expose the dark underlying ocean, allowing more solar radiation to be absorbed. This enhances the warming and leads to further ice melt. Thus, even moderate changes in something like the ocean heat transport associated with AMOC variability could induce a large and rapid retreat of sea ice, in turn amplifying the initial warming. Indeed, a number of studies (e.g., *Dansgaard et al., 1989; Denton et al., 2005; Li et al., 2005*) have suggested that changes in sea-ice extent played an important role in the abrupt climate warming associated with Dansgaard-Oeschger (D-O) oscillations (see Sec. 4.5).

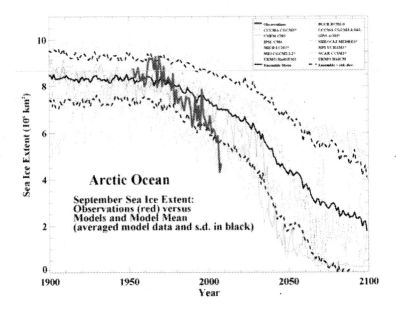

Box 4.1. Figure 1. Arctic September sea ice extent ($\times 10^6$ km^2) from observations (thick red line) and 13 IPCC AR4 climate models, together with the multi-model ensemble mean (solid black line) and standard deviation (dotted black line). Models with more than one ensemble member are indicated with an asterisk. From *Stroeve et al. 2007* (updated to include 2008).

Abrupt, nonlinear behavior in the sea-ice cover has been simulated in simple models. For example, box model studies have shown a "switch-like" behavior in the ice cover (*Gildor and Tziperman, 2001*). Since the ice cover modifies ocean-atmosphere moisture exchange, this in turn affects the source of water for ice sheet growth within these models with possible implications for glacial cycles.

Other simple models, specifically diffusive climate models, also exhibit rapid sea-ice change. These models simulate that an ice cap of sufficiently small size is unstable. This "small ice cap instability" (SICI) (*North, 1984*) leads to an abrupt transition to year-round ice-free conditions under a gradually warming climate. Recently, *Winton (2006)* examined coupled climate model output and found that of two models that simulate a complete loss of Arctic ice cover in response to increased CO_2 forcing, one had SICI-like behavior in which a nonlinear response of surface albedo to the warming climate resulted in an abrupt loss of Arctic ice. The other model showed a more linear response.Perhaps more important for 21st century climate change is the possibility for a rapid transition to seasonally ice-free Arctic conditions. The summer Arctic sea ice cover has undergone dramatic retreat since satellite records began in 1979, amounting to a loss of almost 30% of the September ice cover in 29 years. The late summer ice extent in 2007 was particularly startling and shattered the previous record minimum with an extent that was three standard deviations below the linear trend, as shown in Box 4.1 Figure 1 (from *Stroeve et al., 2007*). Conditions over the 2007-2008 winter have promoted further loss of multi-year ice due to anomalous transport through Fram Strait raising the possibility that rapid and sustained ice loss could result. However, at the time of this writing, it is unclear how this will ultimately affect the 2008 end-of-summer conditions, and there is little scientific consensus that another extreme minimum will occur (http:// www. arcus.org/ search/seaiceoutlook/report_may.php).

Climate model simulations suggest that rapid and sustained September Arctic ice loss is likely in future 21st century climate projections (*Holland et al., 2006*). In one simulation, a transition from conditions similar to pre-2007 levels to a near-ice-free September extent occurred in a decade. Increasing ocean heat transport was implicated in this simulated rapid ice loss, which ultimately resulted from the interaction of large, intrinsic variability and anthropogenically forced change. It is notable that climate models are generally conservative in the modeled rate of Arctic ice loss as compared to observations (*Stroeve et al., 2007*), suggesting that future ice retreat could occur even more abruptly than simulated in almost all current models.

2. What Are the Processes That Control the Overturning Circulation?

We first review our understanding of the fundamental driving processes for the AMOC. We break this discussion into two parts: the main discussion deals with the factors that are thought to be important for the equilibrium state of the AMOC, while the last part (Sec. 2.5) discusses factors of relevance for transient changes in the AMOC.

Like any other steady circulation pattern in the ocean, the flow of the Atlantic Meridional Overturning Circulation (AMOC) must be maintained against the dissipation of energy on the smallest length scales. We wish to determine what processes provide the energy that maintains the steady state AMOC. In general, the energy sources for the ocean are wind stress at the surface, tidal motion, heat fluxes from the atmosphere, and heat fluxes through the ocean bottom.

2.1. Sandström's Experiment

We consider the surface heat fluxes first. They are distributed asymmetrically over the globe. The ocean gains heat in the low latitudes close to the Equator and loses heat in the high latitudes toward the poles. Is this meridional gradient of the surface heat fluxes sufficient for driving a deep overturning circulation? The first one to think about this question was the Swedish researcher *Sandström (1908)*. He conducted a series of tank experiments. His tank was narrow, but long and deep, thus putting the stress on a two- dimensional circulation pattern. He applied heat sources and cooling devices at different depths and observed whether a deep overturning circulation developed. If he applied heating and cooling both at the surface of the fluid, then he could see the water sink under the cooling device. This downward motion was compensated by a slow, broadly distributed upward motion. The resulting overturning circulation ceased once the tank was completely filled with cold water. In addition there developed an extremely shallow overturning circulation in the topmost few centimeters, with warm water flowing toward the cooling device directly at the surface and cooler waters flowing backwards directly underneath. This pattern persisted, but a deep, top-to-bottom overturning circulation did not exist in the equilibrium state.

However, when *Sandström (1908)* put the heat source at depth, then such a deep overturning circulation developed and persisted. Sandström inferred that a heat source at depth is necessary to drive a deep overturning circulation in an equilibrium state. Sources and sinks of heat applied at the surface only can drive vigorous convective overturning for a certain time, but not a steady-state circulation. The tank experiments have been debated and challenged ever since (recently reviewed by *Kuhlbrodt et al., 2007*), but what Sandström inferred for the overturning circulation observed in the ocean remains true.

Thus, if we want to understand the AMOC in a thermodynamical way, we need to determine how heat reaches the deep ocean.

One potential heat source at depth is geothermal heating through the ocean bottom. While it seems to have a stabilizing effect on the AMOC (*Adcroft et al., 2001*), its strength of 0.05 Terawatt (TW, 1 TW = 10^{12} W) is too small to drive the circulation as a whole. Having ruled this out, the only other heat source comes from the surface fluxes. A classical assumption is that vertical mixing in the ocean transports heat downward (*Munk, 1966*). This heat warms the water at depth, decreasing its density and causing it to rise. In other words, vertical advection w of temperature T and its vertical mixing, parameterized as diffusion with strength κ, are in balance:

$$w\frac{\partial T}{\partial z} = \frac{\partial}{\partial z}\kappa\frac{\partial T}{\partial z}$$

(where z denotes the vertical direction). The mixing due to molecular motion is far too small for this purpose: the respective mixing coefficient κ is on the order of 1_{0-7} m2 s-1.

To achieve the observed upwelling of about 30 Sverdrups (Sv, where $1\ Sv = 10_6\ m^3\ s^{-1}$), a vertical mixing with a global average strength of $\kappa = 10\text{-}4\ _{m}2\ s\text{-}1$ is required (*Munk and*

Wunsch, 1998; Ganachaud and Wunsch, 2000). This is presumably accomplished by turbulent mixing.

2.2. Mixing Energy Sources

In order to investigate whether there is enough energy available to drive this mixing, we turn to the schematic overview presented in Figure 4.1. We have already mentioned the heat fluxes through the surface. They are essential because the AMOC is a thermally direct circulation. The other two relevant energy sources of the ocean are winds and tides. The wind stress generates surface waves and acts on the large-scale circulation. Important for vertical mixing at depth are internal waves that are generated in the surface layer and radiate through the ocean. They finally dissipate by turbulence on the smallest length scale, and the water mixes. The interaction of tidal motion with the ocean bottom also generates internal waves, especially where the topography is rough. Again, these internal waves break and dissipate, creating turbulent mixing.

Analysis of the mixing energy budget of the ocean (*Munk and Wunsch, 1998; Wunsch and Ferrari, 2004*) shows that the mixing energy that is available from those energy sources, about 0.4 TW, is just what is needed when one assumes that all 30 Sv of deep water that are globally formed are upwelled from depth by the advection-diffusion balance. However, the estimates of the magnitude of the terms in the mixing budget are highly uncertain. On the one hand, some studies suggest that less than these 0.4 TW are required (e.g., *Hughes and Griffiths, 2006*). On the other hand, the mixing efficiency, a crucial parameter in the computation of this budget, might be smaller than previously thought (*Arneborg, 2002*), which would increase the required energy. Therefore, it cannot be determined whether the mixing energy budget is actually closed. This motivated the search for other possible driving mechanisms for the AMOC.

2.3. Wind-Driven Upwelling in the Southern Ocean

Toggweiler and Samuels (1993a, 1995, 1998) proposed a completely different driving mechanism. The surface wind forcing in the Southern Ocean leads to a northward volume transport. Due to the meridional shear of the winds, this "Ekman" transport is divergent south of 50ºS, and thus water needs to upwell from below the surface to fulfill continuity. The situation is special in the Southern Ocean in that it forms a closed circle around the Earth, with the Drake Passage between South America as the narrowest and shallowest (about 2,500 m) place (outlined dashed in Figure 4.2). No net zonal pressure gradient can be maintained above the sill, and so no net meridional flow balanced by such a large-scale pressure gradient can exist. However, other types of flow are possible—wind-driven for instance. According to *Toggweiler and Samuels (1995)* this Drake Passage effect means that the waters drawn upward by the Ekman divergence must come from below the sill depth, as only from there can they be advected meridionally. Thus we have southward advection at depth, wind-driven upwelling in the Southern Ocean, and northward Ekman transport at the surface. The loop would be closed by the deep water formation in the northern North Atlantic, as that is where deep water of the density found at around 2,500 m depth is formed.

Evidence from observed tracer concentrations supports this picture of the AMOC. A number of studies (e.g., *Toggweiler and Samuels, 1993b; Webb and Suginohara, 2001*) question that deep upwelling occurs in a broad, diffuse manner, and rather point toward substantial upwelling of deep water masses in the Southern Ocean. From model studies it is not clear to what extent wind-driven upwelling is a driver of the AMOC. Recent studies show a weaker sensitivity of the overturning with higher model resolution, casting light on the question as to how strong the regional eddy-driven recirculation is (*Hallberg and Gnanadesikan, 2006*). This could compensate for the northward Ekman transport well above the depth of Drake Passage, short-circuiting the return flow.

As with the mixing energy budget, the estimates of the available energy for wind-driven upwelling are fraught with uncertainty. The work done by the surface winds on that part of the flow that is balanced by the large-scale pressure gradients can be used for wind- driven upwelling from depth. Estimates are between 1 TW (*Wunsch, 1998*) and 2 TW (*Oort et al., 1994*).

2.4. Two Drivers of the Equilibrium Circulation

We define a 'driver' as a process that supplies energy to maintain a steady-state AMOC against dissipation. We find that there are two drivers that are physically quite different from each other. Mixing-driven upwelling (case 1 in Figure 4.3) involves heat flux through the ocean across the surfaces of constant density to depth. The water there expands and then rises to the surface. By contrast, wind-driven upwelling (case 2) means that the waters are pulled to the surface along surfaces of constant density; the water changes its density at the surface when it is in contact with the atmosphere. No interior heat flux is required.

In the real ocean probably both driving processes play a role, as indicated by some recent studies (e.g., *Sloyan and Rintoul, 2001*). If part of the deep water is upwelled by mixing and part by the Ekman divergence in the Southern Ocean, then the tight closure of the energy budget is not a problem anymore (*Webb and Suginohara, 2001*). The question about the drivers is relevant because it implies different sensitivities of the AMOC to changes in the surface forcing, and thus different ways in which climate change can affect it.

2.5. Heat and Freshwater: Relevance for Near-Term Changes

So far we have talked about the equilibrium state of the AMOC to which we applied our energy-based analysis. In models, this equilibrium is reached only after several millennia, owing to the slow time scales of diffusion. However, if we wonder about possible AMOC changes in the next decades or centuries, then model studies show that these are mainly caused by heat and freshwater fluxes at the surface (e.g., *Gregory et al., 2005*), while in principle changes in the wind forcing may also affect the AMOC on short time scales. One can imagine that the drivers ensure that there is an overturning circulation at all, while the distribution of the heat and freshwater fluxes shapes the three-dimensional extent as well as the strength of the overturning circulation. The main influence of these surface fluxes on the AMOC is exerted on its sinking branch, i.e., the formation of deep water masses in the northern North Atlantic. This deep water formation (DWF) occurs in the Nordic and Labrador Seas (see Figure 4.1). Here, strong heat loss of the ocean to the atmosphere leads to a densification and subsequent sinking. Thus, one could see the driving processes as a pump, transporting the waters to the surface, and the DWF processes as the valve through which the waters flow downward (*Samelson, 2004*).

In the Labrador Sea, this heat loss occurs partly in deep convection events, in which the water is mixed vigorously and thoroughly down to 2,000 m or so. These events take place intermittently, each lasting for a few days and covering areas of 50 km to 100 km in width. In the Greenland Sea, the situation is different in that continuous mixing to intermediate depths (around 500 m) prevails. In addition, there is a sill between the Nordic Seas and the rest of the Atlantic (roughly sketched in Figure 4.2). Any water masses from the Nordic Seas that are to join the AMOC must flow over this sill, whose depth is 600 m to 800 m. This implies that deep convection to depths of 2,000 m or 3,000 m is not essential for DWF in the Nordic Seas (*Dickson and Brown, 1994*). Hence the fact that it occurs only rarely is no indication for a weakening of the AMOC. By contrast, deep convection in the Labrador Sea shows strong interannual to decadal variability. This signal can be traced downstream in the deep southward current of North Atlantic Deep Water (*Curry et al., 1999*). This suggests strongly that deep convection in the Labrador Sea can influence the strength of the AMOC.

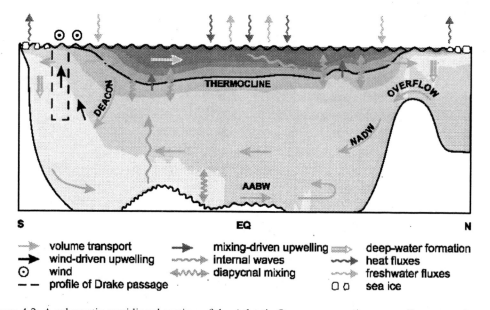

⇢ volume transport	➡ mixing-driven upwelling	⇨ deep-water formation
➡ wind-driven upwelling	⟿ internal waves	⤳ heat fluxes
⊙ wind	⬳ diapycnal mixing	⤳ freshwater fluxes
– – profile of Drake passage		▢ ▢ sea ice

Figure 4.2. A schematic meridional section of the Atlantic Ocean representing a zonally averaged picture (from *Kuhlbrodt et al., 2007*). The AMOC is denoted by straight blue arrows. The background color shading depicts a zonally averaged density profile from observational data. The thermocline lies between the warmer, lighter upper layers and the colder, deeper waters. Short, wavy orange arrows indicate diapycnal mixing, i.e., mixing along the density gradient. This mainly vertical mixing is the consequence of the dissipation of internal waves (long orange arrows). It goes along with warming at depth that leads to upwelling (red arrows). Black arrows denote wind-driven upwelling caused by the divergence of the surface winds in the Southern Ocean together with the Drake Passage effect (explained in the text). The Deacon cell is a wind-driven regional recirculation. The surface fluxes of heat (red wavy arrows) and freshwater (green wavy arrows) are often subsumed as buoyancy fluxes. The heat loss in the northern and southern high latitudes leads to cooling and subsequent sinking, i.e., formation of the deep-water masses North Atlantic Deep Water (NADW) and Antarctic Bottom Water (AABW). The blue double arrows subsume the different deep water formation sites in the North Atlantic (Nordic Seas and Labrador Sea) and in the Southern Ocean (Ross Sea and Weddell Sea)

Both a future warming and increased freshwater input (by more precipitation, more river runoff, enhanced transient export (including sea ice) from the Arctic, and melting inland ice) lead to a diminishing density of the surface waters in the North Atlantic. This hampers the densification of surface waters that is needed for DWF, and thus the overturning slows down. This mechanism can be inferred from data (see Sec. 4) and is reproduced at least qualitatively in the vast majority of climate models (*Stouffer et al., 2006*). However, different climate models show different sensitivities toward an imposed freshwater flux (*Gregory et al., 2005*). Observations of the freshwater budget of the North Atlantic and the Arctic display a strong decadal variability of the freshwater content of these seas, governed by atmospheric circulation modes like the North Atlantic Oscillation (NAO) (*Peterson et al., 2006*). These freshwater transports cause salinity variations (*Curry et al., 2003*). The salinity anomalies affect the amount of deep water formation (*Dickson et al., 1996*). Remarkably though, the strength of crucial parts of the AMOC, such as the sill overflow through Denmark Strait, has been almost constant over many years (*Girton and Sanford, 2003*), with a significant decrease reported only recently (*Macrander et al., 2005*). It is therefore not clear to what degree salinity changes will affect the total overturning rate of the AMOC. In addition, it is hard to assess how strong future freshwater fluxes into the North Atlantic might be. This is due to uncertainties in modeling the hydrological cycle in the atmosphere (*Zhang et al., 2007b*), in modeling the sea-ice dynamics in the Arctic, as well as in estimating the melting rate of the Greenland ice sheet (see Sec. 7).

It is important to distinguish between an AMOC weakening and an AMOC collapse. In global warming scenarios, nearly all coupled General Circulation Model's (GCMs) show a weakening in the overturning strength (*Gregory et al., 2005*). Sometimes this goes along with a termination of deep water formation in one of the main deep water formation sites (Nordic Seas and Labrador Sea; e.g., *Wood et al., 1999; Schaeffer et al., 2002*). This leads to strong regional climate changes, but the AMOC as a whole keeps going. By contrast, in some simpler coupled climate models the AMOC collapses altogether in reaction to increasing atmospheric CO_2 (e.g., *Rahmstorf and Ganopolski, 1999*): the overturning is reduced to a few Sverdrups. Current GCMs do not show this behavior in global warming scenarios, but a transient collapse can always be triggered in models by a large enough freshwater input and has climatic impacts on the global scale (e.g., *Vellinga and Wood, 2007*). In some models, the collapsed state can last for centuries (*Stouffer et al., 2006*) and might be irreversible.

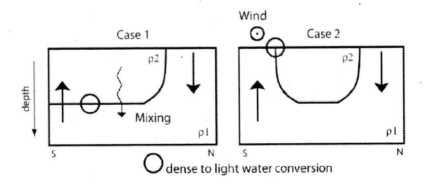

Figure 4.3. Sketch of the two driving mechanisms, mixing (case 1) and wind-driven upwelling (case 2). The sketches are schematic pictures of meridional sections of the Atlantic. Deep water is formed at the right-hand side of the boxes and goes along with heat loss. The curved solid line separates deep dense water ($\rho 1$) from lighter surface water ($\rho 2$). The solid arrows indicate volume flux; the zigzag arrow denotes downward heat flux. Figure from *Kuhlbrodt et al. (2007)*

Finally, it should be mentioned that the driving mechanisms of AMOC's volume flux are not necessarily the drivers of the northward heat transport in the Atlantic (e.g., *Gnanadesikan et al., 2005*). In other words, changes of the AMOC do not necessarily have to affect the heat supply to the northern middle and high latitudes, because other current systems, eddy ocean fluxes, and atmospheric transport mechanisms can to some extent compensate for an AMOC weakening in this respect.

The result of all the mentioned uncertainties is a pronounced discrepancy in experts' opinions about the future of the AMOC. This was seen in a recent elicitation of experts' judgments on the response of the AMOC to climate change (*Zickfeld et al., 2007*). When the twelve experts—paleoclimatologists, observationalists, and modelers—were asked about their individual probability estimates for an AMOC collapse given a 4°C global warming by 2100, their answers lay between 0 and 60% (*Zickfeld et al., 2007*). Enhanced research efforts in the future (see Sec. 8) are required in order to reduce these uncertainties about the future development of the AMOC.

3. What is the Present State of the AMOC?

The concept of a Meridional Overturning Circulation (MOC) involving sinking of cold waters in high-latitude regions and poleward return flow of warmer upper ocean waters can be traced to the early 1 800s (*Rumford, 1800; de Humbolt, 1814*). Since then, the concept has evolved into the modern paradigm of a "global ocean conveyor" connecting a small set of high-latitude sinking regions with more broadly distributed global upwelling patterns via a complex interbasin circulation (*Stommel, 1958; Gordon, 1986*). The general pattern of this circulation has been established for decades based on global hydrographic observations, and continues to be refined. However, measurement of the MOC remains a difficult challenge, and serious efforts toward quantifying the MOC, and monitoring its change, have developed only recently.

Current efforts to quantify the MOC using ocean observations rely on four main approaches:

1. Static ocean "inverse" models utilizing multiple hydrographic sections
2. Analysis of individual transoceanic hydrographic sections
3. Continuous time-series observations along a transoceanic section, and
4. Time-dependent ocean "state estimation" models

We describe, in turn, the fundamentals of these approaches and their assumptions, and the most recent results on the Atlantic MOC that have emerged from each one. In principle the AMOC can also be estimated from ocean models driven by observed atmospheric forcing that are not constrained by ocean observations, or by coupled ocean-atmosphere models. There are many examples of such calculations in the literature, but we will restrict our review to those estimates that are constrained in one way or another by ocean observations.

3.1. Ocean Inverse Models
Ocean "inverse" models combine several (two or more) hydrographic sections bounding a specified oceanic domain to estimate the total ocean circulation through each section. These are often referred to as "box inverse" models because they close off an oceanic "box" defined

by the sections and adjacent continental boundaries, thereby allowing conservation statements to be applied to the domain. The data used in these calculations consist of profiles of temperature and salinity at a number of discrete stations distributed along the sections. The models assume a geostrophic balance for the ocean circulation (apart from the wind-driven surface Ekman layer), and derive the geostrophic velocity profile between each pair of stations, relative to an unknown reference constant, or "reference velocity." The distribution of this reference velocity along each section, and therefore the absolute circulation, is determined by specifying a number of constraints on the circulation within the box and then solving a least-squares (or other mathematical optimization) problem that best fits the constraints, within specified error tolerances. The specified constraints can be many but typically include—above all—overall mass conservation within the box, mass conservation within specified layers, independent observational estimates of mass transports through parts of the sections (e.g., transports derived from current meter arrays), and conservation of property transports (e.g., salt, nutrients, geochemical tracers). Increasingly, the solutions may also be constrained by estimates of surface heat and freshwater fluxes. Once a solution is obtained, the transport profile through each section can be derived, and the AMOC (for zonal basin-spanning sections) can be estimated.

The most comprehensive and up-to-date inverse analyses for the global time-mean ocean include those by *Ganachaud (2003a)* and *Lumpkin and Speer (2007)* (Figure 4.4), based on the World Ocean Circulation Experiment (WOCE) hydrographic data collected during the 1990s. The strength of the Atlantic MOC is given as 18 ± 2.5 Sv by *Lumpkin and Speer (2007)* near 24°N., where it reaches its maximum value. The corresponding estimate from *Ganachaud (2003a)* is 16 ± 2 Sv, in agreement within the error estimates. In both analyses the AMOC strength is nearly uniform throughout the Atlantic from 20°S. to 45°N., ranging from approximately 14 to 18 Sv. These estimates should be taken as being representative of the average strength of the AMOC over the period of the observations.

An implicit assumption in these analyses is that the ocean circulation is in a "steady state" over the time period of the observations, in the above cases over a span of some 10 years. This is likely false, as estimates of relative geostrophic transports across individual repeated sections in the North Atlantic show typical variations of ± 6 Sv (*Ganachaud, 2003a; Lavin et al., 1998*). This variability is accounted for in the inverse models by allowing a relatively generous error tolerance on mass conservation, particularly in upper-ocean layers, which exhibit the strongest temporal variability. While this is an acknowledged weakness of the technique, it is offset by the large number of independent sections included in these (global) analyses, which tend to iron out deviations in individual sections from the time mean. The overall error estimates for the AMOC resulting from these analyses reach about 10-15% of the AMOC magnitude in the mid-latitude North Atlantic, which at the present time can probably be considered as the best constrained available estimate of the "mean" current (1 990s) state of the Atlantic AMOC. However, unless repeated over different time periods, these techniques are unable to provide information on the temporal variability of the AMOC.

3.2. Individual Transoceanic Hydrographic Sections

Historically, analysis of individual transoceanic hydrographic sections has played a prominent role in estimating the strength of the AMOC and the meridional transport of heat of the oceans (*Hall and Bryden, 1982*). The technique is similar to that of the box inverse techniques except that only a single overall mass constraint—the total mass transport across the section—is applied. Other constraints, such as the transports of western boundary currents known from other direct measurements, can also be used where available. The general methodology is summarized in Box 4.2. Determination of the unknown "reference velocity" in the ocean interior is usually accomplished either by assuming that it is uniform across the section or by adjusting it in such a way (subject to overall mass conservation) that it satisfies other *a priori* constraints, such as the expected flow directions of specific water masses. Variability in the reference velocity is only important to the estimation of the AMOC in regions where the topography is much shallower than the mean depth of the section, which is normally confined to narrow continental margins where additional direct observations, if available, are included in the overall calculation.

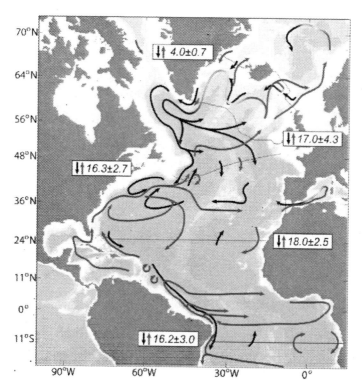

Figure 4.4. Schematic of the Atlantic MOC and major currents involved in the upper (red) and lower (blue) limbs of the AMOC, after *Lumpkin and Speer (2007)*. The boxed numbers indicate the magnitude of the AMOC at several key locations, indicated by gray lines, along with error estimates. The red to green to blue transition on various curves denotes a cooling (red is warm, blue is cold) and sinking of the water mass along its path (figure courtesy of R. Lumpkin, NOAA/AOML)

The best studied location in the North Atlantic, where this methodology has been repeatedly applied to estimate the AMOC strength, is near 24°N., where a total of five transoceanic sections have been acquired between 1957 and 2004. The AMOC estimates derived from these sections range from 14.8 to 22.9 Sv, with a mean value of 18.4 ± 3.1 Sv (Bryden et al., 2005). Individual sections have an estimated error of ± 6 Sv, considerably larger than the error estimates from the above inverse models. Two sections that were acquired during the WOCE period (in 1992 and 1998) yield AMOC estimates of 19.4 and 16.1 Sv, respectively. Therefore these estimates are consistent with the WOCE inverse

AMOC estimates at 24°N. within their quoted uncertainty, as is the mean value of all of the sections (18.4 Sv). Bryden et al. (2005) note a trend in the individual section estimates, with

---

BOX 4.2. HOW DO WE MEASURE THE AMOC?

Observational estimates of the AMOC require the measurement, or inference, of all components of the meridional circulation across a basinwide section. In principle, if direct measurements of the meridional velocity profile are available at all locations across the section, the calculation of the AMOC is straightforward: the velocity is zonally integrated across the section at each depth, and the resulting vertical transport profile is then summed over the northward-moving part of the profile (which is typically the upper ~1,000 m for the Atlantic) to obtain the strength of the AMOC.

In practice, available methods for measuring the absolute velocity across the full width of a transbasin section are either prohibitively expensive or of insufficient accuracy to allow a reliable estimate of the AMOC. Thus, the meridional circulation is typically broken down into several discrete components that can either be measured directly (by current observations), indirectly (by geostrophic calculations based on hydrographic data), or inferred from wind observations (Ekman transports) or mass- balance constraints.

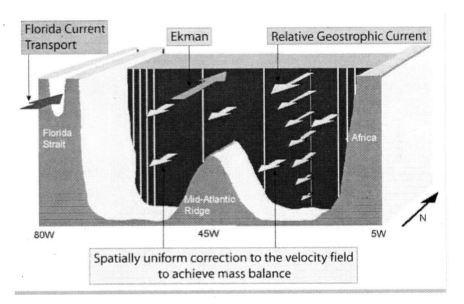

Box 4.2 Figure 1. Circulation components required to estimate the AMOC. The figure depicts the approximate topography along 24-26°N. and the strategy employed by the RAPID monitoring array. The transport of the western boundary current is continuously monitored by a calibrated submarine cable across the Straits of Florida. Hydrographic moorings (depicted by white vertical lines) near the east and west sides of the basin monitor the (relative) geostrophic flow across the basin as well as local flow contributions adjacent to the boundaries. Ekman transport is estimated from satellite wind observations. A uniform velocity correction is included in the interior ocean to conserve mass across the section. (Figure courtesy of J. Hirschi, NOC, Southampton, U.K.)

An illustration of this breakdown is shown in Box 4.2 Figure 1 for the specific situation of the subtropical Atlantic Ocean near 26°N., where the RAPID-MOC array is deployed and where a number of basinwide hydrographic sections have been occupied. The measured transport components include (1) direct measurement of the flow though the Straits of Florida and (2) geostrophic mid-ocean flow derived from density profiles at the eastern and western sides of the ocean, relative to an unknown constant, or "reference velocity." A third component is the ageostrophic flow in the surface layer driven by winds (the Ekman transport), which can be estimated from available wind-stress products. The only remaining unmeasured component is the depth-independent ("barotropic") mid-ocean flow, which is inferred by requiring an overall mass balance across the section. Once combined, these components define the basinwide transport profile and the AMOC strength.

The above breakdown is effective because it takes advantage of the spatially integrating nature of geostrophic computations across the interior of the ocean and limits the need for direct velocity or transport measurements to narrow regions near the coastal boundaries where swift currents may occur (in particular, in the western boundary region). The application is similar for individual hydrographic sections or moored density arrays such as used in RAPID, except that the moored arrays can provide continuous estimates of the interior flow instead of single snapshots in time. Each location where the AMOC is to be measured requires a sampling strategy tuned to the section's topography and known circulation features, but the methodology is essentially the same (*Hall and Bryden, 1982; Bryden et al., 1991; Cunningham et al., 2007*). Inverse models (see Sec. 3.1) follow a similar approach but use a formalized set of constraints with specified error tolerances (e.g., overall mass balance, western boundary current transports, property fluxes) to optimally determine the reference velocity distribution across a section (*Wunsch, 1996*).

the largest AMOC value (22.9 Sv) occurring in 1957 and weakest in 2004 (14.8 Sv), suggesting a nearly 30% decrease in the AMOC over this period (Figure 4.5). Taken at face value, this trend is not significant, as the total change of 8 Sv between 1957 and 2004 falls within the bounds of the error estimates. However, *Bryden et al. (2005)* argue, based upon their finding that the reduced northward transport of upper ocean waters is balanced by a reduction in only the deepest layer of southward NADW, that this change indeed likely reflects a longer term trend rather than random variability. Based upon more recent data collected within the Rapid Climate Change (RAPID) program (see below), it is now believed that the apparent trend could likely have been caused by temporal sampling aliasing.

A similar analysis of available hydrographic sections at 48°N., though less well constrained by western boundary observations than at 24°N., suggests an AMOC variation there of between 9 to 19 Sv, based on three sections acquired between 1957 and 1992 (*Koltermann et al., 1999*). The evidence from individual hydrographic sections therefore points to regional variations in the AMOC of order 4-5 Sv, or about ±25% of its mean value. The time scales associated with this variability cannot be established from these sections, which effectively can only be considered to be "snapshots" in time. Such estimates are, therefore, potentially vulnerable to aliasing by all time scales of AMOC variability.

3.3. Continuous Time-Series Observations

Until recently, there had never been an attempt to continuously measure the AMOC with time-series observations covering the full width and depth of an entire transoceanic section. Motivated by the uncertainty surrounding "snapshot" AMOC estimates derived from hydrographic sections, a joint U.K.-U.S. observational program, referred to as "RAPID–MOC," was mounted in 2004 to begin continuous monitoring of the AMOC at 26°N. in the Atlantic.

The overall strategy consists of the deployment of deep water hydrographic moorings (moorings with temperature and salinity recorders spanning the water column) on either side of the basin to monitor the basin-wide geostrophic shear, combined with observations from clusters of moorings on the western (Bahamas) and eastern (African) continental margins, and direct measurements of the flow though the Straits of Florida by electronic cable (see Box 4.2). Moorings are also included on the flanks of the Mid- Atlantic Ridge to resolve flows in either sub-basin. Ekman transports derived from winds (estimated from satellite measurements) are then combined with the geostrophic and direct current observations and an overall mass conservation constraint to continuously estimate the basin-wide AMOC strength and vertical structure (*Cunningham et al., 2007; Kanzow et al., 2007*).

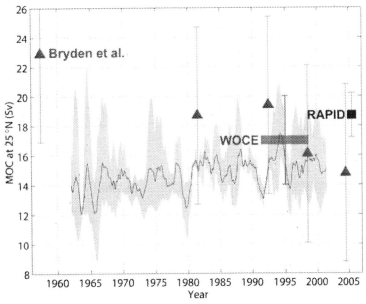

Figure 4.5. Strength of the Atlantic MOC at 25°N. derived from an ensemble average of three state estimation models (solid curve), and the model spread (shaded), for the period 1962-2002 (courtesy of the CLIVAR Global Synthesis and Observations Panel, GSOP). The estimates from individual hydrographic sections at 24°N. (from *Bryden et al., 2005*), from the WOCE inverse model estimates at 24°N. (*Ganachaud, 2003a; Lumpkin and Speer, 2007*), and from the 2004-05 RAPID–MOC Array at 26°N (*Cunningham et al., 2007*) are also indicated, with respective uncertainties.

Although only the first year of results is presently available from this program, these results provide a unique new look at AMOC variability (Figure 4.6) and provide new insights on estimates derived from one-time hydrographic sections. The annual mean strength and standard deviation of the AMOC, from March 2004 to March 2005, was 18.7 ± 5.6 Sv, with instantaneous (daily) values varying over a range of nearly 10-30 Sv. The Florida Current,

Ekman, and mid-ocean geostrophic transport were found to contribute about equally to the variability in the upper ocean limb of the AMOC. The compensating southward flow in the deep ocean (identical to the red curve in Figure 4.6 but opposite in sign), also shows substantial changes in the vertical structure of the deep flow, including several brief periods where the transport of lower NADW across the entire section (associated with source waters originating in the Norwegian-Greenland Sea dense overflows) is nearly, or totally, interrupted.

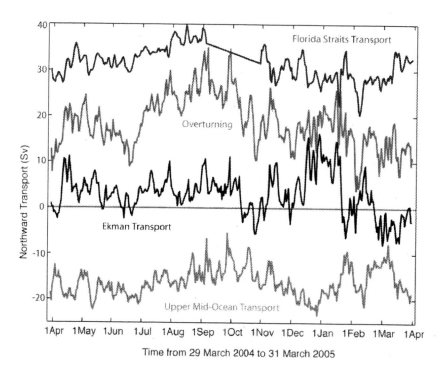

Figure 4.6. Time series of AMOC variability at 26°N. ("overturning", red curve), derived from the 2004-05 RAPID Array (from *Cunningham et al., 2007*). Individual contributions to the total upper ocean flow across the section by the Florida Current (blue), Ekman transport (black), and the mid-ocean geostrophic flow (magenta) are also shown. A 2- month gap in the Florida current transport record during September to November 2004 was caused by hurricane damage to the electromagnetic cable monitoring station on the Bahamas side of the Straits of Florida

These result show that the AMOC can, and does, vary substantially on relatively short time scales and that AMOC estimates derived from one-time hydrographic sections are likely to be seriously aliased by short-term variability. Although the short-term variability of the AMOC is large, the standard error in the 1-year RAPID estimate derived from the autocorrelation statistics of the time series is approximately 1.5 Sv (*Cunningham et al., 2007*). Thus, this technique should be capable of resolving year-to-tear changes in the annual mean AMOC strength of the order of 1-2 Sv. The one year (2004-05) estimate of the AMOC strength of 18.7 ±1.5 Sv is consistent, within error estimates, with the corresponding values near 26°N. determined from the WOCE inverse analysis (16-18 ±2.5 Sv). It is also consistent with the 2004 hydrographic section estimate of 14.8 ±6 Sv, which took place during the first month of the RAPID time series (April 2004), during a period when the AMOC was weaker than its year-long average value (Figure 4.5).

3.4. Time-Varying Ocean State Estimation

With recent advances in computing capabilities and global observations from both satellites and autonomous in-situ platforms, the field of oceanography is rapidly evolving toward operational applications of ocean state estimation analogous to that of atmospheric reanalysis activities. A large number of these activities are now underway that are beginning to provide first estimates of the time-evolving ocean "state" over the last 50+ years, during which sufficient observations are available to constrain the models.

There are two basic types of methods, (1) variational adjoint methods based on control theory and (2) sequential estimation based on stochastic estimation theory. Both methods involve numerical ocean circulation models forced by global atmospheric fields (typically derived from atmospheric reanalyses) but differ in how the models are adjusted to fit ocean data. Sequential estimation methods use specified atmospheric forcing fields to drive the models, and progressively correct the model fields in time to fit (within error tolerances) the data as they become available (e.g., *Carton et al., 2000*). Adjoint methods use an iterative process to minimize differences between the model fields and available data over the entire duration of the model run (up to 50 years), through adjustment of the atmospheric forcing fields and model initial conditions, as well as internal model parameters (e.g., *Wunsch, 1996*). Except for the simplest of the sequential estimation techniques, both approaches are computationally expensive, and capabilities for running global models for relatively long periods of time and at a desirable level of spatial resolution are currently limited. However, in principle these models are able to extract the maximum amount of information from available ocean observations and provide an optimum, and dynamically self-consistent, estimate of the time-varying ocean circulation. Many of these models now incorporate a full suite of global observations, including satellite altimetry and sea surface temperature observations, hydrographic stations, autonomous profiling floats, subsurface temperature profiles derived from bathythermographs, surface drifters, tide stations, and moored buoys.

Progress in this area is fostered by the International Climate Variability and Predictability (CLIVAR) Global Synthesis and Observations Panel (GSOP) through synthesis intercomparison and verification studies (http://www.clivar.org/organization/ gsop/ reference. php). A time series of the Atlantic AMOC at 25°N. derived from an ensemble average of three of these state estimation models, covering the 40-year period from 1962 to 2002, is shown in Figure 4.5. The average AMOC strength over this period is about 15 Sv, with a typical model spread of ±3 Sv. The models suggest interannual AMOC variations of 2-4 Sv with a slight increasing (though insignificant) trend over the four decades of the analysis. The mean estimate for the WOCE period (1990-2000) is 15.5 Sv, and agrees within errors with the 16-18 Sv mean AMOC estimates from the foregoing WOCE inverse analyses.

In comparing these results with the individual hydrographic section estimates, it is notable that only the 1998 (and presumably also the more recent 2004) estimates fall within the spread of the model values. However, owing to the large error bars on the individual section estimates, this disagreement cannot be considered statistically significant. The limited number of models presently available for these long analyses may also underestimate the model spread that will occur when more models are included. It should be noted that these models are formally capable of providing error bars on their own AMOC estimates, although as yet this task has generally been beyond the available computing resources. This should become a priority within climate science once feasible.

A noteworthy feature of Figure 4.5 is the apparent increase in the AMOC strength between the end of the model analysis period in 2002 and the 2004-05 RAPID estimate, an increase of some 4 Sv. The RAPID estimate lies near the top of the model spread of the preceding four decades. Whether this represents a temporary interannual increase in the AMOC that will also be captured by the synthesis models when they are extended through this period, or will represent an ultimate disagreement between the estimates, awaits determination.

3.5. Conclusions and Outlook

The main findings of this chapter concerning the present state of the Atlantic MOC can be summarized as follows:

The WOCE inverse model results (e.g., *Ganachaud, 2003b; Lumpkin and Speer, 2007*) provide, at this time, our most robust estimates of the recent "mean state" of the AMOC, in the sense that they cover an analysis period of about a decade (1990-2000) and have quantifiable (and reasonably small) uncertainties. These analyses indicate an average AMOC strength in the mid-latitude North Atlantic of 16-18 Sv.

Individual hydrographic sections widely spaced in time are not a viable tool for monitoring the AMOC. However, these sections, especially when combined with geochemical observations, still have considerable value in documenting longer-term property changes that may accompany changes in the AMOC, and in the estimation of meridional property fluxes including heat, freshwater, carbon, and nutrients.

Continuous estimates of the AMOC from programs such as RAPID are able to provide accurate estimates of annual AMOC strength and interannual variability, with uncertainties on the annually averaged AMOC of 1-2 Sv, comparable to uncertainties available from the WOCE inverse analyses. RAPID is planned to continue through at least 2014 and should provide a critical benchmark for ocean synthesis models.

Time-varying ocean state estimation models are still in a development phase but are now providing first estimates of AMOC variability, with ongoing intercomparison efforts between different techniques. While there is still considerable research required to further refine and validate these models, including specification of uncertainties, this approach should ultimately lead to our best estimates of the large-scale ocean circulation and AMOC variability.

Our assessment of the state of the Atlantic MOC has been focused on 24°N., owing to the concentration of observational estimates there, which, in turn, is historically related to the availability of long-term, high-quality western boundary current observations at this location. The extent to which AMOC variability at this latitude, apart from that due to local wind-driven (Ekman) variability, is linked to other latitudes in the Atlantic remains an important research question. Also important are changes in the structure of the AMOC, which could have long-term consequences for climate independent of changes in overall AMOC strength. For example, changes in the relative contributions of Southern Hemisphere water masses that supply the upper ocean return flow of the cell (i.e., relatively warm and salty Indian Ocean thermocline water vs. cooler and fresher Subantarctic Mode Waters and Antarctic Intermediate Waters) could significantly impact the temperature and salinity of of the North Atlantic over time and feed back on the deep water formation process.

. Natural variability of the AMOC is driven by processes acting on a wide range of time scales. On intraseasonal to intrannual time scales, the dominant processes are wind- driven Ekman variability and internal changes due to Rossby or Kelvin (boundary) waves. On interannual to decadal time scales, both variability in Labrador Sea convection related to NAO forcing and wind-driven baroclinic adjustment of the ocean circulation are implicated in models (e.g., *Boning et al., 2006*). Finally, on multidecadal time scales, there is growing model evidence that large-scale observed interhemispheric SST anomalies are linked to AMOC variations (*Knight et al., 2005; Zhang and Delworth, 2006*). Our ability to detect future changes and trends in the AMOC depends critically on our knowledge of the spectrum of AMOC variability arising from these natural causes. The identification, and future detection, of AMOC changes will ultimately rely on building a better understanding of the natural variability of the AMOC on the interannual to multidecadal time scales that make up the lower frequency end of this spectrum.

4. What Is the Evidence for Past Changes in the Overturning Circulation?

Our knowledge of the mean state and variability of the AMOC is limited by the short duration of the instrumental record. Thus, in order to gain a longer term perspective on AMOC variability and change, we turn to geologic records from past climates that can yield important insights on past changes in the AMOC and how they relate to climate changes. In particular, we focus on records from the last glacial period, for which there is evidence of changes in the AMOC that can be linked to a rich spectrum of climate variability and change. Improving our ability to characterize and understand past AMOC changes will increase confidence in our ability to predict any future changes in the AMOC, as well as the global impact of these changes on the Earth's natural systems.

The last glacial period was characterized by large, widespread and often abrupt climate changes at millennial time scales, many of which have been attributed to changes in the AMOC and its attendant feedbacks (*Broecker et al., 1985; Clark et al., 2002a, 2007; Alley, 2007*). In the following, we first summarize various types of evidence (commonly referred to as proxy records, in that they provide an indirect measure of the physical property of interest) used to infer changes in the AMOC. We then discuss the current understanding of changes in the AMOC during the following four time windows (Figure 4.7):

1. The Last Glacial Maximum (19,000-23,000 years ago), when ice sheets covered large parts of North America and Eurasia, and the concentration of atmospheric CO_2 was approximately 30% lower than during pre-industrial times. Although the Last Glacial Maximum (LGM) was characterized by relatively low climate variability at millennial time scales, it had a different AMOC than the modern AMOC, which provides a good target for the coupled climate models that are used to predict future changes.
2. The last deglaciation (11,500-19,000 years ago), which was a time of natural global warming associated with large changes in insolation, rising atmospheric CO_2, and melting ice sheets, but included several abrupt climate changes which likely involved changes in the AMOC.

3. Marine Isotope Stage (MIS) 3 (30,000-65,000 years ago), which was a time of pronounced millennial-scale climate variability characterized by abrupt transitions that occurred over large parts of the globe in spite of relatively small changes in insolation, atmospheric CO_2 concentration, and ice-sheet size. Just how these signals originated and were transmitted and modified around the globe, and the extent to which they are associated with changes in the AMOC, remains controversial.

4. The Holocene (0-11,500 years ago), which was a time of relative climate stability (compared to glacial climates) in spite of large changes in insolation. This period of time is characterized by atmospheric CO_2 levels similar to pre-industrial times. Although AMOC changes during the Holocene were smaller than during glacial times, our knowledge of them extends the record of natural variability under near modern boundary conditions beyond the instrumental record.

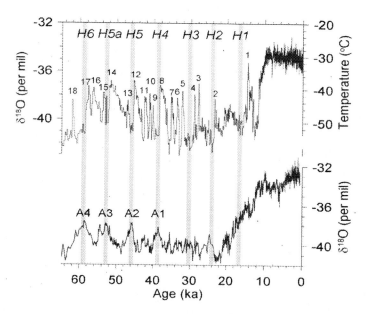

Figure 4.7. Records showing characteristic climate changes for the interval from 65,000 years ago to the present. (Top) The Greenland Ice Sheet Project (GISP2) O record (*Grootes et al., 1993; Stuiver and Grootes, 2000*), which is a proxy for air temperature, with more positive values corresponding to warmer temperatures (*Cuffey and Clow, 1997*). Numbers 1-18 correspond to conventional numbering of warm peaks of Dansgaard-Oeschger oscillations. (Bottom) The Byrd O record (*Johnsen et al., 1972; Hammer et al., 1994*), with the time scale synchronized to the GISP2 time scale by methane correlation (*Blunier and Brook, 2001*). Antarctic warm events identified as A1, etc. Vertical gray bars correspond to times of Heinrich events, with each Heinrich event labeled by conventional numbering (H6, H5, etc.)

4.1. Proxy Records Used to Infer Past Changes in the AMOC

4.1.1. Water mass tracers

The most widely used proxy of millennial-scale changes in the AMOC is C of dissolved inorganic carbon, as recorded in the shells of bottom-dwelling (benthic) foraminifera, which differentiates the location, depth, and volume of nutrient-depleted North Atlantic Deep Water (NADW) relative to underlying nutrient-enriched Antarctic Bottom Water (AABW) (*Boyle and Keigwin, 1982; Curry and Lohmann, 1982; Duplessy et al., 1988*). Millennial-scale water mass variability is also seen in the distribution of other elements linked to nutrients such as Cd and Zn in foraminifera shells (*Boyle and Keigwin, 1982; Marchitto et al., 1998*). The radiocarbon content of deep waters (high in NADW that has recently exchanged carbon with the radiocarbon-rich atmosphere, and low in the older AABW) is recorded both in foraminifera and deep-sea corals (*Keigwin and Schlegel, 2002; Robinson et al., 2005*) and has also been used as a water mass tracer. The deep water masses also carry a distinct Nd isotope signature, which can serve as a tracer that is independent of carbon and nutrient cycles (*Rutberg et al., 2000; Piotrowski et al., 2005*).

4.1.2. Dynamic tracers

While the water mass tracers provide information on water mass geometry, they cannot be used alone to infer the rates of flow. Variations in the grain size of deep-sea sediments can provide information on the vigor of flow at the sediment-water interface, with stronger flows capable of transporting larger particle sizes (*McCave and Hall, 2006*). The magnetic properties of sediments related to particle size have also been used to infer information about the vigor of near-bottom flows (*Kissel et al., 1999*).

The contrasting residence times of the particle-reactive decay products of dissolved uranium (Pa and Th) provide an integrated measure of the residence time of water in the overlying water column. Today, the relatively vigorous renewal of waters in the deep Atlantic results in low ratios of Pa/Th in the underlying sediments, but this ratio should increase if NADW production slows (*Bacon and Anderson, 1982; Yu et al., 1996*). While radiocarbon has been used most successfully as a tracer of water masses in the deep Atlantic, the *in situ* decay of radiocarbon within the Atlantic could potentially be used to infer flow rates, given a sufficiently large number of precise measurements (*Adkins and Boyle, 1997; Wunsch, 2003*).

Finally, as for the modern ocean, we can use the fact that the large-scale oceanic flows are largely in geostrophic balance and infer flows from the distribution of density in the ocean. For paleoclimate reconstructions, the distribution of seawater density can be estimated from oxygen isotope ratios in foraminifera (*Lynch-Stieglitz et al., 1999*) as well as other proxies for temperature and salinity (*Adkins et al., 2002; Elderfield et al., 2006*).

Most of the proxies for water mass properties and flow described above are imperfect recorders of the quantity of interest. They can also be affected to varying degrees by biological, physical, and chemical processes that are not necessarily related to deep water properties and flows. These proxies are most useful for identifying relatively large changes, and the confidence in our inferences based on them increases when there is consistency between more than one independent line of evidence.

4.3. Evidence for state of the AMOC during the last glacial maximum

Although the interval corresponding to the LGM (19,000 to 23,000 years ago) does not correspond to an abrupt climate change, a large body of evidence points to a significantly different AMOC at that time (*Lynch-Stieglitz et al., 2007*), providing an important target for coupled climate model simulations that are used to predict future changes. Among these indicators of a different AMOC, the geographic distribution of different species of surface-dwelling (planktonic) organisms can be used to suggest latitudinal shifts in sites of deep water formation. Accordingly, while warm currents extend far into the North Atlantic today, compensating the export of deep waters from the polar seas, during the LGM, planktonic species indicate that the North Atlantic was marked by a strong east- west trending polar front separating the warm subtropical waters from the cold waters which dominated the North

Atlantic during glacial times, suggesting a southward displacement of deep water formation (*CLIMAP, 1981; Ruddiman and McIntyre, 1981; Paul and Schafer-Neth, 2003; Kucera et al., 2005*).

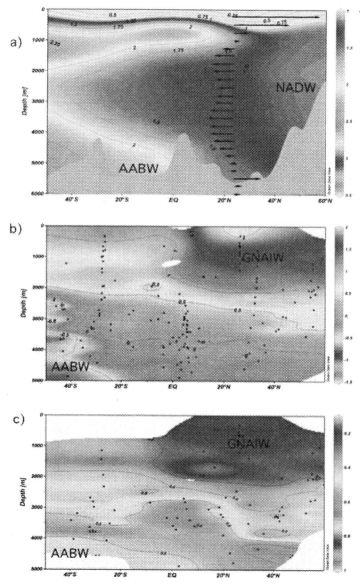

Figure 4.8. (a) The modern distribution of dissolved phosphate (PO4, mmol liter^{-1})—a biological nutrient—in the western Atlantic (*Conkright et al., 2002*). Also indicated is the southward flow of North Atlantic Deep Water (NADW), which is compensated by the northward flow of warmer waters above 1 km, and the Antarctic Bottom Water (AABW) below. (b) The distribution of the carbon isotopic composition ($^{13}C/^{12}C$, expressed as ,C, Vienna Pee Dee belemnite standard) of the shells of benthic foraminifera in the western and central Atlantic during the Last Glacial Maximum (LGM) (*Bickert and Mackensen, 2004; Curry and Oppo, 2005*). Data (dots) from different longitudes are collapsed in the same meridional plane. GNAIW, Glacial North Atlantic Intermediate Water. (c) Estimates of the Cd (nmol kg^{-1}) concentration for LGM from the ratio of Cd/Ca in the shells of benthic foraminifera, from *Marchitto and Broecker (2006)*. Today, the isotopic composition of dissolved inorganic carbon and the concentration of dissolved Cd in seawater both show "nutrient"-type distributions similar to that of PO4

The chemical and isotopic compositions of benthic organisms suggest that low-nutrient NADW dominates the modern deep North Atlantic (Figure 4.8). During the LGM, however, these proxies indicate that the deep water masses below 2 kilometers (km) depth appear to be older (*Keigwin, 2004*) and more nutrient rich (*Duplessy et al., 1988; Sarnthein et al., 1994; Bickert and Mackensen, 2004; Curry and Oppo, 2005; Marchitto and Broecker, 2006*) than the waters above 2 km, suggesting a northward expansion of AABW and corresponding shoaling of NADW to form Glacial North Atlantic Intermediate Water (GNAIW) (Figure 4.8). Finally, pore-water chloride data from deep-sea sediments in the Southern Ocean indicate that the north-south salinity gradient in the deep Atlantic was reversed relative to today, with the deep Southern Ocean being much saltier than the North Atlantic (*Adkins et al., 2002*).

The accumulation of the decay products of uranium in ocean sediments (Pa/Th ratio) is consistent with an overall residence time of deep waters in the Atlantic that was slightly longer than today (*Vu et al., 1996; Marchal et al., 2000; McManus et al., 2004*). Reconstructions of seawater density based on the isotopic composition of benthic shells suggest a reduced density contrast across the South Atlantic basin, implying a weakened AMOC in the upper 2 km of the South Atlantic (*Lynch-Stieglitz et al., 2006*). Inverse modeling (*Winguth et al., 1999*) of the carbon isotope data is also consistent with a slightly weaker AMOC during the LGM.

4.4. Evidence for changes in the AMOC during the last deglaciation

Multiple proxies indicate that the AMOC underwent several large and abrupt changes during the last deglaciation (11,500 to 19,000 years ago). Proxies of temperature and precipitation suggest corresponding changes in climate (Figure 4.7) that can be attributed to these changes in the AMOC and its attendant feedbacks (*Broecker et al., 1985; Clark et al., 2002a; Alley, 2007*). Many of the AMOC proxy records from marine sediments show that the changes in deep water properties and flow were quite abrupt, but due to slow sedimentation rates and mixing of the sediments at the sea floor these records can only provide an upper bound on the transition time between one circulation state and another. Radiocarbon data from fossil deep-sea corals, however, show that deep water properties can change substantially in a matter of decades (*Adkins et al., 1998*). Several possible freshwater forcing mechanisms have been identified that may explain this variability, although there are still large uncertainties in understanding the relation between these mechanisms and changes in the AMOC (Box 4.3).

Early in the deglaciation, starting at ~19,000 years ago, water mass tracers (^{14}C and .C) suggest that low-nutrient, radiocarbon-enriched GNAIW began to contract and shoal from its LGM distribution so that by ~17.5 ka, a significant fraction of the North Atlantic basin was filled with high-nutrient, radiocarbon-depleted AABW (Figure 4.9) (*Sarnthein et al., 1994; Zahn et al., 1997; Curry et al., 1999; Willamowski and Zahn, 2000; Rickaby and Elderfield, 2005; Robinson et al., 2005*). Dynamic tracers of the AMOC (grain size and Pa/Th ratios of deep-sea sediments) similarly show a shift starting at ~1 9 ka toward values that indicate a reduction in the rate of the AMOC (Figure 4.9) (*Manighetti and McCave, 1995; McManus et al., 2004*). By ~17.5 ka, the Pa/Th ratios almost reach the ratio at which they are produced in the water column, requiring a slowdown or shutdown of deep water renewal in the deep Atlantic (*Siddall et al., 2007*), thus explaining the extreme contraction of GNAIW inferred from the water mass tracers. At the same time, radiocarbon data from the Atlantic basin not only support a reduced flux of GNAIW but also indicate a vigorous circulation of AABW in the North Atlantic basin (*Robinson et al., 2005*).

BOX 4.3. PAST MECHANISMS FOR FRESHWATER FORCING OF THE AMOC

Ice sheets represent the largest readily exchangeable reservoir of freshwater on Earth. Given the proximity of modern and former ice sheets to critical sites of intermediate and deep water formation (Figure 4.1), variations in their freshwater fluxes thus have the potential to induce changes in the AMOC. In this regard, the paleorecord has suggested four specific mechanisms by which ice sheets may rapidly discharge freshwater to the surrounding oceans and cause abrupt changes in the AMOC: (1) Heinrich events, (2) meltwater pulses, (3) routing events, and (4) floods.

1. . Heinrich events are generally thought to represent an ice-sheet instability resulting in abrupt release of icebergs that triggers a large reduction in the AMOC. Paleoclimate records, however, indicate that Heinrich events occur after the AMOC has slowed down or largely collapsed. An alternative explanation is that Heinrich events are triggered by an ice-shelf collapse induced by subsurface oceanic warming that develops when the AMOC collapses, with the resulting flux of icebergs acting to sustain the reduced AMOC.

2. The ~20-m sea-level rise ~14,500 years ago, commonly referred to as meltwater pulse (MWP) 1A, indicates an extraordinary episode of ice-sheet collapse, with an associated freshwater flux to the ocean of ~0.5 Sv over several hundred years (see Chapter 2). Nevertheless, the timing, source, and the effect on climate of MWP-1A remain unclear. In one scenario, the event was triggered by an abrupt warming (start of the Bølling warm interval) in the North Atlantic region, causing widespread melting of Northern Hemisphere ice sheets. Although this event represents the largest freshwater forcing yet identified from paleo-sea-level records, there was little response by the AMOC, leading to the conclusion that the meltwater entered the ocean as a sediment-laden, very dense bottom flow, thus reducing its impact on the AMOC. In another scenario, MWP-1A largely originated from the Antarctic Ice Sheet, possibly in response to the prolonged interval of warming in the Southern Hemisphere that preceded the event. In this case, climate model simulations indicate that the freshwater perturbation in the Southern Ocean may have triggered the resumption of the AMOC that caused the Bølling warm interval.

3. The most well-known hypothesis for a routing event involves retreat of the Laurentide Ice Sheet (LIS) that redirected continental runoff from the Mississippi to the St. Lawrence River, triggering the Younger Dryas cold interval. There is clear paleoceanographic evidence for routing of freshwater away from the Mississippi River at the start of the Younger Dryas, and recent paleoceanographic evidence now clearly shows a large salinity decrease in the St. Lawrence estuary at the start of the Younger Dryas associated with an increased freshwater flux derived from western Canada.

4. The most well-known flood is the final sudden drainage of glacial Lake Agassiz that is generally considered to be the cause of an abrupt climate change ~8400 years ago. For this event, the freshwater forcing was likely large but short; the best current estimate suggests a freshwater flux of 4-9 Sv over 0.5 year. This event was unique to the last stages of the LIS, however, and similar such events should only be expected in association with similar such ice-sheet configurations. Other floods have been inferred at other times, but they would have been much smaller (~0.3 Sv in 1 year), and model simulations suggest they would have had a negligible impact on the AMOC.

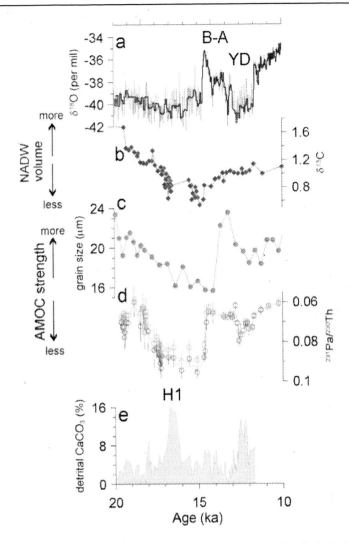

Figure 4.9. Proxy records of changes in climate and the AMOC during the last deglaciation. Ka, thousand years. (a) The GISP2 .O record (*Grootes et al., 1993; Stuiver and Grootes, 2000*). B-A is the Bølling-Allerød warm interval, YD is the Younger Dryas cold interval, and H1 is Heinrich event 1. (b) The .C record from core SO75-26KL in the eastern North Atlantic (*Zahn et al., 1997*). (c) Record of changes in grain size ("sortable silt") from core BOFS 10k in the eastern North Atlantic (*Manighetti and McCave, 1995*). (d) The record of ^{231}Pa/^{230}Th in marine sediments from the Bermuda Rise, western North Atlantic (*McManus et al., 2004*). Purple symbols are values based on total 238U activity, green symbols are based on total 232Th activity. (e) Record of changes in detrital carbonate from core VM23-81 from the North Atlantic (*Bond et al., 1997*)

The cause of this extreme slowdown of the AMOC is often attributed to Heinrich event 1, which represented a massive release of icebergs from the Laurentide Ice Sheet into the North Atlantic Ocean (Box 4.3) (*Broecker, 1994; McManus et al., 2004; Timmermann et al., 2005b*). The best estimate for the age of Heinrich event 1 (~17.5 ka), however, indicates the decrease in the AMOC began ~1,500 years earlier, with the event only coinciding with the final near-cessation of the AMOC ~17.5 ka (Figure 4.9) (*Bond et al., 1993; Bond and Lotti, 1995; Hemming, 2004*). These relations thus suggest that some other mechanism was responsible for the decline and eventual near-collapse of the AMOC prior to the event (Box 4.3).

This interval of a collapsed AMOC continued until ~14.6 ka, when dynamic tracers indicate a rapid resumption of the AMOC to near-interglacial rates (Figure 4.9). This rapid change in the AMOC was accompanied by an abrupt warming throughout much of the Northern Hemisphere associated with the onset of the Bølling-Allerød warm interval (*Clark et al., 2002b*). The renewed overturning filled the North Atlantic basin with NADW, as shown by Cd/Ca ratios (*Boyle and Keigwin, 1987*) and Nd isotopes (*Piotrowski et al., 2004*) from the North and South Atlantic, respectively. Moreover, the distribution of radiocarbon in the North Atlantic was similar to the modern ocean, with the entire water column filled by radiocarbon-enriched water (*Robinson et al., 2005*).

An abrupt reduction in the AMOC occurred again at ~12.9 ka, corresponding to the start of the ~1200-year Younger Dryas cold interval. During this time period, many of the paleoceanographic proxies suggest a return to a circulation state similar to the LGM. Unlike the near-collapse earlier in the deglaciation at ~17.5 ka, for example, Pa/Th ratios suggest only a partial reduction in the AMOC during the Younger Dryas (Figure 4.9). Sediment grain size (*Manighetti and McCave, 1995*) also shows evidence for reduced NADW input into the North Atlantic during the Younger Dryas event (Figure 4.9). Radiocarbon concentration in the atmosphere rises at the start of the Younger Dryas, which is thought to reflect decreased ocean uptake due to a slowdown of the AMOC (*Hughen et al., 2000*). Radiocarbon-depleted AABW replaced radiocarbon-enriched NADW below ~2500 m, suggesting a shoaling of NADW coincident with a reduction of the AMOC (*Keigwin, 2004*). The ,C values also suggest a return to the LGM water mass configuration (*Sarnthein et al., 1994; Keigwin, 2004*), as do other nutrient tracers (*Boyle and Keigwin, 1987*) and the Nd isotope water mass tracer (*Piotrowski et al., 2005*).

The cause of the reduced AMOC during the Younger Dryas has commonly been attributed to the routing of North American runoff with a resulting increase in freshwater flux draining eastward through the St. Lawrence River (*Johnson and McClure, 1976; Rooth, 1982; Broecker et al., 1989*), which is supported by recent paleoceanographic evidence (*Flower et al., 2004; Carlson et al., 2007*) (Box 4.3).

4.5. Evidence for changes in the AMOC during stage 3

Marine isotope stage 3 (30,000—65,000 years ago) was a period of intermediate ice volume that occurred prior to the LGM. This period of time is characterized by the Dansgaard-Oeschger (D-O) oscillations, which were first identified from Greenland ice- core records (*Johnsen et al., 1992; Grootes et al., 1993*) (Figure 4.7). These oscillations are similar to the abrupt climate changes during the last deglaciation and are characterized by alternating warm (interstadial) and cold (stadial) states lasting for millennia, with abrupt transitions between states of up to 16°C occurring over decades or less (*Cuffey and Clow, 1997; Huber et al., 2006*). Bond et al. (1993) recognized that several successive D-O oscillations of decreasing amplitude represented a longer term (~7,000-yr) climate oscillation which culminates in a massive release of icebergs from the Laurentide Ice Sheet, known as a Heinrich event (Figure 4.7) (Box 4.3). The D-O signal seems largely confined to the Northern Hemisphere, while the Southern Hemisphere often exhibits less abrupt, smaller amplitude millennial climate changes (*Clark et al., 2007*), best represented by A-events seen in Antarctic ice core records (Figure 4.7). Synchronization of Greenland and Antarctic ice core records (*Sowers and Bender, 1995; Bender et al., 1994, 1999; Blunier et al., 1998; Blunier and Brook, 2001; EPICA Community Members, 2006*) suggests an out-of-phase "seesaw" relationship between temperatures of the Northern and Southern Hemispheres, and that the thermal contrast between hemispheres is greatest at the time of Heinrich events (Figure 4.7).

By comparison to the deglaciation, there are fewer proxy records constraining millennial-scale changes in the AMOC during stage 3. Most inferences of these changes are based on C as a proxy for water-mass nutrient content. A depth transect of well-correlated C records is required in order to capture temporal changes in the vertical distribution of any given water mass, since the C values at any given depth may not change significantly if the core site remains within the same water mass.

Figure 4.10 illustrates one such time-depth transect of C records from the eastern North Atlantic that represent changes in the depth and volume (but not rate) of the AMOC during an interval (3 5-48 ka) of pronounced millennial-scale climate variability (Figure 4.7). We emphasize this interval only because it encompasses a highly resolved and well-dated array of C records. The distinguishing feature of these records is a minimum in C at the same time as Heinrich events 4 and 5, indicating the near- complete replacement of nutrient-poor, high C NADW with nutrient-rich, low C AABW in this part of the Atlantic basin. The inference of a much reduced rate of the AMOC from these data is supported by the proxy records during the last deglaciation (Figure 4.9), which indicate a similar distribution of C at a time when Pa/Th ratios suggest the AMOC had nearly collapsed by the time of Heinrich event 1 (see above). Insofar as we understand the interhemispheric seesaw relationship established by ice core records (Figure 4.7) to reflect changes in the AMOC and corresponding ocean heat transport (*Broecker, 1998; Stocker and Johnsen, 2003*), the fact that Heinrich events during stage 3 only occur at times of maximum thermal contrast between hemispheres (cold north, warm south) further indicates that some other mechanism was responsible for causing the large reduction in the AMOC by the time a Heinrich event occurred.

While many of the Heinrich stadials show up clearly in these and other $\delta^{13}C$ records, there is often no clear distinction between D-O interstadials and non-Heinrich D-O stadials (Figure 4.10) (*Boyle, 2000; Shackleton et al., 2000; Elliot et al., 2002*). While some $\delta^{13}C$ and Nd records do show millennial-scale variability not associated with the Heinrich events (*Charles et al., 1996; Curry et al., 1999; Hagen and Keigwin, 2002; Piotrowski et al., 2005*), there are many challenges that have impeded the ability to firmly establish the presence or absence of coherent changes in the North Atlantic water masses (and by inference the AMOC) during the D-O oscillations. These challenges include accurately dating and correlating sediment records beyond the reach of radiocarbon, and having low abundances of the appropriate species of benthic foraminifera in cores with high-enough resolution to distinguish the D-O oscillations.

In contrast to these difficulties in distinguishing and resolving D-O oscillations with water-mass tracers, the relative amount of magnetic minerals in deep-sea sediments in the path of the deep Atlantic overflows shows contemporaneous changes with all of the D-O oscillations (*Kissel et al., 1999*). These magnetic minerals are derived from Tertiary basaltic provinces underlying the Norwegian Sea and are interpreted to record an increase (or decrease) in the velocity of the overflows from the Nordic Seas during D-O interstadials (or stadials). Taken at face value, the C and magnetic records may indicate that latitudinal shifts in the AMOC occurred, but with little commensurate change in the depth of deep water formation. The corresponding changes in the relative amount of magnetic minerals then reflect times when NADW formation occurred either in the Norwegian Sea, thus entraining magnetic minerals from the sea floor there, or in the open North Atlantic, at sites to the south of the source of the magnetic minerals. What remains unclear is whether changes in the overall strength of the AMOC accompanied these latitudinal shifts in NADW formation.

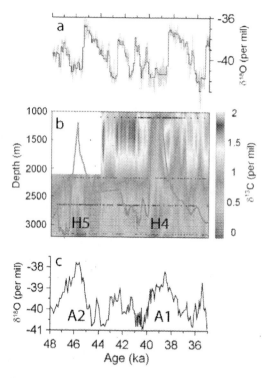

Figure 4.10. (a) The GISP2 ä ^{18}O record (*Grootes et al., 1993; Stuiver and Grootes, 2000*). Times of Heinrich events 4 and 5 identified (H4 and H5). (b) Time-varying ä ^{13}C, a proxy for distribution of deep-water masses, as a function of depth in the eastern North Atlantic based on four ä ^{13}C records at water depths of 1,099 m (*Zahn et al., 1997*), 2,161 m (*Elliot et al., 2002*), 2,637 m (*Skinner and Elderfield, 2007*), and 3,146 m (*Shackleton et al., 2000*). Control points from four cores used for interpolation are shown (black dots). More negative ä ^{13}C values correspond to nutrient-rich Antarctic Bottom Water (AABW), whereas more positive ä ^{13}C values correspond to nutrient-poor North Atlantic Deep Water (see Figure 4.8). Also shown by the thick gray line is a proxy for Heinrich events, with peak values corresponding to Heinrich events H5 and H4 (*Stoner et al., 2000*) (note that scale for this proxy is not shown). During Heinrich events H5 and H4, nutrient-rich AABW displaces NADW to shallow depths in the eastern North Atlantic basin. (c) The Byrd ä ^{18}O record (*Johnsen et al., 1972*), with the time scale synchronized to the GISP2 time scale by methane correlation (*Blunier and Brook, 2001*). ka, thousand years. A1, A2, Antarctic warm events

The fact that the global pattern of millennial-scale climate changes is consistent with that predicted from a weaker AMOC (see Sec. 6) has been taken as a strong indirect confirmation that the stage 3 D-O oscillations are caused by AMOC changes (*Alley, 2007; Clark et al., 2007*). However, care must be taken to separate the climate impacts of a much-reduced AMOC during Heinrich stadials, for which there is good evidence, from the non-Heinrich stadials, for which evidence of changes in the AMOC remains uncertain. This is often difficult in all but the highest resolution climate records. It has also been shown that changes in sea-ice concentrations in the North Atlantic can have a significant impact (*Barnett et al., 1989; Douville and Royer, 1996; Chiang et al., 2003*) and were likely involved in some of the millennial-scale climate variability during the deglaciation and stage 3 (*Denton et al., 2005; Li et al., 2005; Masson-Delmotte et al., 2005*). Sea-ice changes may be a mechanism to amplify the impact of small changes in AMOC strength or location, but they may also result from changes in atmospheric circulation (*Seager and Battisti, 2007*).

4.6. Evidence for changes in the AMOC during the holocene

The proxy evidence for the state of the AMOC during the Holocene (0-11,500 years ago) is scarce and sometimes contradictory but clearly points to a more stable AMOC on millennial time scales than during the deglaciation or glacial times. Some $ä^{13}C$ reconstructions suggest relatively dramatic changes in deep Atlantic water-mass properties on millennial time scales, but these changes are not always coherent between different sites (*Oppo et al., 2003; Keigwin et al., 2005*). Similarly, the $ä^{13}C$ and tracemetal-based nutrient reconstructions on the same cores may disagree (*Keigwin and Boyle, 2000*). There is some indication from sediment grain size for variability in the strength of the overflows (*Hall et al., 2004*), but the relatively constant flux of Pa/Th to the Atlantic sediments suggests only small changes in the AMOC (*McManus et al., 2004*). The geostrophic reconstructions of the flow in the Florida Straits also suggest that small changes in the strength of the AMOC are possible over the last 1,000 years (*Lund et al., 2006*).

There was a brief (about 150 year) cold snap in parts of the Northern Hemisphere at ~8,200 years ago, and it was proposed that this event may have resulted from a meltwater-induced reduction in the AMOC (*Alley and Agustdottir, 2005*). There is now evidence of a weakening of the overflows in the North Atlantic from sediment grain size and magnetic properties (*Ellison et al., 2006; Kleiven et al., 2008*), and also a replacement of NADW (with high $ä^{13}C$ ratios) by AABW (with low $ä^{13}C$ ratios) in the deep North Atlantic (*Kleiven et al., 2008*).

While many of the deep-sea sediment records are only able to resolve changes on millennial to centennial time scales, a recent study (*Boessenkool et al., 2007*) reconstructs the strength of the Iceland-Scotland overflow on sub-decadal time scales over the last 230 years. This grain-size based study suggests that the recent weakening over the last decades falls mostly within the range of its variability over the period of study. This work shows that paleoceanographic data may, in some locations, be used to extend the instrumental record of decadal- and centennial-scale variability.

4.7. Summary

We now have compelling evidence from a variety of paleoclimate proxies that the AMOC existed in a different state during the LGM, providing concrete evidence that the AMOC changed in association with the lower CO_2 and presence of the continental ice sheets. The LGM can be used to test the response of AMOC in coupled ocean atmosphere models to these changes (Sec. 5). We also have strong evidence for abrupt changes in the AMOC during the last deglaciation and during the Heinrich events, although the relation between these changes and known freshwater forcings is not always clear (Box 4.3). Better constraining both the magnitude and location of the freshwater perturbations that may have caused these changes in the AMOC will help to further refine the models, enabling better predictions of future abrupt changes in the AMOC. The relatively modest AMOC variability during the Holocene presents a challenge for the paleoclimate proxies and archives, but further progress in this area is important, as it will help establish the range of natural variability from which to compare any ongoing changes in the AMOC.

5. How Well Do the Current Coupled Ocean-Atmosphere Models Simulate the Overturning Circulation?

Coupled ocean-atmosphere models are commonly used to make projections of how the AMOC might change in future decades. Confidence in these models can be improved by making comparisons of the AMOC both between models and between models and observational data. Even though the scarcity of observations presents a major challenge, it is apparent that significant mismatches are present and that continued efforts are needed to improve the skill of coupled models. This section reviews simulations of the present-day (Sec. 5.1), Last Glacial Maximum (Sec. 5.2), and transient events of the past (Sec. 5.3). Model projections of future changes in the AMOC are presented in Section 7.

5.1. Present-Day Simulations

A common model-model and model-data comparison uses the mean strength of the AMOC. Observational estimates are derived from either hydrographic data (Sec. 3.3; *Ganachaud, 2003a; Talley et al., 2003; Lumpkin and Speer, 2007*) or inventories of chlorofluorocarbon tracers in the ocean (*Smethie and Fine, 2001*). The estimates are consistent with each other and suggest a mean overturning of about 15-18 Sv with errors of about 2-5 Sv.

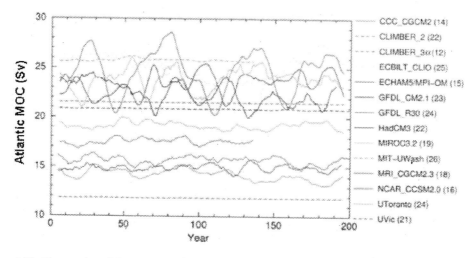

Figure 4.11. Time series of the strength of the Atlantic meridional overturning as simulated by a suite of coupled ocean-atmosphere models using present-day boundary conditions, from *Stouffer et al. (2006)*. The strength is listed along the y-axis in Sverdrups (Sv; 1 Sv = 10^6 m^{3}s^{-1}). Curves were smoothed with a 10-yr running mean to reduce high-frequency fluctuations. The numbers after the model names indicate the long-term mean of the Atlantic MOC

Coupled atmosphere-ocean models using modern boundary conditions yield a wide range of values for overturning strength, which is usually defined as the maximum meridional overturning streamfunction value in the North Atlantic excluding the surface circulation. While the maximum overturning streamfunction is not directly observable, it is a very useful metric for model intercomparisons. Present-day control (i.e., fixed forcing) simulations yield average AMOC intensities from model to model between 12 and 26 Sv (Figure 4.11; *Stouffer et al., 2006*), while simulations of the 20th century that include historical variations in forcing have a range from 10 to 30 Sv (*Randall et al., 2007*; see also Figure 4.17). In addition, some of the 20th century simulations show substantial drifts that might hinder predictions of future AMOC strength (*Randall et al., 2007*).

There are also substantial differences among models in AMOC variability, which tends to scale with the mean strength of the overturning. Models with a more vigorous overturning tend to produce pronounced multidecadal variations, while variability in models with a weaker AMOC is more damped (*Stouffer et al., 2006*). Time series of the AMOC are too incomplete to give an indication of which mode is more accurate, although recent observations suggest that the AMOC is highly variable on sub-annual time scales (Sec. 3.3; *Cunningham et al., 2007*).

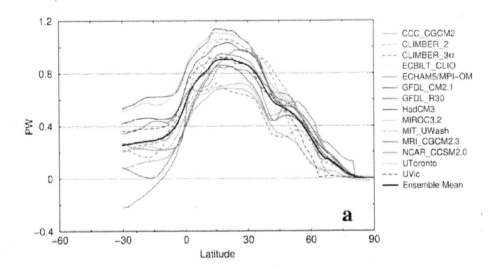

Figure 4.12. Northward heat transport in the Atlantic Ocean in an ensemble of coupled ocean-atmosphere models, from *Stouffer et al. (2006)*. For comparison, observational estimates at 20-25°N. are about 1.3 ± 0.2 Petawatts (PW; 1 PW = 1,015 Watts) (*Ganachaud and Wunsch, 2000; Stammer et al., 2003*).

Another useful model-data comparison can be made for ocean heat transport in the Atlantic. A significant fraction of the northward heat transport in the Atlantic is due to the AMOC, with additional contributions from horizontal circulations (e.g., *Roemmich and Wunsch, 1985*). In the absence of variations in radiative forcing, changes in ocean heat storage are small when averaged over long periods. Under these conditions, ocean heat transport must balance surface heat fluxes, and the heat transport therefore provides an indication of how well surface fluxes are simulated. There are several calculations of heat transport at 20-25° N. in the Atlantic derived by combining hydrographic observations in inverse models. These methods yield estimates of about 1.3 Petawatts (PW; 1 PW = 1,015

Watts) with errors on the order of about 0.2 PW (*Ganachaud and Wunsch, 2000; Stammer et al., 2003*). While all models agree that heat transport in the Atlantic is northward at 20°N., the modeled magnitude varies greatly (Figure 4.12). Most models tend to underestimate the ocean heat transport, with ranges generally between 0.5 to 1.1 PW (*Jia, 2003; Stouffer et al., 2006*). The mismatch is believed to result from two factors: (1) smaller than observed temperature differences between the upper and lower branches of the AMOC, with surface waters too cold and deep waters too warm, and (2) overturning that is too weak (*Jia, 2003*). The source of these model errors will be discussed further.

Schmittner et al. (2005) and *Schneider et al. (2007)* have proposed that the skill of a model in producing the climatological spatial patterns of temperature, salinity, and pycnocline depth in the North Atlantic is another useful measure of model ability to simulate the overturning circulation. These authors found that models simulate temperature better than salinity; they attribute errors in the latter to biases in the hydrologic cycle in the atmosphere (*Schneider et al., 2007*). Large errors in pycnocline depth are probably the result of compounded errors from both temperature and salinity fields. Also, errors over the North Atlantic alone tend to be significantly larger than those for the global field (*Schneider et al., 2007*). Large cold biases of up to several degrees Celsius in the North Atlantic, seen in most coupled models, are attributed partly to misplacement of the Gulf Stream and North Atlantic Current and the large SST gradients associated with them (*Randall et al., 2007*). Cold surface biases commonly contrast with temperatures that are about 2° C too warm at depth in the region of North Atlantic Deep Water (*Randall et al., 2007*).

Some of these model errors, particularly in temperature and heat transport, are related to the representation of western boundary currents (Gulf Stream and North Atlantic Current) and deep-water overflow across the Greenland-Iceland-Scotland ridge. Two common model biases in the western boundary current are (1) a separation of the Gulf Stream from the coast of North America that occurs too far north of Cape Hatteras (*Dengg et al., 1996*) and (2) a North Atlantic Current whose path does not penetrate the southern Labrador Sea, and is instead too zonal with too few meanders (*Rossby, 1996*). The effect of the first bias is to prohibit northward meanders and warm core eddies, negatively affecting heat transport and water mass transformation, while the second bias results in SSTs that are too cold. Both of these biases have been improved in standalone ocean models by increasing the resolution to about 0.1° so that mesoscale eddies may be resolved (e.g., *Smith et al., 2000; Bryan et al., 2007*). The resolution of current coupled ocean-atmosphere models is typically on the order of 1° or more, requiring an increase in computing power of an order of magnitude before coupled ocean eddy-resolving simulations become routine. Initial results from coupling a high-resolution ocean model to an atmospheric model indicate that a corresponding increase in atmospheric resolution may also be necessary (*Roberts et al., 2004*).

Ocean model resolution is also one of the issues involved in the representation of ocean convection, which can occur on very small spatial scales (Wadhams et al., 2002), and in deep-water overflows. Deep-water masses in the North Atlantic are formed in marginal seas and enter the open ocean through overflows such as the Denmark Strait and the Faroe Bank Channel. Model simulations of overflows are unrealistic in several aspects, including (1) the specification of sill bathymetry, which is made difficult because the resolution is often too coarse to represent the proper widths and depths (*Roberts and Wood, 1997*), and (2) the representation of mixing of dense overflow waters with ambient waters downstream of the sill

(*Winton et al., 1998*). In many ocean models, topography is specified as discrete levels, which leads to a "stepped" profile descending from sills. Mixing of overflow waters with ambient waters occurs at each step, leading to excessive entrainment. As a result, deep waters in the lower branch of the AMOC are too warm and too fresh (e.g., *Tang and Roberts, 2005*). Efforts are being made to improve this model deficiency through new parameterizations (*Thorpe et al., 2004; Tang and Roberts, 2005*) or by using isopycnal or terrain-following vertical coordinate systems (*Willebrand et al., 2001*).

Realistic simulation of sea ice is also important for the AMOC due to the effects of sea ice on the surface energy and freshwater budgets of the North Atlantic. The representation of dynamical and thermodynamical processes has become more sophisticated in the current generation of sea-ice models. Nevertheless, when coupled to atmosphere-ocean general circulation models, sea-ice models tend to yield unrealistically large sea-ice extents in the Northern Hemisphere, a poor simulation of regional distributions, and a large range in ice thickness (e.g., *Arzel et al., 2006; Zhang and Walsh, 2006*). These tendencies are the result of biases in winds, ocean mixing, and surface heat fluxes (*Randall et al., 2007*).

5.2. Last Glacial Maximum Simulations

Characteristics of the overturning circulation at the LGM were reviewed in Section 3. Those that are the most robust and, therefore, the most useful for evaluating model performance are (1) a shallower boundary, at a level of about 2,000-2,500 m, between Glacial North Atlantic Intermediate Water and Antarctic Bottom Water (*Duplessy et al., 1988; Boyle, 1992; Curry and Oppo, 2005; Marchitto and Broecker, 2006*); (2) a reverse in the north-south salinity gradient in the deep ocean to the Southern Ocean being much saltier than the North Atlantic (*Adkins et al., 2002*); and (3) formation of Glacial North Atlantic Intermediate Water south of Iceland (*Duplessy et al., 1988; Sarnthein et al., 1994; Pflaumann et al., 2003*).

It is more difficult to compare model results to inferred flow speeds, due to the lack of agreement among proxy records for this variable. Some studies suggest a vigorous circulation with transports not too different from today (*McCave et al., 1995; Yu et al., 1996*), while others suggest a decreased flow speed (*Lynch-Stieglitz et al., 1999; McManus et al., 2004*). All that can be said confidently is that there is no evidence for a significant strengthening of the overturning circulation at the LGM.

Results from LGM simulations are strongly dependent on the specified boundary conditions. In order to facilitate model-model and model-data comparisons, the second phase of the Paleoclimate Modelling Intercomparison Project (PMIP2; *Braconnot et al., 2007*) coordinated a suite of coupled atmosphere-ocean model experiments using common boundary conditions. Models involved in this project include both General Circulation Models (GCMs) and Earth System Models of Intermediate Complexity (EMICs). LGM boundary conditions are known with varying degrees of certainty. Some are known well, including past insolation, atmospheric concentrations of greenhouse gases, and sea level. Others are known with less certainty, including the topography of the ice sheets, vegetation and other land-surface characteristics, and freshwater fluxes from land. For these, PMIP2 simulations used best estimates (see *Braconnot et al., 2007*). More work is necessary to narrow the uncertainty of these boundary conditions, particularly since some could have important effects on the AMOC.

PMIP2 simulations using LGM boundary conditions were completed with five models, three coupled atmosphere-ocean models and two EMICs. Only one of the models, the ECBilt-

CLIO EMIC, employs flux adjustments. Although EMICs generally have not been included in future climate projections using multimodel ensembles, considering them within the context of model evaluation may yield additional understanding about how various model parameterizations and formulations affect the simulated AMOC.

The resulting AMOC in the the LGM simulations varies widely between the models, and several of the simulations are clearly not in agreement with the paleodata (Figures 4.7, 4.13). A shoaling of the circulation is clear in only one of the models (the NCAR CCSM3); all other models show either a deepening or little change (*Weber et al., 2007;*

Otto-Bliesner et al., 2007). Also, the north-south salinity gradient of the LGM deep ocean is not consistently reversed in these model simulations (*Otto-Bliesner et al., 2007*). All models do show a southward shift of GNAIW formation, however. In general, the better the model matches one of these criteria, the better it matches the others as well (*Weber et al., 2007*).

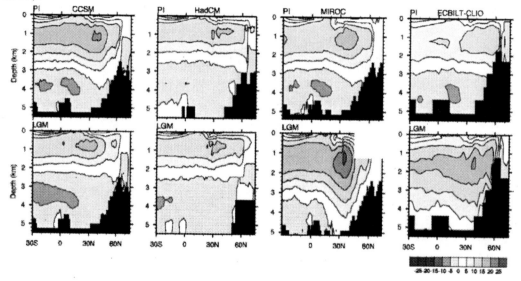

Figure 4.13. Atlantic meridional overturning (in Sverdrups) simulated by four PMIP2 coupled ocean-atmosphere models for modern (top) and the Last Glacial Maximum (bottom). From *Otto-Bliesner et al. (2007)*

There is a particularly large spread among the models in terms of overturning strength (Figure 4.13). Some models show a significantly increased AMOC streamfunction for the LGM compared to the modern control (by ~25-40%). Others have a significantly decreased streamfunction (by ~20-30%), while another shows very little change (*Weber et al., 2007*). Again, the overturning strength is not constrained well enough from the paleodata to make this a rigorous test of the models. It is likely, though, that simulations with a significantly strengthened AMOC are not realistic, and this tempers the credibility of their projections of future AMOC change. A more complete understanding of past AMOC changes and our ability to simulate those in models will lead to increased confidence in the projection of future changes.

Several factors control the AMOC response to LGM boundary conditions. These include changes in the freshwater budget of the North Atlantic, the density gradient between the North and South Atlantic, and the density gradient between GNAIW and AABW (*Schmittner et al., 2002; Weber et al., 2007*). The density gradient between GNAIW and AABW appears to be particularly important, and sea-ice concentrations have been shown to play a central role in determining this gradient (*Otto-Bliesner et al., 2007*). The AMOC response also has some dependence on the accuracy of the control state. For example, models with an unrealistically shallow overturning circulation in the control simulation do not yield a shoaled circulation for LGM conditions (*Weber et al., 2007*).

5.3. Transient Simulations of Past AMOC Variability

In addition to the equilibrium simulations discussed thus far, transient simulations of past meltwater pulses to the North Atlantic (see Sec. 4) may offer another test of model skill in simulating the AMOC. Such a test requires quantitative reconstructions of the freshwater pulse, including its volume, duration and location, plus the magnitude and duration of the resulting reduction in the AMOC. This information is not easy to obtain; coupled GCM simulations of most events, including the Younger Dryas and Heinrich events, have been forced with idealized freshwater pulses and compared with qualitative reconstructions of the AMOC (e.g., *Peltier et al., 2006; Hewitt et al., 2006*). There is somewhat more information about the freshwater pulse associated with the 8.2 ka event, though important uncertainties remain (*Clarke et al., 2004; Meissner and Clark, 2006*). A significant problem, however, is the scarcity of data about the AMOC during the 8.2 ka event. New ocean sediment records suggest the AMOC weakened following the freshwater pulse, but a quantitative reconstruction is lacking (*Ellison et al., 2006; Kleiven et al., 2008*). Thus, while simulations forced with the inferred freshwater pulse at 8.2 ka have produced results in quantitative agreement with reconstructed climate anomalies (e.g., *LeGrande et al., 2006; Wiersma et al., 2006*), the 8.2 ka event is currently limited as a test of a model's ability to reproduce changes in the AMOC itself.

6. What Are the Global and Regional Impacts of a Change in the Overturning Circulation?

In this section we review some of the climatic impacts of the AMOC over a range of time scales. While all of the impacts are not necessarily abrupt, they indicate consistent physical relationships that might be anticipated with any abrupt change in the AMOC. We start with evidence of the climatic impact of AMOC changes during glacial periods. While AMOC changes are not hypothesized to cause Ice Ages, there are indications of large AMOC changes within glacial periods, and these offer excellent opportunities to evaluate the global-scale climatic impact of large AMOC changes. We then move on to possible impacts of AMOC changes during the instrumental era. All of these results point to global-scale, robust impacts of AMOC changes on the climate system. In particular, a central impact of AMOC changes is to alter the interhemispheric temperature gradient, thereby moving the position of the Intertropical Convergence Zone (ITCZ). Such ITCZ changes induce a host of regional climate impacts.

6.1. Extra-tropical impacts during the last ice age

During the last glacial period, records indicate there were significant abrupt climate change events, such as the D-O oscillations and Heinrich events discussed in details in Section 4. These are thought to be associated with changes in the AMOC, and thus offer important insights into the climatic impacts of large changes in the AMOC. The paleoproxies from the Bermuda Rise (*McManus et al., 2004*) further indicate that the AMOC was substantially weakened during the Younger Dryas cooling event and was almost shut down during the latest Heinrich event—H1. The AMOC transports a substantial amount of heat northward. A rapid shutdown of the AMOC causes a cooling in the North Atlantic and a warming in the South Atlantic, associated with the reduction of the northward ocean heat transport, as simulated by many climate models (*Vellinga and Wood, 2002; Dahl et al., 2005; Zhang and Delworth, 2005; Stouffer et al., 2006*).

The cooling stadials of the Greenland D-O oscillations were also synchronous with higher oxygen levels off the California coast (indicating reduced upwelling and reduced California Current) (*Behl and Kennett, 1996*), enhanced North Pacific intermediate-water formation, and the strengthening of the Aleutian Low (*Hendy and Kennett, 2000*). This teleconnection is seen in coupled modeling simulations in which the AMOC is suppressed in response to massive freshwater inputs (*Mikolajewicz et al., 1997; Zhang and Delworth, 2005*), i.e., cooling in the North Atlantic induced by a weakened AMOC can lead to the strengthening of the Aleutian Low and large-scale cooling in the central North Pacific.

The millennial-scale abrupt climate change events found in Greenland ice cores have been linked to the millennial-scale signal seen in Antarctic ice cores (*Blunier et al., 1998; Bender et al., 1999; Blunier and Brook, 2001*). A very recent high resolution glacial climate record derived from the first deep ice core in the Atlantic sector of the Southern Ocean region (Dronning Maud Land, Antarctica) shows a one-to-one coupling between all Antarctic warm events (i.e., the A events discussed in detail in Sec. 3) and Greenland D-O oscillations during the last ice age (*EPICA Community Members, 2006*). The amplitude of the Antarctic warm events is found to be linearly dependent on the duration of the concurrent Greenland cooling events. Such a bipolar seesaw pattern was explained by changes in the heat flux connected to the reduction of the AMOC (*Manabe and Stouffer, 1988; Stocker and Johnsen, 2003; EPICA Community Members, 2006*).

6.2. Tropical impacts during the last ice age and Holocene

Recently, many paleorecords from different tropical regions have revealed abrupt changes that are remarkably coherent with the millennial-scale abrupt climate changes recorded in the Greenland ice cores during the glacial period, indicating that changes in the AMOC might have significant global-scale impacts on the tropics. A paleoproxy from the Cariaco basin off Venezuela suggests that the ITCZ shifted southward during cooling stadials of the Greenland D-O oscillations (*Peterson et al., 2000*). Stott et al. (2002) suggest that Greenland cooling events were related to an El Niño–like pattern of sea surface temperature (SST) change, a weakened Walker circulation, and a southward shift of the ITCZ in the tropical Pacific. The tropical Pacific east-west SST contrast was further reduced during the latest Heinrich event (H1) and Younger Dryas event (*Lea et al., 2000; Koutavas et al., 2002*). Drying conditions in the northeastern tropical Pacific west of Central America were synchronous with the Younger Dryas and the latest Heinrich event—H1 (*Benway et al.,*

2006). When Greenland was in cooling condition, the summer Asian monsoon was reduced, as indicated by a record from Hulu Cave in eastern China (*Wang et al., 2001*). Wet periods in northeastern Brazil are synchronous with Heinrich events, cold periods in Greenland, and periods of weak east Asian summer monsoons and decreased river runoff to the Cariaco basin (*Wang et al., 2004*). Sediment records from the Oman margin in the Arabian Sea indicate that weakened Indian summer monsoon upwelling occurred during Greenland stadials (*Altabet et al., 2002*).

The global synchronization of abrupt climate changes as indicated by these paleorecords, especially the anti-phase relationship of precipitation changes between the Northern Hemisphere (Hulu Cave in China, Cariaco basin) and the Southern Hemisphere (northeastern Brazil), is thought to be induced by changes in the AMOC. Global coupled climate models are employed to test this hypothesis. Figure 4.14 compares paleorecords with simulated changes in response to the weakening of the AMOC using the Geophysical Fluid Dynamics Laboratory (GFDL) coupled climate model (CM2.0). In the numerical experiment, the AMOC was substantially weakened by freshening the high latitudes of the North Atlantic (*Zhang and Delworth, 2005*). This leads to a southward shift of the ITCZ over the tropical Atlantic (Figure 4.14, upper right), similar to that found in many modeling studies (*Vellinga and Wood, 2002; Dahl et al., 2005; Stouffer et al., 2006*). This southward shift of the Atlantic ITCZ is consistent with paleorecords of drier conditions over the Cariaco basin (*Peterson et al., 2000*) and wetter conditions over northeastern Brazil during Heinrich events (*Wang et al., 2004*) (Figure 4.14, lower right). Beyond the typical responses in the Atlantic, this experiment also shows many significant remote responses outside the Atlantic, such as a southward shift of the ITCZ in the tropical Pacific (Figure 4.14, upper right), consistent with drying conditions over the northeastern tropical Pacific during the Younger Dryas and Heinrich events (*Benway et al., 2006*). The modeled weakening of the Indian and East Asian summer monsoon in response to the weakening of the AMOC (Figure 4.14, upper left) is also consistent with paleoproxies from the Indian Ocean (*Altabet et al., 2002*; Figure 4.14, lower left) and the Hulu Cave in eastern China (*Wang et al., 2001, 2004*; Figure 4.14, lower right). The simulated weakening of the AMOC also led to reduced cross-equatorial and east-west SST contrasts in the tropical Pacific, an El Niño-like condition, and a weakened Walker circulation in the southern tropical Pacific, a La Niña-like condition, and a stronger Walker circulation in the northern tropical Pacific. Coupled air-sea interactions and ocean dynamics in the tropical Pacific are important for connecting the Atlantic changes with the Asian monsoon variations (*Zhang and Delworth, 2005*). Thus, both atmospheric teleconnections and coupled air-sea interactions play crucial roles for the global-scale impacts of the AMOC.

Similar global-scale synchronous changes on a multidecadal to centennial time scale have also been found during the Holocene. For example, the Atlantic ITCZ shifted southward during the Little Ice Age and northward during the Medieval Warm Period (*Haug et al., 2001*). Sediment records in the anoxic Arabian Sea show that centennial-scale Indian summer monsoon variability coincided with changes in the North Atlantic region during the Holocene, including a weaker summer monsoon during the Little Ice Age and an enhanced summer monsoon during the Medieval Warm Period (*Gupta et al., 2003*). These changes might also be associated with a reduction of the AMOC during the Little Ice Age (*Lund et al., 2006*).

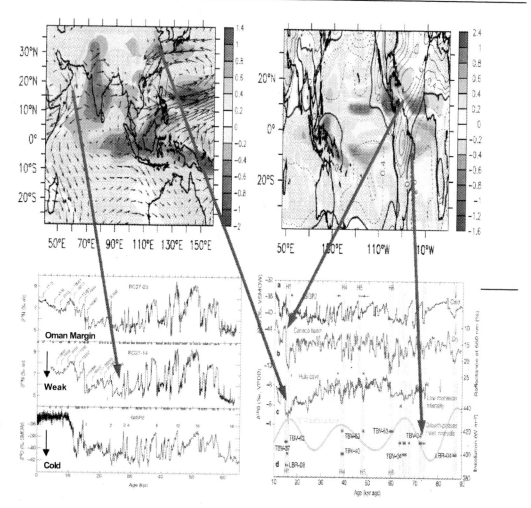

Figure 4.14. Comparison of simulated changes in response to the weakening of the AMOC using the Geophysical Fluid Dynamics Laboratory (GFDL) coupled model (CM2.0) with paleorecords. Upper left (*Zhang and Delworth, 2005*): Simulated summer precipitation change (color shading, units are m yr^{-1}) and surface wind change (black vectors) over the Indian and eastern China regions. Upper right (*Zhang and Delworth, 2005*): Simulated annual mean precipitation change (color shading, units are m yr^{-1}) and sea-level pressure change (contour, units are hPa). Negative values correspond to a reduction of precipitation. Lower left (*Altabet et al., 2002*): The δ^{15}N records for denitrification from sediment cores from the Oman margin in the Arabian Sea were synchronous with D-O oscillations recorded in Greenland ice cores (GISP2) during the last glacial period, i.e., the reduced denitrification, indicating weakened Indian summer monsoon upwelling, occurred during cold Greenland stadials. Lower right (*Wang et al., 2004*): Comparison of the growth patterns of speleothems from northeastern Brazil (d) with (a) δ^{18}O values of Greenland ice cores (GISP2), (b) Reflectance of the Cariaco basin sediments from ODP Hole 1002C (*Peterson et al., 2000*), (c) δ^{18}O values of Hulu cave stalagmites (*Wang et al., 2001*). The modeled global response to the weakening of the AMOC (*Zhang and Delworth, 2005*) is consistent with all these synchronous abrupt climate changes found from the Oman margin, Hulu Cave, Cariaco basin, and northeastern Brazil during cold Greenland stadials, i.e., drying at the Cariaco basin, weakening of the Indian and Asian summer monsoon, and wetting in northeastern Brazil (red arrows). Abbreviations: %, percent; ‰, per mil; SMOW, Standard Mean Ocean Water; kyr, thousand years ago; H1, H4, H5, H6, Heinrich events; W m^{-2}, watts per square meter; nm, nanometer; m yr^{-1}, meters per year; hPa, hectoPascals

6.3. Possible impacts during the 20th century

Instrumental records in the 20th century can also provide clues about possible AMOC impacts. Instrumental records show significant large-scale multidecadal variations in the Atlantic SST. The observed detrended 20th century multidecadal SST anomaly averaged over the North Atlantic, often called the Atlantic Multidecadal Oscillation (AMO) (*Enfield et al., 2001; Knight et al., 2005*), has significant regional and hemispheric climate impacts (*Enfield et al., 2001; Knight et al., 2006; Zhang and Delworth, 2006; Zhang et al., 2007a*). The warm AMO phases occurred during 1925–65 and the recent decade since 1995, and cold phases occurred during 1900–25 and 1965–95. The AMO index is highly correlated with multidecadal variations of the tropical North Atlantic (TNA) SST and Atlantic hurricane activity (*Goldenberg et al., 2001; Landsea, 2005; Knight et al., 2006; Zhang and Delworth, 2006; Sutton and Hodson, 2007*). The observed TNA surface warming is correlated with above-normal Atlantic hurricane activity during the 1950-60s and the recent decade since 1995.

While the origin of these multidecadal SST variations is not certain, one leading hypothesis involves fluctuations of the AMOC. Models provide some support for this (*Delworth and Mann, 2000; Knight et al., 2005*), with typical AMOC variability of several Sverdrups on multidecadal time scales, corresponding to 5-10% of the mean in these models. Another hypothesis is that they are forced by changes in radiative forcing (*Mann and Emanuel, 2006*). *Delworth et al. (2007)* suggest that both processes— radiative forcing changes, along with internal variability, possibly associated with the AMOC—may be important. A very recent study (*Zhang, 2007*) lends support to the hypothesis that AMOC fluctuations are important for the multidecadal variations of observed TNA SSTs. *Zhang (2007)* finds that observed TNA SST is strongly anticorrelated with TNA subsurface ocean temperature (after removing long-term trends). This anticorrelation is a distinctive signature of the AMOC variations in coupled climate models; in contrast, simulations driven by external radiative forcing changes do not generate anticorrelated surface and subsurface TNA variations, lending support to the idea that the observed TNA SST fluctuations may be AMOC-induced.

6.3.1. Tropical impacts

Empirical analyses have demonstrated a link between multidecadal fluctuations of Atlantic sea surface temperatures and Sahelian (African) summer rainfall variations (*Folland et al., 1986*), in which an unusually warm North Atlantic is associated with increased summer rainfall over the Sahel. Studies with atmospheric general circulation models (e.g., *Giannini et al., 2003; Lu and Delworth, 2005*) have shown that models, when given the observed multidecadal SST variations, are able to reproduce much of the observed Sahelian rainfall variations. However, these studies do not identify the source of the SST fluctuations. Recent work (*Held et al., 2005*) suggests that increasing greenhouse gases and aerosols may also be important factors in the late 20th century Sahelian drying.

The source of the observed Atlantic multidecadal SST variations has not been firmly established. One leading candidate mechanism involves fluctuations of the AMOC. *Knight et al. (2006)* have analyzed a 1,400-year control integration of the coupled climate model HADCM3 and found a clear relationship between AMO-like SST fluctuations and surface air temperature over North America and Eurasia, modulation of the vertical shear of the zonal wind in the tropical Atlantic, and large-scale changes in Sahel and Brazil rainfall. Linkages between the AMO and these tropical variations were often based on statistical analyses. Linkages between AMOC changes and tropical conditions, emphasizing the importance of changes in the atmospheric and oceanic energy budgets, are emphasized in *Cheng et al. (2007)*. To investigate the causal link between the AMO and other multidecadal variability, *Zhang and Delworth (2006)* simulated the impact of AMO-like SST variations on climate with a hybrid coupled model. They demonstrated that many features of observed multidecadal climate variability in the 20th century may be interpreted—at least partially—as a response to the AMO. A warm phase of the AMO leads to a northward shift of the Atlantic ITCZ, and thus an increase in the Sahelian and Indian summer monsoonal rainfall, as well as a reduction in the vertical shear of the zonal wind in the tropical Atlantic region that is important for the development of Atlantic major Hurricanes (Figure 4.15). Thus, the AMO creates large-scale atmospheric circulation anomalies that would be favorable for enhanced tropical storm activity. The study of *Black et al. (1999)* using Caribbean sediment records suggests that a southward shift of the Atlantic ITCZ when the North Atlantic is cold—similar to what is seen in the models—has been a robust feature of the climate system for more than 800 years, and is similar to results from the last ice age.

6.3.2. Impacts on North America and Western Europe

The recent modeling studies (*Sutton and Hodson, 2005, 2007*) provide a clear assessment of the impact of the AMO over the Atlantic, North America, and Western Europe (Figure 4.16). In response to a warm phase of the AMO, a broad area of low pressure develops over the Atlantic, extending westward into the Caribbean and Southern United States. The pressure anomaly pattern denotes weakened easterly trade winds, potentially reinforcing the positive SST anomalies in the tropical North Atlantic Ocean by reducing the latent heat flux. Precipitation is generally enhanced over the warmer Atlantic waters and is reduced over a broad expanse of the United States. The summer temperature response is clear, with substantial warming over the United States and Mexico, with weaker warming over Western Europe.

Observational analyses (*Enfield et al., 2001*) suggest that the AMO has a strong impact on the multidecadal variability of U.S. rainfall and river flows. *McCabe et al. (2004)* further suggest that there is significant positive correlation between the AMO and the Central U.S. multidecadal drought frequency, and the positive AMO phase contributes to the droughts observed over the continental U.S. in the decade since 1995.

6.3.3. Impacts on Northern Hemisphere mean temperature

Knight et al. (2005) find in the 1,400-year control integration of the HADCM3 climate model that variations in the AMOC are correlated with variations in the Northern Hemisphere mean surface temperature on decadal and longer time scales. *Zhang et al. (2007a)* demonstrate that AMO-like SST variations can contribute to the Northern Hemispheric mean surface temperature fluctuations, such as the early 20th century warming, the pause in hemispheric-scale warming in the mid-20th century, and the late 20th century rapid warming, in addition to the long-term warming trend induced by increasing greenhouse gases.

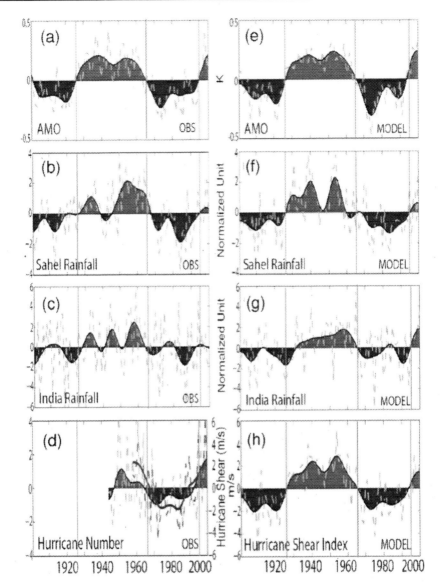

Figure 4.15. Left: various observed (OBS) quantities with an apparent association with the AMO. Right: Simulated responses of various quantities to AMO-like fluctuations in the Atlantic Ocean from a hybrid coupled model (adapted from *Zhang and Delworth, 2006*). Dashed green lines are unfiltered values, while the red and blue color-shaded values denote low-pass filtered values. Blue shaded regions indicate values below their long-term mean, while red shading denotes values above their long-term mean. The vertical blue lines denote transitions between warm and cold phases of the AMO. Time in calendar years is along the bottom axis. (a), (e) AMO Index, a measure of SST over the North Atlantic. Positive values denote an unusually warm North Atlantic. (b), (f) Normalized summer rainfall anomalies over the Sahel (20°W.-40°E.,10-20°N.). (c), (g) Normalized summer rainfall over west-central India (65-80°E.,15-25°N.). (d) Number of major Atlantic Hurricanes from the NOAA HURDAT data set. The brown lines denote the vertical shear of the zonal (westerly) wind (multiplied by -1) derived from the ERA- 40 reanalysis, i.e., the difference in the zonal wind between 850 and 200 hectopascals (hPa) over the south-central part of the main development region (MDR) for tropical storms (10-14°N.,70-20°W.). (h) Vertical shear of the simulated zonal wind (multiplied by -1), calculated as in (d).

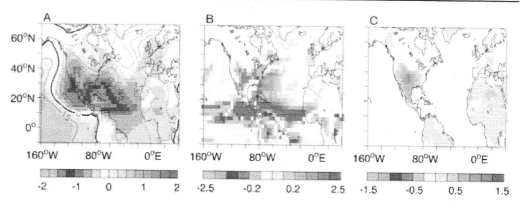

Figure 4.16. These panels (adapted from Sutton and Hodson, 2005) show the simulated response of various fields to an idealized AMO SST anomaly using the HADAM3 Atmospheric General Circulation Model. Results are time means for the August-October period. (a) Sea level pressure, units are pascals (Pa), with an interval of 15 Pa. (b) Precipitation, units are millimeters per day. (c) Surface air temperature, units are kelvin.

6.4. Simulated Impacts on ENSO Variability

Modeling studies suggest that changes in the AMOC can modulate the characteristics of El-Niño Southern Oscillation (ENSO). *Timmermann et al. (2005a)* found that the simulated weakening of the AMOC leads to a deepening of the tropical Pacific thermocline, and a weakening of ENSO, through the propagation of oceanic waves from the Atlantic to the tropical Pacific. Very recent modeling studies (*Dong and Sutton, 2007; Timmermann et al., 2007*) found opposite results, i.e., the weakening of the AMOC leads to an enhanced ENSO variability through atmospheric teleconnections. *Dong et al. (2006)* also show that a negative phase of the AMO leads to an enhancement of ENSO variability.

6.5. Impacts on Ecosystems

Recent coupled climate–ecosytem model simulations (*Schmittner, 2005*) find that a collapse of the AMOC leads to a reduction of North Atlantic plankton stocks by more than 50%, and a reduction of global productivity by about 20% due to reduced upwelling of nutrient-rich deep water and depletion of upper ocean nutrient concentrations. The model results are consistent with paleorecords during the last ice age indicating low productivity during Greenland cold stadials and high productivity during Greenland warm interstadials (*Rasmussen et al., 2002*). Multidecadal variations in abundance of Norwegian spring-spawning herring (a huge pelagic fish stock in the northeast Atlantic) have been found during the 20th century. These variations of the Atlantic herring are in phase with the AMO index and are mainly caused by variations in the inflowing Atlantic water temperature (*Toresen and Østvedt, 2000*). Model simulations show that the stocks of Arcto-Norwegian cod could decrease substantially in reaction to a weakened AMOC (*Vikebø et al., 2007*). Further, *Schmittner et al. (2007)* show that changes in Atlantic circulation can have large effects on marine ecosystems and biogeochemical cycles, even in areas remote from the Atlantic, such as the Indian and North Pacific oceans.

6.6. Summary and Discussion

A variety of observational and modeling studies demonstrate that changes in the AMOC induce a near-global-scale suite of climate system changes. A weakened AMOC cools the North Atlantic, leading to a southward shift of the ITCZ, with associated drying in the Caribbean, Sahel region of Africa, and the Indian and Asian monsoon regions. Other near-global-scale impacts include modulation of the Walker circulation and associated air-sea interactions in the Pacific basin, possible impacts on North American drought, and an imprint on hemispheric mean surface air temperatures. These relationships appear robust across a wide range of time scales, from observed changes in the 20th century to changes inferred from paleoclimate indicators from the last ice age climate.

In addition to the above impacts, regional changes in sea level would accompany a substantial change in the AMOC. For example, in simulations of a collapse of the AMOC (*Levermann et al., 2005; Vellinga and Wood, 2007*) there is a sea level rise of up to 80 cm in the North Atlantic. This sea level rise is a dynamic effect associated with changes in ocean circulation. This would be in addition to other global warming induced changes in sea level arising from large-scale warming of the global ocean and melting of land- based ice sheets induced by increasing CO_2. This additional sea leve rise could affect the coastlines of the United States, Canada, and Europe.

7. What Factors That Influence the Overturning Circulation Are Likely to Change in the Future, and What Is the Probability That the Overturning Circulation Will Change?

As noted in the Intergovernmental Panel for Climate Change (IPCC) Fourth Assessment Report (AR4), all climate model projections under increasing greenhouse gases lead to an increase in high-latitude temperature as well as an increase in high-latitude precipitation (*Meehl et al., 2007*). Both warming and freshening tend to make the high-latitude surface waters less dense, thereby increasing their stability and inhibiting convection.

In the IPCC AR4, 19 coupled atmosphere-ocean models contributed projections of future climate change under the SRES A1B scenario (*Meehl et al., 2007*). Of these, 16 models did not use flux adjustments (all except CGCM3.1, INM-CM3.0, and MRI-CGCM2.3.2). In making their assessment, *Meehl et al. (2007)* noted that several of the models simulated a late 20th century AMOC strength that was inconsistent with present-day estimates: 14-18 Sv at 24°N. (*Ganachaud and Wunsch, 2000; Lumpkin and Speer, 2003*); 13-19 Sv at 48° N. (*Ganachaud, 2003a*); maximum values of 17.2 Sv (*Smethie and Fine, 2001*) and 18 Sv (*Talley et al., 2003*) with an error of ± 3-5 Sv. As a consequence of their poor 20th century simulations, these models were not used in their assessment.

The full range of late 20th century estimates of the Atlantic MOC strength (12-23 Sv) is spanned by the model simulations (Figure 4.17; *Schmittner et al., 2005; Meehl et al., 2007*). The models further project a decrease in the AMOC strength of between 0% and 50%, with a multimodel average of 25%, over the course of the 21st century. None of the models simulated an abrupt shutdown of the AMOC during the 21st century.

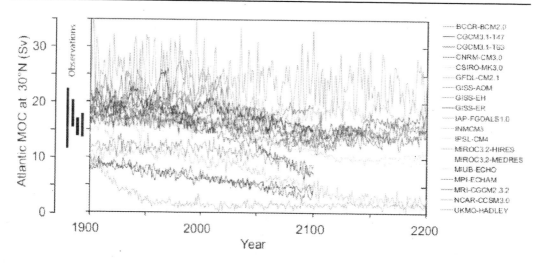

Figure 4.17. The Atlantic meridional overturning circulation (AMOC) at 30°N from the 19 coupled atmosphere-ocean models assessed in the IPCC AR4. The SRES A1B emissions scenario was used from 1999 to 2100. Those model projections that continued to 2200 retained the year 2100 radiative forcing for the remainder of the integration. Observationally based estimates of the late 20th century AMOC strength are also shown on the left as black bars. Taken from *Meehl et al. (2007)* as originally adapted from *Schmittner et al. (2005)*.

Schneider et al. (2007) extended the analysis of *Meehl et al. (2007)* by developing a multimodel average in which the individual model simulations were weighted a number of ways. The various weighting estimates were based on an individual model's simulation of the contemporary ocean climate, and in particular its simulated fields of temperature, salinity, pycnocline depth, as well as its simulated Atlantic MOC strength. Their resulting best estimate 21st century AMOC weakening of 25-30% was invariant to the weighting scheme used and is consistent with the simple multimodel mean of 25% obtained in the IPCC AR4.

In early versions of some coupled atmosphere-ocean models, (e.g., *Dixon et al.,* 1999), increased high-latitude precipitation dominated over increased high-latitude warming in causing the projected weakening of the AMOC under increasing greenhouse gases, while in others (e.g., *Mikolajewicz and Voss,* 2000), the opposite was found. However, *Gregory et al.* (2005) undertook a recent model intercomparison project in which, in all 11 models analyzed, the AMOC reduction was caused more by changes in surface heat flux than changes in surface freshwater flux. Weaver et al. (2007) extended this analysis by showing that, in one model, this conclusion was independent of the initial mean climate state.

A number of stabilization scenarios have been examined using both coupled Atmosphere-Ocean General Circulation Models (AOGCMs) (*Stouffer and Manabe, 1999; Voss and Mikolajewicz, 2001; Stouffer and Manabe, 2003; Wood et al., 2003; Yoshida et al., 2005; Bryan et al., 2006*) as well as Earth System Models of Intermediate Complexity (EMICs) (*Meehl et al.,* 2007). Typically the atmospheric CO_2 concentration in these models is increased at a rate of 1%/year to either two times or four times the preindustrial level of atmospheric CO_2, and held fixed thereafter. In virtually every simulation, the AMOC reduces but recovers to its initial strength when the radiative forcing is stabilized at two times or four times the preindustrial levels of CO_2. Only one early flux-adjusted model simulated a complete shutdown, and even this was not permanent (*Manabe and Stouffer,* 1994; *Stouffer and Manabe,* 2003). The only model to exhibit a permanent cessation of the AMOC in response to increasing greenhouse gases was an intermediate complexity model which incorporates a zonally averaged ocean component (*Meehl et al.,* 2007).

Historically, coupled models that eventually lead to a collapse of the AMOC under global warming conditions were of lower resolution, used less complete physics, used flux adjustments, or were models of intermediate complexity with zonally averaged ocean components (wherein convection and sinking of water masses are coupled). The newer models assessed in the IPCC AR4 typically do not involve flux adjustments and have more stable projections of the future evolution of the AMOC.

One of the most misunderstood issues concerning the future of the AMOC under anthropogenic climate change is its often cited potential to cause the onset of the next ice age (see Box 4.4). A relatively solid understanding of glacial inception exists wherein a change in seasonal incoming solar radiation (warmer winters and colder summers), which is associated with changes in the Earth's axial tilt, longitude of perihelion, and the precession of its elliptical orbit around the sun, is required. This small change must then be amplified by albedo feedbacks associated with enhanced snow and ice cover, vegetation feedbacks associated with the expansion of tundra, and greenhouse gas feedbacks associated with the uptake (not release) of carbon dioxide and reduced release or increased destruction rate of methane. As discussed by *Berger and Loutre (2002)* and *Weaver and Hillaire-Marcel (2004a,b),* it is not possible for global warming to cause an ice age.

BOX 4.4. WOULD A COLLAPSE OF THE AMOC LEAD TO COOLING OF EUROPE AND NORTH AMERICA?

One of the motivations behind the study of abrupt change in the AMOC is its potential influence on the climates of North America and western Europe. Some reports, particularly in the media, have suggested that a shutdown of the AMOC in response to global warming could plunge western Europe and even North America into conditions much colder than our current climate. On the basis of our current understanding of the climate system, such a scenario appears very unlikely. On the multidecadal to century time scale, it is very likely that Europe and North America will warm in response to increasing greenhouse gases (although natural variability and regional shifts could lead to periods of decadal-scale cooling in some regions). A significant weakening of the AMOC in response to global warming would moderate that long-term warming trend. If a complete shutdown of the AMOC were to occur (viewed as very unlikely, as described in this assessment), the reduced ocean heat transport could lead to a net cooling of the ocean by several degrees in parts of the North Atlantic, and possibly 1 to 2 degrees Celsius over portions of extreme western and northwestern Europe. However, even in such an extreme (and very unlikely) scenario, a multidecadal to century-scale warming trend in response to increasing greenhouse gases would still be anticipated over most of North America, eastern and southern Europe, and Asia.

Wood et al. (1999), using HADCM3 with sufficient resolution to resolve Denmark Strait overflow, performed two transient simulations starting with a preindustrial level of atmospheric CO_2 and subsequently increasing it at a rate of 1% or 2% per year. Convection and overturning in the Labrador Sea ceased in both these experiments, while deep water formation persisted in the Nordic seas. As the climate warmed, the Denmark Strait overflow water became warmer and hence lighter, so that the density contrast between it and the deep Labrador Sea water (LSW) was reduced. This made the deep circulation of the Labrador Sea collapse, while Denmark Strait overflow remained unchanged, a behavior suggested from the paleoreconstructions of *Hillaire-Marcel et al. (2001)* for the Last Interglacial (Eemian). The results of *Hillaire-Marcel et al. (2001)* suggest that the modern situation, with active LSW formation, has apparently no analog throughout the last glacial cycle, and thus appears a feature exclusive to the present interglacial.

Results similar to those of *Wood et al.* (1999) were found by *Hu et al.* (2004), although *Hu et al.* (2004) also noted a significant increase in Greenland–Iceland–Norwegian (GIN) Sea convection as a result of enhanced inflow of saline North Atlantic water, and reduced outflow of sea ice from the Arctic. Some coupled models, on the other hand, found significant reductions in convection in the GIN Sea in response to increasing atmospheric greenhouse gases (*Bryan et al.*, 2006; *Stouffer et al.*, 2006). A cessation of LSW formation by 2030 was also found in high-resolution ocean model simulations of the Atlantic Ocean driven by surface fluxes from two coupled atmosphere-ocean climate models (*Schweckendiek and Willebrand, 2005*). *Cottet-Puinel et al. (2004)* obtained similar results to *Wood et al. (1999)* concerning the transient cessation of LSW formation and further showed that LSW formation eventually reestablished upon stabilization of anthropogenic greenhouse gas levels. The same model experiments of *Wood et al. (1999)* suggest that the freshening North Atlantic surface waters presently observed (*Curry et al., 2003*) is associated with a transient increase of the AMOC (*Wu et al., 2004*). Such an increase would be consistent with findings of *Latif et al. (2006)*, who argued that their analysis of ocean observations and model simulations supported the notion of a slight AMOC strengthening since the 1980s.

The best estimate of sea level rise from 1993 to 2003 associated with mass loss from the Greenland ice sheet is 0.21 ± 0.07 mm yr^{-1} (*Bindoff et al., 2007*). This converts to only 0.0015 to 0.0029 Sv of freshwater forcing, an amount that is too small to affect the AMOC in models (see *Weaver and Hillaire-Marcel*, 2004a; *Jungclaus et al.*, 2006). Recently, *Velicogna and Wahr (2006)* analyzed the Gravity Recovery and Climate Experiment (GRACE) satellite data to infer an acceleration of Greenland ice loss from April 2002 to April 2006 corresponding to 0.5 ± 0.1 mm/yr of global sea level rise. The equivalent 0.004–0.006 Sv of freshwater forcing is, once more, too small to affect the AMOC in models. *Stouffer et al. (2006)* undertook an intercomparison of 14 coupled models subject to a 0.1-Sv freshwater perturbation (17 times the upper estimate from GRACE data) applied for 100 years to the northern North Atlantic Ocean. A simple scaling analysis (conducted by the authors of this assessment report) shows that if over a 10-year period Arctic sea ice were to completely melt away in all seasons, North Atlantic freshwater input would be about half this rate (see Box 4.1 for a discussion of observed and projected Arctic sea ice change). In all cases, the models exhibited a weakening of the AMOC (by a multimodel mean of 30% after 100 years), and none of the models simulated a shutdown. *Ridley et al.* (2005) elevated greenhouse gas levels to four times preindustrial values and retained them fixed thereafter to investigate the evolution of the Greenland Ice sheet in their coupled model. They found a peak melting rate of about 0.1 Sv, which occurred early in the simulation, and noted that this perturbation had little effect on the AMOC. *Jungclaus et al. (2006)* independently applied 0.09 Sv freshwater forcing along the boundary of Greenland as an upper-bound estimate of potential external freshwater forcing from the melting of the Greenland ice sheet. Under the SRES A1B scenario they, too, only found a weakening of the AMOC with a subsequent recovery in its strength. They concluded that Greenland ice sheet melting would not cause abrupt climate change in the 21st century.

Based on our analysis, we conclude that it is very likely that the strength of the AMOC will decrease over the course of the 21st century. Both weighted and unweighted multimodel ensemble averages under an SRES A1B future emission scenario suggest a best estimate of 25-30% reduction in the overall AMOC strength. Associated with this reduction is the possible cessation of LSW water formation. In models where the AMOC weakens, warming still occurs downstream over Europe due to the radiative forcing associated with increasing greenhouse gases (*Gregory et al., 2005; Stouffer et al., 2006*). No model under idealized (1%/year or 2%/year increase) or SRES scenario forcing exhibits an abrupt collapse of the AMOC during the 21st century, even accounting for estimates of accelerated Greenland ice sheet melting. We conclude that it is very unlikely that the AMOC will undergo an abrupt transition during the course of the 21st century. Based on available model simulations and sensitivity analyses, estimates of maximum Greenland ice sheet melting rates, and our understanding of mechanisms of abrupt climate change from the paleoclimate record, we further conclude it is unlikely that the AMOC will collapse beyond the end of the 21st century as a consequence of global warming, although the possibility cannot be entirely excluded.

8. What Are the Observational and Modeling Requirements Necessary to Understand the Overturning Circulation and Evaluate Future Change?

It has been shown in this chapter that the AMOC plays a vital role in the climate system. In order to more confidently predict future changes—especially the possibility of abrupt change—we need to better understand the AMOC and the mechanisms governing its variability and sensitivity to forcing changes. Improved understanding of the AMOC comes at the interface between observational and theoretical studies. In that context, theories can be tested, oftentimes using numerical models, against the best available observational data. The observational data can come from the modern era or from proxy indicators of past climates.

We describe in this section a suite of activities that are necessary to increase our understanding of the AMOC and to more confidently predict its future behavior. While the activities are noted in separate categories, the true advances in understanding— leading to a predictive capability—come in the synthesis of the various activities described below, particularly in the synthesis of modeling and observational analyses.

8.1. Sustained modern observing system

We currently lack a long-term, sustained observing system for the AMOC. Without this in place, our ability to detect and predict future changes of the AMOC—and their impacts—is very limited. The RAPID project may be viewed as a prototype for such an observing system. The following set of activities is therefore needed:

- Research to delineate what would constitute an efficient, robust observational network for the AMOC. This could include studies in which model results are sampled according to differing observational networks, thereby evaluating the utility of those networks for observing the AMOC and guiding the development of new observational networks and the enhancement of existing observational networks.

- Sustained deployment over decades of the observational network identified above to robustly measure the AMOC. This would likely include observations of key processes involved in deep water formation in the Labrador and Norwegian Seas, and their communication with the rest of the Atlantic (e.g., *Lozier et al., 2007*).

- Focused observational programs as part of process studies to improve understanding of physical processes of importance to the AMOC, such as ocean- atmosphere coupling, mixing processes, and deep overflows. These should lead to improved representation of such processes in numerical models.

8.2. Acquisition and interpretation of paleoclimate data

While the above stresses current observations, much can be learned from the study of ancient climates that provide insight into the past behavior of the AMOC. We need to develop paleoclimate datasets that allow robust, quantitative reconstructions of past ocean circulations and their climatic impacts. Therefore, the following set of activities is needed:

- Acquisition and analysis of high-resolution records from the Holocene that can provide insight on decadal to centennial time scales of AMOC-related climate variability. This is an important baseline against which to judge future change.

- Acquisition and analysis of paleoclimate records to document past changes in the AMOC, including both glacial and nonglacial conditions. These will provide a more robust measure of the response of the AMOC to changing radiative forcing and will allow new tests of models. Our confidence in predictions of future AMOC changes is enhanced to the extent that models faithfully simulate such past AMOC changes.

- More detailed assessment of the past relationship between AMOC and climate, especially the role of AMOC changes in abrupt climate change.

- Acquisition and analysis of paleoclimate records that can provide improved estimates of past changes in meltwater forcing. This information can lead to improved understanding of the AMOC response to freshwater input and can help to better constrain models.

8.3. Improvement and use of models

Models provide our best tools for predicting future changes in the AMOC and are an important pathway toward increasing our understanding of the AMOC, its variability, and its sensitivity to change. Such insights are limited, however, by the fidelity of the models employed. There is an urgent need both to (1) improve the models we use and (2) use models in innovative ways to increase our understanding of the AMOC. Therefore, the following set of activities is needed:

- Development of models with increased resolution in order to more faithfully represent the small-scale processes that are important for the AMOC. The models used for the IPCC AR4 assessment had oceanic resolution on the order of 50-100 km in the horizontal, with 30-50 levels in the vertical. In reality, processes with spatial scales of several kilometers (or less) are important for the AMOC.

- Development of models with improved numerics and physics, especially those that appear to influence the AMOC. In particular, there is a need for improved representation of small-scale processes that significantly impact the AMOC. For example, overflows of dense water over sills in the North Atlantic are an important feature for the AMOC, and their representation in models needs to be improved.
- Development of advanced models of land-based ice sheets, and their incorporation in climate models. This is particulary crucial in light of uncertainties in the interaction between the AMOC and land-based ice sheets on long time scales.
- Design and execution of innovative numerical experiments in order to (1) shed light on the mechanisms governing variability and change of the AMOC, (2) estimate the inherent predictability of the AMOC, and (3) develop methods to realize that predictability. The use of multimodel ensembles is particularly important.
- Development and use of improved data assimilation systems for providing estimates of the current and past states of the AMOC, as well as initial conditions for prediction of the future evolution of the AMOC.
- Development of prototype prediction systems for the AMOC. These prediction systems will start from the observed state of the AMOC and use the best possible models, together with projections of future changes in atmospheric greenhouse gases and aerosols, to make the best possible projections for the future behavior of the AMOC. Such a prediction system could serve as a warning system for an abrupt change in the AMOC.

8.4. Projections of future changes in radiative forcing and related impacts

One of the motivating factors for the study of AMOC behavior is the possibility of abrupt change in the future driven by increasing greenhouse gas concentrations. In order to evaluate the likelihood of such an abrupt change, it is crucial to have available the best possible projections for future changes in radiative forcing, especially those changes in radiative forcing due to human activity. This includes not only greenhouse gases, which tend to be well mixed and long lived in the atmosphere, but also aerosols, which tend to be shorter lived with more localized spatial patterns. Thus, realistic projections of aerosol concentrations and their climatic effects are crucial for AMOC projections.

One of the important controls on the AMOC is the freshwater flux into the Atlantic, including the inflow of freshwater from rivers surrounding the Arctic. For example, observations (*Peterson et al., 2002*) have shown an increase during the 20th century of Eurasian river discharge into the Arctic. For the prediction of AMOC changes it is crucial to have complete observations of changes in the high-latitude hydrologic cycle, including precipitation, evaporation, and river discharge, as well as water released into the Atlantic from the Greenland ice sheet and from glaciers. This topic is discussed more extensively in Chapter 2.

5. POTENTIAL FOR ABRUPT CHANGES IN ATMOSPHERIC METHANE

Lead Author: Edward Brook,[*] Oregon State University, Corvallis, OR

---

[*] SAP 3.4 Federal Advisory Committee Member

Contributing Authors: David Archer, University of Chicago, Chicago, IL
Ed Dlugokencky, NOAA Earth System Research Laboratory, Boulder, CO
Steve Frolking, University of New Hampshire, Durham, NH
David Lawrence, National Center for Atmospheric Research, Boulder, CO

Key Findings

- The main concerns about abrupt changes in atmospheric methane (CH_4) stem from (1) the large quantity of methane believed to be stored as methane hydrate in the sea floor and permafrost soils and (2) climate-driven changes in methane emissions from northern high-latitude and tropical wetlands.

- The size of the methane hydrate reservoir is uncertain, perhaps by up to a factor of 10. Because the size of the reservoir is directly related to the perceived risks, it is difficult to make certain judgment about those risks.

- There are a number of suggestions in the scientific literature about the possibility of catastrophic release of methane to the atmosphere based on both the size of the hydrate reservoir and indirect evidence from paleoclimatological studies. However, modeling and detailed studies of ice core methane so far do not support catastrophic methane releases to the atmosphere in the last 650,000 years or in the near future. A very large release of methane may have occurred at the Paleocene- Eocene boundary (about 55 million years ago), but other explanations for the evidence have been offered.

- The current network of atmospheric methane monitoring sites is sufficient for capturing large-scale changes in emissions, but it is insufficient for attributing changes in emissions to one specific type of source.

- Observations show that there have not yet been significant increases in methane emissions from northern terrestrial high-latitude hydrates and wetlands resulting from increasing Arctic temperatures.

- Catastrophic release of methane to the atmosphere appears very unlikely in the near term (e.g., this century). However, it is very likely that climate change will accelerate the pace of chronic emissions from both hydrate sources and wetlands. The magnitude of these releases is difficult to estimate with existing data. Methane release from the hydrate reservoir will likely have a significant influence on global warming over the next 1,000 to 100,000 years.

Recommendations

- Monitoring of the abundance of atmospheric methane and its isotopic composition sufficient to allow detection of change in emissions from northern and tropical wetland regions should be prioritized. Specifically, systematic measurements of CH_4 from tall towers and aircraft in the Arctic and subarctic regions and expanded surface flux measurements and continued observation of CH_4 abundance in the tropics and subtropics would allow detection of changes in emissions from sparsely monitored but important regions.

- The feasibility of monitoring methane in the ocean water column near marine hydrate deposits, or in the atmosphere near terrestrial hydrate deposits, to detect changes in emissions from those sources, should be investigated, and if feasible, this monitoring should be implemented.

- Efforts should be made to increase certainty in the size of the global methane hydrate reservoirs. The level of concern about catastrophic release of methane to the atmosphere is directly linked to the size of these reservoirs.

- The size and location of hydrate reservoirs that are most vulnerable to climate change (for example shallow-water deposits, shallow sub-surface deposits on land, or regions of potential large submarine landslides) should be identified accurately and their potential impact on future methane concentrations should be evaluated.

- Improvement in process-based modeling of methane release from marine hydrates is needed. The transport of bubbles is particularly important, as are the migration of gas through the stability zone and the mechanisms controlling methane release from submarine landslides.

- Modeling efforts should establish the current and future climate-driven acceleration of chronic release of methane from wetlands and terrestrial hydrate deposits. These efforts should include development of improved representations of wetland hydrology and biogeochemistry, and permafrost dynamics, in earth system and global climate models.

- Further work on the ice core record of atmospheric methane is needed to fully understand the implications of past abrupt changes in atmospheric methane. This work should include high-resolution and high-precision measurements of methane mixing ratios and isotopic ratios, and biogeochemical modeling of past methane emissions and relevant atmospheric chemical cycles. Further understanding of the history of wetland regions is also needed.

1. Background: Why Are Abrupt Changes in Methane of Potential Concern?

1.1. Introduction

Methane (CH_4) is the second most important greenhouse gas that humans directly influence, carbon dioxide (CO_2) being first. Concerns about methane's role in abrupt climate change stem primarily from (1) the large quantities of methane stored as solid methane hydrate on the sea floor and to a lesser degree in terrestrial sediments, and the possibility that these reservoirs could become unstable in the face of future global warming, and (2) the possibility of large-scale conversion of frozen soil in the high- latitude Northern Hemisphere to methane producing wetland, due to accelerated warming at high latitudes. This chapter summarizes the current state of knowledge about these reservoirs and their potential for forcing abrupt climate change.

1.2. Methane and Climate

A spectral window exists between ~7 and 12 micrometers (μm) where the atmosphere is somewhat transparent to terrestrial infrared (IR) radiation. Increases in the atmospheric abundance of molecules that absorb IR radiation in this spectral region contribute to the greenhouse effect. Methane is a potent greenhouse gas because it strongly absorbs terrestrial IR radiation near 7.66 μm, and its atmospheric abundance has more than doubled since the start of the Industrial Revolution. Radiative forcing (RF) is used to assess the contribution of a perturbation (in this case, the increase in CH_4 since 1750 A.D.) to the net irradiance at the top of the tropopause (that area of the atmosphere between the troposphere and the stratosphere) after allowing the stratosphere to adjust to radiative equilibrium. The direct radiative forcing of atmospheric methane determined from an increase in its abundance from its pre-industrial value of 700 parts per billion (ppb) (*MacFarling-Meure et al., 2006; Etheridge et al., 1998*) to its globally averaged abundance of 1,775 ppb in 2006 is 0.49±0.05 watts per square meter (W m^{-2}) (*Hofmann et al., 2006*). Methane oxidation products, stratospheric water (H_2O) vapor and tropospheric ozone (O3), contribute indirectly to radiative forcing, increasing methane's total contribution to ~0.7 W m^{-2} (e.g., *Hansen and Sato, 2001*), nearly half of that for carbon dioxide (CO2) . Increases in methane emissions can also increase the methane lifetime and the lifetimes of other gases oxidized by the hydroxyl radical (OH). Assuming the abundances of all other parameters that affect OH stay the same, the lifetime for an additional pulse of CH_4 (e.g., 1 teragram, Tg; 1 Tg = 10^{12}g = 0.001Gt, gigaton) added to the atmosphere would be ~40% larger than the current value. Additionally, CH_4 is oxidized to CO_2; CO_2 produced by CH_4 oxidation is equivalent to ~6% of CO_2 emissions from fossil fuel combustion. Over a 100-year time horizon, the direct and indirect effects on RF of emission of 1 kilogram (kg) CH_4 are 25 times greater than for emission of 1 kg CO_2 (*Forster et al., 2007*).

The atmospheric abundance of CH_4 increased with human population because of increased demand for energy and food. Beginning in the 1970s, as CH_4 emissions from natural gas venting and flaring at oil production sites declined and rice agriculture stabilized, the growth rate of atmospheric CH_4 decoupled from population growth. Since 1999, the global atmospheric CH_4 abundance has been nearly stable; globally averaged CH_4 in 1999 was only 3 ppb less than the 2006 global average of 1775 ppb. Potential contributors to this stability are decreased emissions from the Former Soviet Union after their economy collapsed in 1992 (*Dlugokencky et al., 2003*), decreased emissions from natural wetlands because of widespread drought (*Bousquet et al., 2006*), decreased emissions from rice paddies due to changes in water management (*Li et al., 2002*), and an increase in the chemical sink (removal terms in the methane budget are referred to as "sinks") because of changing climate (*Fiore et al., 2006*). Despite attempts to explain the plateau in methane levels, the exact causes remain unknown, making predictions of future methane levels difficult. *Hansen et al. (2000)* have suggested that, because methane has a relatively short atmospheric lifetime (see below) and reductions in emissions are often cost effective, it is an excellent gas to target to counter increasing RF of CO_2 in the short term.

1.3. The Modern Methane Budget

The largest individual term in the global methane budget is removal from the atmosphere by oxidation of methane initiated by reaction with hydroxyl radical (OH; OH + CH_4 → CH_3 + H_2O) in the troposphere.

Approximately 90% of atmospheric CH_4 is removed by this reaction, so estimates of OH concentrations as a function of time can be used to establish how much methane is removed from the atmosphere. When combined with measurements of the current trends in atmospheric methane concentrations, these estimates provide a powerful constraint on the total source. OH is too variable for its large-scale, time-averaged concentration to be determined by direct measurements, so measurements of 1,1,1 -trichloroethane (methyl chloroform), an anthropogenic compound with relatively well-known emissions and predominant OH sink, are most commonly used as a proxy. As assessed by the Intergovernmental Panel on Climate Change (IPCC) Fourth Assessment Report *(Forster et al., 2007)*, the globally averaged OH concentration is ~10^6 per cubic centimeter (cm^{-3}), and there was no detectable change from 1979 to 2004. Reaction with OH is also the major CH_4 loss process in the stratosphere. Smaller atmospheric sinks include oxidation by chlorine in the troposphere and stratosphere and oxidation by electronically excited oxygen atoms [O(1D)] in the stratosphere. Atmospheric CH_4 is also oxidized by bacteria (methanotrophs) in soils, a term which is usually included in budgets as a negative source. These sink terms result in an atmospheric CH_4 lifetime of ~9 years ($\pm$ 10%). In other words, at steady state, each year one ninth of the total amount of methane in the atmosphere is removed by oxidation, and replaced by emissions to the atmosphere.

When an estimate of the lifetime is combined with global observations in a one-box mass balance model of the atmosphere (that is, considering the entire atmosphere to be a well-mixed uniform box), total global emissions can be calculated with reasonable certainty. Using a lifetime of 8.9 years and National Oceanic and Atmospheric Administration (NOAA) Earth System Research Laboratory (ESRL) global observations of CH4 and its trend gives average emissions of 556±10 teragrams (Tg) CH_4 per year (yr^{-1}), with no significant trend for 1984-2006 (Figures 5.1 and 5.7). The uncertainty on total emissions is 1 standard deviation (s.d.) of the interannual variability; total uncertainty is on order of ±10%. The total amount of methane in the atmosphere (often referred to as the atmospheric "burden") is ~5,000 Tg, or 5 Gt CH_4.

Methane is produced by a variety of natural and anthropogenic sources. Estimates of emissions from individual sources are made using bottom-up and top-down methods. Bottom-up inventories use emission factors (e.g., average emissions of CH_4 per unit area for a specific wetland type) and activity levels (e.g., total area of that wetland type) to calculate emissions. Because the relatively few measurements of emission factors are typically extrapolated to large spatial scales, uncertainties in emissions estimated with the bottom-up approach are typically quite large. An example of the top-down method applied to the global scale using a simple 1-box model is shown in Figure 5.1 and described above, but the method can also be applied using a three-dimensional chemical transport model to optimize emissions from regional to continental scales based on a comparison between model-derived mixing ratios and observations. Bottom-up inventories are normally used as initial guesses in this approach. This approach is used to estimate emissions by source and region. Table 5.1 shows optimized CH_4 emissions calculated from an inverse modeling study *(Bergamaschi et al., 2007*, scenario 3) that was constrained by *in situ* surface observations and satellite-based estimates of column- averaged CH_4 mixing ratios. It should be noted that optimized emissions from inverse model studies depend on the *a priori* estimates of emissions and the observational constraints, and realistic estimates of uncertainties are still a challenge. For example, despite the small uncertainties given in the table for termite emissions, emissions from this sector varied from ~3 1 to 67 Tg yr^{-1} over the range of scenarios tested, which is a larger range than the uncertainties in the table would imply. While total global emissions are fairly well constrained by this combination of measurements and lifetime, individual source terms still have relatively large uncertainties.

The constraint on the total modern source strength is important because any new proposed source (for example, a larger than previously identified steady-state marine hydrate source) would have to be balanced by a decrease in the estimated magnitude of another source. The budget presented in Table 5.1 refers to net fluxes to the atmosphere only. The gross production of methane is very likely to be significantly larger, but substantial quantities of methane are consumed in soils, oxic freshwater, and the ocean before reaching the atmosphere (*Reeburgh, 2004*). (The soil sink in Table 5.1 refers only to removal of atmospheric methane by oxidation in soils.)

Given the short CH_4 lifetime (~9 yr), short-term changes in methane emissions from climatically sensitive sources such as biomass burning and wetlands, or in sinks, are seen immediately in surface observations of atmospheric methane. As implied above, reaction with methane is one of the major sinks for the OH radical (the main methane sink), and therefore increases in methane levels should cause an increase in the lifetimes of methane and other long-lived greenhouse gases consumed by OH. Higher methane emissions therefore mean increased methane lifetimes, which in turn means that the impact of any short-term increase in methane emissions will last longer.

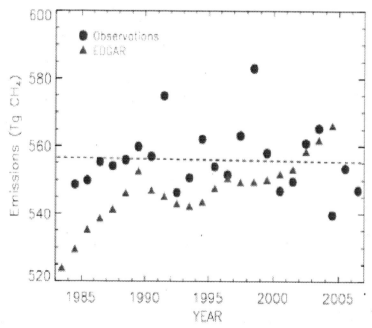

Figure 5.1. Methane emissions as function of time calculated with constant lifetime; emissions from EDGAR inventory with constant natural emissions shown as red triangles. The dashed line is a linear least-squares fit to the calculated emissions; its slope is -0.05+/-0.31 Tg CH4 yr^{-2}. EDGAR is Emission Database for Global Atmospheric Research (described in *Olivier and Berdowski, 2001*); 2001 to 2004 emissions are preliminary (source: http://www.milieuennatuurcompendium.nl/indicatoren/nl0167-Broeikasgasemissies%2C-mondiaal.html?i=9-20). Tg, teragrams; 1 Tg = 10^{12} g.

**Table 5.1. Annual CH4 emissions for 2003 by source type
(from scenario 3 of *Bergamaschi et al., 2007*); chemical sinks are scaled to total
emissions based on *Lelieveld et al. (1998)*. Tg/yr, teragrams per year; 1 Tg = 10^{12} g**

Source	Emissions (Tg/yr)	Fraction of total (%)
Coal	35.6±4.4	6.7
Oil and gas	41.8±5.5	7.9
Enteric fermentation	82.0±9.6	15.4
Rice agriculture	48.7±5.1	9.2
Biomass burning	21.9±2.6	4.1
Waste	67.0±10.7	12.6
Wetlands	208.5±7.6	39.2
Wild animals	6.8±2.0	1.3
Termites	42.0±6.7	7.9
Soil	-21.3±5.8	-4.0
Oceans	-1.3±2.9	-0.2
Total	**531.6±3.7**	
Chemical Sinks	**Loss (Tg/yr)**	
Troposphere	490±50	92.5
Stratosphere	40±10	7.5
Total	**530**	

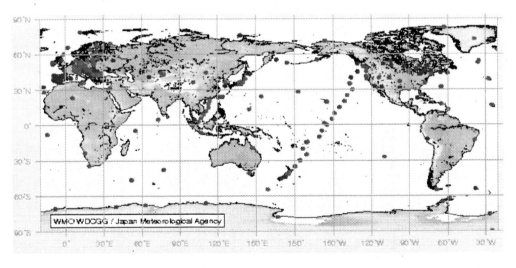

Figure 5.2. World Meteorological Organization global network of monitoring sites (blue dots) for long-term observation of atmospheric methane as of this date (http://gaw.kishou.go.jp/wdcgg/).

1.4. Observational network and its current limitations, particularly relative to the hydrate, permafrost, and Arctic Wetland Sources

The network of air sampling sites where atmospheric methane mixing ratios are measured can be viewed on the World Meteorological Organization (WMO) World Data Centre for Greenhouse Gases (WDCGG) Web site (http://gaw.kishou.go.jp/wdcgg/) and is reproduced in Figure 5.2. Methane data have been reported to the WDCGG for ~130 sites. Relatively few measurements are reported for the Arctic, and sites are typically far from potential permafrost, hydrate, and wetland sources. Existing Arctic sites have been used to infer decreased emissions from the fossil-fuel sector of the Former Soviet Union (*Dlugokencky et al., 2003*) and provide boundary conditions for model studies of emissions, but they are too remote from source regions to accurately quantify emissions, so uncertainties on northern emissions will remain large until more continuous measurement sites are added close to sources. The optimal strategy would include continuous measurements from tall towers and vertical profiles collected from aircraft. Measurements from tall towers are influenced by emissions from much larger areas than those derived from eddy-correlation flux techniques, which have footprints on the order of 1 square kilometer (km^2). When combined with global- or regional-scale models, these measurements can be used to quantify fluxes; the vertical profiles would be used to assess the quality of the model results through the troposphere. To properly constrain CH_4 emissions in the tropics, retrievals of CH_4 column-averaged mixing ratios must be continued to complement surface observations.

BOX 5.1. CHEMISTRY, PHYSICS, AND OCCURRENCE OF METHANE HYDRATE

A clathrate is a substance in which a chemical lattice or cage of one type of molecule traps another type of molecule. Gas hydrates are substances in which gas molecules are trapped in a lattice of water molecules (Figure 5.3). The potential importance of methane hydrate to abrupt climate change results from the fact that large amounts of methane can be stored in a relatively small volume of solid hydrate. For example, 1 cubic meter (m3) of methane hydrate is equivalent to 164 m3 of free gas (and 0.8 m3 of water) at standard temperature and pressure (Kvenvolden, 1993). Naturally occurring gas hydrate on Earth is primarily methane hydrate and forms under high pressure – low temperature conditions in the presence of sufficient methane. These conditions are most often found in relatively shallow marine sediments on continental margins, but also in some high-latitude terrestrial sediments (Figure 5.4). Although the amount of methane stored as hydrate in geological reservoirs is not well quantified, it is very likely that very large amounts are sequestered in comparison to the present total atmospheric methane burden.

The right combination of pressure and temperature conditions forms what is known as the hydrate stability zone, shown schematically in Figure 5.5. In marine sediments, pressure and temperature both increase with depth, creating a relatively narrow region where methane hydrate is stable. Whether or not methane hydrate forms depends not only on temperature and pressure but also on the amount of methane present. The latter constraint limits methane hydrate formation to locations of significant biogenic or thermogenic methane (Kvenvolden, 1993). When ocean bottom water temperatures are near $0°C$, hydrates can form at shallow depths, below ~200 m water depth, if sufficient methane is present. The upper limit of the hydrate stability zone can therefore be at the sediment surface, or deeper in the sediment, depending on pressure and temperature. The thickness of the stability zone increases with water depth in typical ocean sediments. It is important to note, however, that most marine methane hydrates are found in shallow water near continental margins, in areas where the organic carbon content of the sediment is sufficient to fuel methanogenesis. In terrestrial sediments, hydrate can form at depths of ~200 m and deeper, in regions where surface temperatures are cold enough that temperatures at 200 m are within the hydrate stability zone.

1.5. Abrupt changes in atmospheric methane?

Concern about abrupt changes in atmospheric methane stems largely from the large amounts of methane present as solid methane hydrate in ocean sediments and terrestrial sediments, which may become unstable in the face of future warming. Methane hydrate is a solid substance that forms at low temperatures and high pressures in the presence of sufficient methane, and is found primarily in marine continental margin sediments and some arctic terrestrial sedimentary deposits (see Box 5.1). Warming or release of pressure can destabilize methane hydrate, forming free gas that may ultimately be released to the atmosphere. The processes controlling hydrate stability and gas transport are complex and only partly understood. Estimates of the total amount of methane hydrate vary widely, from 500 to 10,000 gigatons of carbon (GtC) stored as methane in hydrates in marine sediments, and 7.5 to 400 GtC in permafrost (both figures are uncertain, see Sec. 4). The total amount of carbon in the modern atmosphere is ~810 GtC, but the total methane content of the atmosphere is only ~4 GtC (*Dlugokencky et al., 1998*). Therefore, even a release of a small portion of the methane hydrate reservoir to the atmosphere could have a substantial impact on radiative forcing.

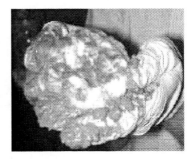

Figure 5.3. Photographs of methane hydrate as nodules, veins, and laminae in sediment. Courtesy of USGS (http://geology

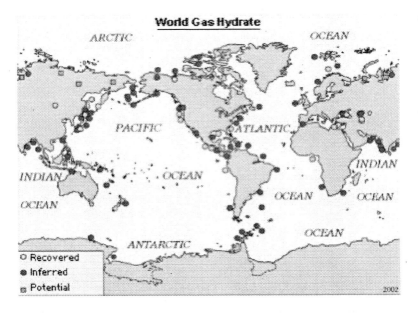

Figure 5.4. Map of methane hydrate deposit locations. Courtesy of USGS (http://geology

Massive releases of methane from marine or terrestrial hydrates have not been observed. Evidence from the ice core record indicates that abrupt shifts in methane concentration have occurred in the past 110,000 years (*Chappellaz et al., 1993a; Brook et al., 1996, 2000*), although the concentration changes during these events were relatively small. Farther back in geologic time, an abrupt warming at the Paleocene-Eocene boundary (about 55 million years ago) has been attributed to a large release of methane to the atmosphere, although alternate carbon sources such as oxidation of sedimentary organic carbon or peats have also been proposed (see discussion in Sec. 4). These past abrupt changes are discussed in detail below, and their existence provides further motivation for considering the potential for future abrupt changes in methane.

The large impact of a substantial release of methane hydrates to the atmosphere, if it were to occur, coupled with the potential for a more steady increase in methane production from melting hydrates and from wetlands in a warming climate, motivates several questions this chapter attempts to address:

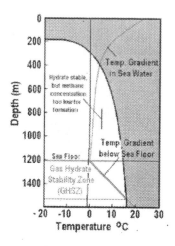
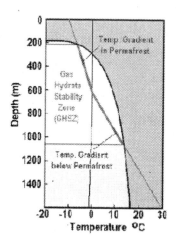

Figure 5.5. Schematic diagram of hydrate stability zone for typical continental margin (left) and permafrost (right) settings. The red line shows the temperature gradient with depth. The hydrate stability zone is technically the depth interval where the *in situ* temperature is lower than the temperature of the phase transition between hydrate and free gas. In the ocean this can occur above the sea floor, but generally there is not sufficient methane in the water column for methane hydrate to form. For this reason the stability zone in the left figure terminates at the sea floor. From National Energy Technology Laboratory (http://204. 154.137. 1 4/technologies/oil-gas/Future Supply/MethaneHydrates/about-hydrates/conditions.htm).

1. What is the volume of methane in terrestrial and marine sources and how much of it is likely to be released if the climate warms in the near future?
2. What is the impact on the climate system of the release of varying quantities of methane over varying intervals of time?
3. What is the evidence in the past for abrupt climate change caused by massive methane release?
4. What conditions (in terms of sea-level rise and warming of bottom waters) would allow methane release from hydrates locked up in sea-floor sediments?
5. How much methane is likely to be released by warming of northern high-latitude soils, sea-level rise, and other climate-driven changes in wetlands?

6. What are the observational and modeling requirements necessary to understand methane storage and its release under various future scenarios of abrupt climate change?

2. History of Atmospheric Methane

Over the last ~300 years the atmospheric methane mixing ratio increased from ~700-750 ppb in 1700 A.D. to a global average of ~1,775 ppb in 2006. Direct atmospheric monitoring has been conducted in a systematic way only since the late 1970s, and data for previous times come primarily from ice cores (Figure 5.6). Current levels of methane are anomalous with respect to the long-term ice core record, which now extends back to 800,000 years (*Spahni et al., 2005; Loulergue et al., 2008*). New international plans to drill at a site of very low accumulation rate in Antarctica may in the future extend the record to 1.5 million years (*Brook and Wolff, 2005*).

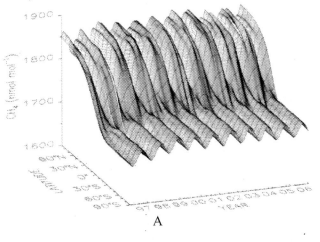

A

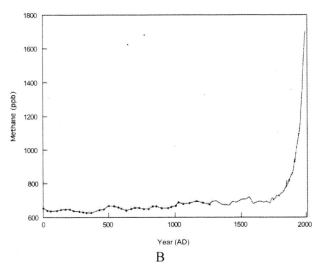

B

Figure 5.6. Continued

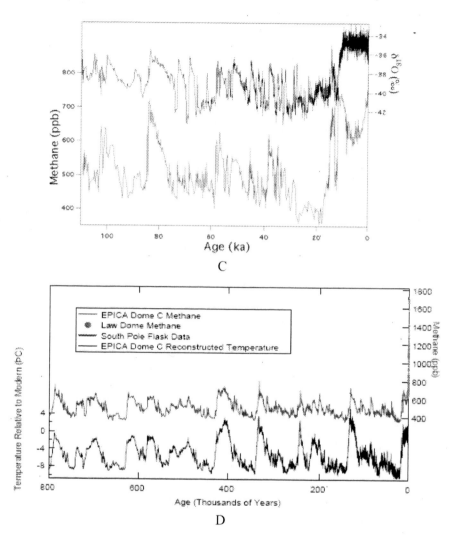

C

D

Figure 5.6. The history of atmospheric methane as derived from ice cores and direct measurements. A, Zonally averaged representation of seasonal and interannual trends in tropospheric methane and interhemispheric gradient over the last decade from NOAA Earth System Research Laboratory (ESRL) data. B, The last 1,000 years from ice cores and direct measurements (*MacFarling-Meure et al., 2006*) and NOAA ESRL data. NOAA ESRL data are updated from Dlugokencky et al., 1994. Unprocessed data and additional figures are available from NOAA ESRL web pages: http://www.esrl.noaa.gov/_gmd/ Photo_Gallery/GMD_Figures/ccgg_figures/ and ftp://ftp.cmdl.noaa.gov/ccg/ch4/flask/, C, The last 100,000 years of methane history from the Greenland Ice Sheet Project 2 (GISP2) ice core in Greenland (Brook et al., 2006; Grachev et al., 2007; Brook and Mitchell, 2007). $\delta^{18}O$ is the stable isotope composition of the ice, a proxy for temperature, with more positive values indicating warmer temperatures. The amplitude of abrupt methane variations appears positively correlated with Northern Hemisphere summer insolation (*Brook et al., 1996*). D, Ice core data from the EPICA Dome C ice cores for the last 800,000 years from *Loulergue et al. (2008)* with additional data for the late Holocene from *MacFarling-Meure et al. (2006)* and NOAA ESRL. Temperature reconstruction is based on the D/H ratio of ice at Dome C. Abbreviations: nmol mol[-1], nanomoles per mole; ppb, parts per billion by mole (same as nanomoles per mole); ‰, per mil.

2.1. Direct Observations

Early systematic measurements of the global distribution of atmospheric CH_4 established a rate of increase of ~16 ppb yr^{-1} in the late 1970s and early 1980s and a strong gradient between high northern and high southern latitudes of ~150 ppb (*Blake and Rowland, 1988*). By the early 1990s it was clear that the CH_4 growth rate was decreasing (*Steele et al., 1992*) and that, if the CH_4 lifetime were constant, atmospheric CH_4 was approaching steady state where emissions were approximately constant (*Dlugokencky et al., 1998*).

Significant variations are superimposed on this declining growth rate and have been attributed to climate-induced variations in emissions from biomass burning (*van der Werf et al., 2004*) and wetlands (*Walter et al., 2001*), and changes in the chemical sink after the eruption of Mount Pinatubo (*Dlugokencky et al., 1996*). Recent measurements show that the global atmospheric CH4 burden has been nearly constant since 1999 (Figure 5.7). This observation is not well understood, underscoring our lack of understanding of how individual methane sources are changing.

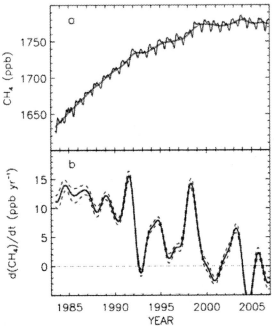

Figure 5.7. Recent trends in atmospheric methane from global monitoring data (NOAA Earth System Research Laboratory, ESRL). NOAA ESRL data are updated from Dlugokencky et al., 1994. Unprocessed data and additional figures are available from NOAA ESRL web pages: http://www.esrl.noaa.gov/gmd/Photo_Gallery/GMD_Figures/ccgg_figures/ and ftp://ftp.cmdl.noaa.gov/ccg/ch4/flask/. A, Global average atmospheric methane mixing ratios (blue line) determined using measurements from the ESRL cooperative air sampling network. The red line represents the long-term trend. B, Solid line is the instantaneous global average growth rate for methane; dashed lines are uncertainties (1 standard deviation) calculated with a Monte Carlo method that assesses uncertainty in the distribution of sampling sites (*Dlugokencky et al., 2003*).

Recently published column-averaged CH_4 mixing ratios determined from a satellite sensor greatly enhance the spatial coverage of CH4 observations (*Frankenberg et al., 2006*). Coverage in the tropics greatly increases measurements there, but coverage in the Arctic remains poor because of the adverse impact of clouds on the retrievals. Use of these satellite data in inverse model studies will reduce uncertainties in emissions estimates, particularly in the tropics.

2.2. The Ice Core Record

The long-term record shows changes in methane on glacial-interglacial time scales of ~300-400 ppb (Figure 5.6D), dominated by a strong ~100,000-year periodicity, with higher levels during warm interglacial periods and lower levels during ice ages. Periodicity of ~40,000 and 20,000 years is also apparent, associated with Earth's cycles of obliquity and precession (*Delmotte et al., 2004*). Methane is believed to provide a positive feedback to warming ultimately caused by changes in the Earth's orbital parameters on these time scales. The cyclicity is widely attributed to processes affecting both northern high- latitude and tropical wetlands, including growth and decay of Northern Hemisphere ice sheets, and variations in the strength of the monsoon circulation and associated rainfall patterns in Asia, Africa, and South America (*Delmotte et al., 2004; Spahni et al., 2005; Loulergue et al., 2008*).

The ice core record also clearly shows another scale of variability, abrupt shifts in methane on millennial time scales that are coincident with abrupt changes in temperature observed in Greenland ice cores (Figure 5.6C). These abrupt shifts have been studied in detail in three deep ice cores from Greenland and in several Antarctic ice cores (*Chappellaz et al., 1993a; Brook et al., 1996; Brook et al., 2000; Severinghaus et al., 1998; Severinghaus and Brook, 1999; Huber et al., 2006; Grachev et al., 2007*). Detailed work using nitrogen and argon isotope ratios as gas phase indicators of warming in the ice core record shows clearly that the increase in methane associated with the onset of abrupt warming in Greenland is coincident with, or slightly lags (by a few decades at most), the warming (*Severinghaus et al., 1998; Severinghaus and Brook, 1999; Huber et al., 2006; Grachev et al., 2007*). Methane closely follow the Greenland ice isotopic record (Figure 5.6C), and the amplitude of methane variations associated with abrupt warming in Greenland appears to vary with time. *Brook et al. (1996)* suggested a longterm modulation of the atmospheric methane response to abrupt climate change related to global hydrologic changes on orbital time scales, an issue further quantified by *Flückiger et al. (2004)*.

2.3. What caused the abrupt changes in methane in the ice core record?

Because the modern natural methane budget is dominated by emissions from wetlands, it is logical to interpret the ice core record in this context. The so-called "wetland hypothesis" postulates that abrupt warming in Greenland is associated with warmer and wetter climate in terrestrial wetland regions, which results in greater emissions of methane from wetlands. Probable sources include tropical wetlands (including regions now below sea level) and high-latitude wetlands in regions that remained ice free or were south of the major ice sheets. Cave deposits in China, as well as marine and lake sediment records, indicate that enhanced monsoon rainfall in the Northern Hemisphere's tropics and subtropics was closely linked to abrupt warming in Greenland (e.g., *Kelly et al., 2006; Wang et al., 2004; Yuan et al., 2004; Dykoski et al., 2005; Peterson et al., 2000*). The cave records in particular are important because they are extremely well dated using uranium series isotopic techniques, and high-resolution oxygen isotope records from caves, interpreted as rainfall indicators, convincingly match large parts of the Greenland ice core methane record.

The wetland hypothesis is based on climate-driven changes in methane sources, but it is also possible that changes in methane sinks, primarily the OH radical, played a role in the variations observed in ice cores. Both *Kaplan et al. (2006)* and *Valdes et al. (2005)* proposed that the glacial-interglacial methane change cannot be explained entirely by changes in emissions from wetlands, because in their global climate-biosphere models the difference between Last Glacial Maximum (LGM) and early Holocene methane emissions is not large enough to explain the observed changes in the ice core record. Both studies explain this apparent paradox by invoking increased production of volatile organic carbon (VOC) from the terrestrial biosphere in warmer climates. VOCs compete with methane for reaction with OH, increasing the methane lifetime and the steady-state methane concentration that can be maintained at a given emission rate. Neither of these studies is directly relevant to the abrupt changes in the ice core record, and there are considerable uncertainties in the modeling. Nonetheless, further work on the role of changes in the methane sink on time scales relevant to abrupt methane changes is warranted.

The wetland hypothesis has been challenged by authors calling attention to the large marine and terrestrial hydrate reservoirs. The challenge was most extensively developed by *Kennett et al. (2003)*, who postulated that the abrupt shifts in methane in the ice core record were caused by abrupt release of methane from methane hydrates in sea-floor sediments on continental margins. This hypothesis originated from observations of negative carbon isotope excursions in marine sediment records in the Santa Barbara basin, which apparently coincided with the onset of abrupt warming in Greenland and increases in atmospheric methane in the ice core record. The "clathrate gun hypothesis" postulates that millennial-scale abrupt warming during the last ice age was actually driven by atmospheric methane from hydrate release, and further speculates on a central role for methane in causing late Quaternary climate change (*Kennett et al., 2003*).

Some proponents of the clathrate gun hypothesis further maintain that wetlands were not extensive enough during the ice age to be the source of the abrupt variations in methane in the ice core record. For example, *Kennett et al. (2003)* maintain that large accumulations of carbon in wetland ecosystems are a prerequisite for significant methanogenesis and that these established wetlands are exclusively a Holocene phenomenon. Process-based studies of methane emissions from wetlands, on the other hand, emphasize the relationship between annual productivity and emissions (e.g., *Christensen et al., 1996*). In this view methane production is closely tied to the production of labile carbon (*Schlesinger, 1997*) in the annual productivity cycle *(Christensen et al., 1996)*. From this perspective it has been postulated that the ice core record reflects changes in rainfall patterns and temperature that could quickly influence the development of anoxic conditions, plant productivity, and methane emissions in regions where the landscape is appropriate for development of water-saturated soil (e.g., *Brook et al., 2000; von Huissteten, 2004*).

The hypothesis that there was very little methane emission from wetlands prior to the onset of the Holocene is at odds with models of both wetland distribution and emissions for pre-Holocene times, the latter indicating emissions consistent with, or exceeding, those inferred from the ice core record (e.g., *Valdes et al., 2005; Kaplan, 2002; Kaplan et al., 2006; Chappellaz et al., 1993b; von Huissteten, 2004*). V*on Huissteten (2004)* specifically considered methane emissions during the stadial and interstadial phases of Marine Isotope Stage 3 (~30,000-60,000 years ago), when ice core data indicate that several rapid changes in atmospheric methane occurred (Figure 5.6C). *Von Huissteten* describes wetland sedimentary

---

BOX 5.2. THE ICE CORE RECORD AND ITS FIDELITY IN CAPTURING ABRUPT EVENTS

Around the time of discovery of the abrupt, but small, changes in methane in the late Quaternary ice core records (Figure 5.6C) (*Chappellaz et al., 1993a*) some authors suggested that very large releases of methane to the atmosphere might be consistent with the ice core record, given the limits of time resolution of ice core data at that time, and the smoothing of atmospheric records due to diffusion in the snowpack (*e.g., Thorpe et al., 1996*). Since that time a large number of abrupt changes in methane in the Greenland ice core record (which extends to ~120,000 years before present) have been sampled in great detail, and no changes greatly exceeding those shown in Figure 5.6C have been discovered (*Brook et al., 1996; 2000; 2005; Blunier and Brook, 2001; Chappellaz et al., 1997; Severinghaus et al. 1998; Severinghaus and Brook, 1999; Huber et al. 2006; EPICA Members, 2006; Grachev et al., 2007*).

Could diffusion in the snowpack mask much larger changes? Air is trapped in polar ice at the base of the firn (snowpack) where the weight of the overlying snow transforms snow to ice, and air between the snow grains is trapped in bubbles (Figure 5.8). The trapped air is therefore younger than the ice it is trapped in (this offset is referred to as the gas age-ice age difference). It is also mixed by diffusion, such that the air trapped at an individual depth interval is a mixture of air of different ages. In addition, bubbles do not all close off all the same depth, so there is additional mixing of air of different ages due to this variable bubble close-off effect. The overall smoothing depends on the parameters that control firn thickness, densification, and diffusion – primarily temperature and snow accumulation rate.

Spahni et al. (2003) used the firn model of *Schwander et al. (1993)* to study the impact of smoothing on methane data from the Greenland Ice Core Project (GRIP) ice core in Greenland for the late Holocene. They examined the impact of smoothing on abrupt changes in methane in the Greenland ice core record. *Brook et al. (2000)* investigated a variety of scenarios for abrupt changes in methane, including those proposed by *Thorpe et al. (1996)*, and compared what the ice core record would record of those events with high-resolution data for several abrupt shifts in methane (Figure 5.9).

Two aspects of the ice core data examined by Brook *et al.* argue against abrupt, catastrophic releases of methane to the atmosphere as an explanation of the ice core record. First, the abrupt shifts in methane concentration take place on time scales of centuries, whereas essentially instantaneous releases would be recorded in the Greenland ice core record as more abrupt events (Figure 5.9). While this observation says nothing about the source of the methane, it does indicate that the ice core record is not recording an essentially instantaneous atmospheric change (*Brook et al., 2000*). Second, the maximum levels of methane reached in the ice core record are not high enough to indicate extremely large changes in the atmospheric methane concentration (Figure 5.9).

---

deposits in northern Europe dating from this period and used a process-based model to estimate methane emissions for the cold and warm intervals. The results suggest that emissions from Northern Hemisphere wetlands could be sufficient to cause emissions variations inferred from ice core data. *MacDonald et al. (2006)* presented a compilation of basal peat ages for the circumarctic and showed that peat accumulation started early in the deglaciation (at about 16,000 years before present), and therefore emissions of methane from Northern Hemisphere peat ecosystems very likely played a role in the methane increase at the end of the last ice age. The coincidence of peatland development and the higher Northern Hemisphere summer insolation of late glacial and early Holocene time supports the hypothesis that such wetlands were methane sources at previous times of higher Northern Hemisphere summer insolation (*MacDonald et al., 2006),* for example during insolation and methane peaks in the last ice age or at previous glacial-interglacial transitions (*Brook et al., 1996; 2000*). In summary, although the sedimentary record of wetlands and the factors controlling methane production in wetlands are imperfectly known, it appears likely that wetlands were important in the pre- Holocene methane budget.

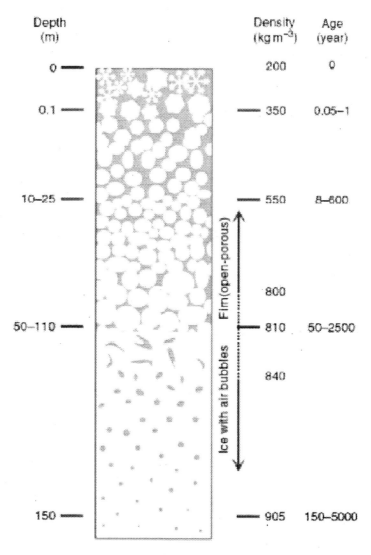

Figure 5.8. The firn column of a typical site on a polar ice sheet, from *Schwander (2006)*. Abbreviations: m, meter; kg m^{-3}, kilograms per cubic meter

The clathrate gun hypothesis is important for understanding the future potential for abrupt changes in methane – concern for the near future is warranted if the clathrate reservoir was unstable on the time scale of abrupt late Quaternary climate change. However, as an explanation for late Quaternary methane cycles, the clathrate gun hypothesis faces several challenges, elaborated upon further in Section 4. First, the radiative forcing of the small variations in atmospheric methane burden during the ice age should have been quite small (*Brook et al., 2000*), although it has been suggested that impacts on stratospheric water vapor may have increased the greenhouse power of these small methane variations (*Kennett et al., 2003*). Second, the ice core record clearly shows that the abrupt changes in methane lagged the abrupt temperature changes in the Greenland ice core record, albeit by only decades (*Severinghaus et al., 1998; Severinghaus and Brook, 1999; Huber et al., 2006; Grachev et al., 2007*). These observations imply that methane is a feedback to rather than a cause of warming, ruling out one aspect of the clathrate gun hypothesis (hydrates as trigger), but they do not constrain the cause of the abrupt shifts in methane. Third, isotopic studies of ice core methane do not support methane hydrates as a source for abrupt changes in methane (*Sowers, 2006; Schaefer et al., 2006*). The strongest constraints come from hydrogen isotopes (*Sowers, 2006)* and are described further in Section 4.

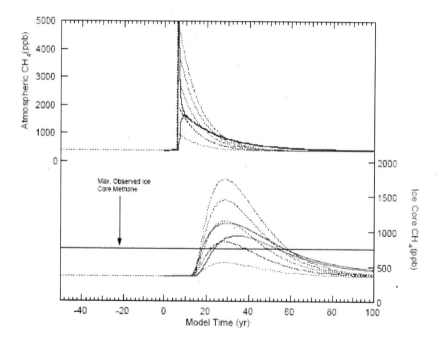

Figure 5.9. Model simulations of smoothing instantaneous release of methane from clathrates to the atmosphere, and the ice core response to those events. The ice core response was calculated by convolving the atmospheric histories in the top panel with a smoothing function appropriate for the GISP2 ice core. The solid lines are the atmospheric history and smoothed result for the model of a 4,000 teragram release of methane from *Thorpe et al. (1994)*. The blue solid line represents how an Arctic ice core would record a release in the Northern Hemisphere, and the red solid line represents how an Antarctic ice core would record that event (from *Brook et al., 2000*). The dashed lines represent instantaneous arbitrary increases of atmospheric methane to values of 1,000, 2,000, 3,000, 4,000, or 5,000 ppb (colored dashed lines in top panel) and the ice core response (bottom panel, same color scheme)

3. Potential Mechanisms for Future Abrupt Changes in Atmospheric Methane

Three general mechanisms are considered in this chapter as potential causes of abrupt changes in atmospheric methane in the near future large enough to cause abrupt climate change. These are outlined briefly in this section, and discussed in detail in Sections 4-6.

3.1. Destabilization of marine methane hydrates

This issue is probably the most well known due to extensive research on the occurrence of methane hydrates in marine sediments, and the large quantities of methane apparently present in this solid phase in continental-margin marine sediments. Destabilization of this solid phase requires mechanisms for warming the deposits and/or reducing pressure on the appropriate time scale, transport of free methane gas to the sediment-water interface, and transport to the atmosphere (see Box 5.1). There are a number of physical impediments to abrupt release, in addition to the fact that bacterial methanotrophy consumes methane in oxic sediments and the ocean water column. Warming of bottom waters, slope failure, and their interaction are the most commonly discussed mechanisms for abrupt release.

3.2. Destabilization of permafrost hydrates

Hydrate deposits at depth in permafrost are known to exist, and although their extent is uncertain, the total amount of methane in permafrost hydrates is very likely much smaller than in marine sediments. Surface warming eventually would increase melting rates of permafrost hydrates. Inundation of some deposits by warmer seawater and lateral invasion of the coastline are also concerns and may be mechanisms for more rapid change.

3.3. Changes in wetland extent and methane productivity

Although a destabilization of either the marine or terrestrial methane hydrate reservoirs is the most probable pathway for a truly abrupt change in atmospheric methane concentration, the potential exists for a more chronic, but substantial, increase in natural methane emissions in association with projected changes in climate. The most likely region to experience a dramatic change in natural methane emission is the northern high latitudes, where there is increasing evidence for accelerated warming, enhanced precipitation, and widespread permafrost thaw which could lead to an expansion of wetland areas into organic-rich soils that, given the right environmental conditions, would be fertile areas for methane production.

In addition, although northern high-latitude wetlands seem particularly sensitive to climate change, the largest natural source of methane to the atmosphere is from tropical wetlands, and methane emissions there may also be sensitive to future changes in temperature and precipitation. Modeling studies addressing this issue are therefore also included in our discussion.

4. Potential for Abrupt Methane Change from Marine Hydrate Sources

4.1. Impact of temperature change on marine methane hydrates

A prominent concern about marine methane hydrates is that warming at the Earth's surface will ultimately propagate to hydrate deposits and melt them, releasing methane to the ocean-atmosphere system. The likelihood of this type of methane release depends on the propagation of heat through the sea floor, the migration of methane released from hydrate deposits through sediments, and the fate of this methane in the water column.

4.1.1. Propagation of temperature change to the hydrate stability zone

The time dependence of changes in the inventory of methane in the hydrate reservoir depends on the time scale of warming and chemical diffusion. There is evidence from paleotracers (*Martin et al., 2005*) and from modeling (*Archer et al., 2004*) that the temperature of the deep sea is sensitive to the climate of the Earth's surface. In general, the time scale for changing the temperature of the ocean increases with water depth, reaching a maximum of about 1,000 years for the abyssal ocean. This means that abrupt changes in temperature at the surface ocean would not be transmitted immediately to the deep sea. There are significant regional variations in the ventilation time of the ocean, and in the amount of warming that might be expected in the future. The Arctic is expected to warm particularly strongly because of the albedo feedback from the melting Arctic ice cap. Temperatures in the North Atlantic appear to be sensitive to changes in ocean circulation such as during rapid climate change during the last ice age (*Dansgaard et al., 1989*).

The top of the hydrate stability zone is at 200 to 600 m water depth, depending mainly on the temperature of the water column. Within the sediment column, temperature increases with depth along the geothermal temperature gradient, 30-50°C km^{-1} (*Harvey and Huang, 1995*). The shallowest sediments that could contain hydrate only have a thin hydrate stability zone, and the stability zone thickness increases with water depth. A change in the temperature of the deep ocean will act as a change in the upper boundary condition of the sediment temperature profile. Warming of the overlying ocean may not put surface sediments into undersaturation, but the warmer overlying temperature propagates downward until a new profile with the same geothermal temperature gradient can be established. How long this takes is a strong (second order) function of the thickness of the stability zone, but the time scales are in general long. In 1,000 years the temperature signal should have propagated about 180 m in the sediment. In steady state, an increase in ocean temperature will decrease the thickness of the stability zone. *Dickens (2001b)* calculated that the volume of the stability zone ought to decrease by about half with a temperature increase of 5°C.

4.1.2. Impact on stratigraphic-type deposits

Hydrate deposits formed within sedimentary layers are referred to as stratigraphic-type deposits. After an increase in temperature of the overlying water causes hydrate to melt at the base of the stability zone, the fate of the released methane is difficult to predict. The increase in pore volume and pressure could provoke gas migration through the stability zone or a landslide, or the bubbles could remain enmeshed in the sediment matrix. Hydrate moves down to the base of the stability zone by the accumulation of overlying sediment at the sea floor, so melting of hydrate at the stability zone takes place continuously, not just in association with ocean warming.

When hydrate melts, most of the released methane goes into the gas phase to form bubbles, assuming that the porewaters were already saturated in dissolved methane. The fate of the new bubbles could be to remain in place, to migrate, or to diffuse away and react chemically (*Hinrichs et al., 1999; Wakeham et al., 2003*), and it is difficult to predict which will occur. The potential for gas migration through the stability zone is one of the more significant uncertainties in forecasting the ocean hydrate response to anthropogenic warming (*Harvey and Huang, 1995*).

In cohesive sediments, bubbles expand by fracturing the sediment matrix, resulting in elongated shapes (*Boudreau et al., 2005*). Bubbles tend to rise because they are less dense than the water they are surrounded by, even at the 200+ atmosphere pressures in sediments of the deep sea. If the pressure in the gas phase exceeds the lithostatic pressure in the sediment, fracture and gas escape can occur (*Flemings et al., 2003*). Modeled and measured (*Dickens et al., 1995*) porewater pressures in the sediment column at Blake Ridge approach lithostatic pressures, indicating that new gas bubbles added to the sediment might be able to escape to the overlying water by this mechanism.

A differential-pressure mechanism begins to operate when the bubbles occupy more than about 10% of the volume of the pore spaces (*Hornbach et al., 2004*). If a connected bubble spans a large enough depth range, the pressure of the porewater will be higher at the bottom of the bubble than it is at the top, because of the weight of the porewater over that depth span. The pressure inside the bubble will be more nearly constant over the depth span, because the compressed gas is not as dense as the porewater is. This will result in a pressure gradient at the top and the bottom of the bubble, tending to push the bubble upward. *Hornbach et al. (2004)* postulated that this mechanism might be responsible for allowing methane to escape from the sediment column, and they calculated the maximum thickness of an interconnected bubble zone required, before the bubbles would break through the overlying sediment column. In their calculations, and in stratigraphic deposits (they refer to them as "basin settings"), the thickness of the bubble column increases as the stability zone gets thicker. It takes more pressure to break through a thicker stability zone, so a taller column of gas is required. In compressional settings, where the dominant force is directed sideways by tectonics, rather than downward by gravity, the bubble layer is never as thick, reflecting an easier path to methane escape.

Multiple lines of evidence indicate that gas can be transported through the hydrate stability zone without freezing into hydrate. Seismic studies at Blake Ridge have observed the presence of bubbles along faults in the sediment matrix (*Taylor et al., 2000*). Faults have been correlated with sites of methane gas emission from the sea floor (*Aoki et al., 2000; Zuhlsdorff et al., 2000; Zuhlsdorff and Spiess, 2004*). Seismic studies often show "wipeout zones" where the bubble zone beneath the hydrate stability zone is missing, and all of the layered structure of the sediment column within the stability zone is smoothed out. These are interpreted to be areas where gas has broken through the structure of the sediment to escape to the ocean (*Riedel et al., 2002; Wood et al., 2002;*

Hill et al., 2004). Bubbles associated with seismic wipeout zones are observed within the depth range that should be within the hydrate stability zone, assuming that the temperature of the sediment column is the steady-state expression of the local average geothermal gradient

(*Gorman et al., 2002*). This observation has been explained by assuming that upward migration of the fluid carries with it heat, maintaining a warm channel where gas can be transported through what would otherwise be thermodynamically hostile territory (*Taylor et al., 2000; Wood et al., 2002*).

The sediment surface of the world's ocean has holes in it called pockmarks (*Hovland and Judd, 1988; Hill et al., 2004*), interpreted to be the result of catastrophic or continuous escape of gas to the ocean. Pockmarks off Norway are accompanied by authigenic carbonate deposits associated with anaerobic oxidation of methane (*Hovland et al., 2005*). Pockmarks range in size from meters to kilometers (*Hovland et al., 2005*), with one 700-km^2 example on the Blake Ridge (*Kvenvolden, 1999*). If the Blake Ridge pockmark is the result of a catastrophic explosion, it might have released less than 1GtC as methane (assuming a 500-m-thick layer of 4% methane yields 1 GtC). Since each individual pockmark releases a small amount of methane relative to the atmospheric inventory, pockmark methane release could impact climate as part of the ongoing "chronic" methane source to the atmosphere, if the frequency of pockmark eruptions increased. In this sense pockmarks do not represent "catastrophic" methane releases. However, *Kennett et al. (2003)* hypothesized that some apparently inactive pockmark fields may have formed during the last deglaciation and are evidence of active methane discharge at that time.

Another mechanism for releasing methane from the sediment column is by submarine landslides. These are a normal, integral part of the ocean sedimentary system (*Hampton et al., 1996; Nisbet and Piper, 1998*). Submarine landslides are especially prevalent in river deltas because of the high rate of sediment delivery and because of the presence of submarine canyons. The tendency for slope failure can be amplified if the sediment accumulates more quickly than the excess porosity can be squeezed out. This accumulation can lead to instability of the sediment column, causing periodic Storeggatype landslides off the coast of Norway (see section below on Storegga Landslide), in the Mediterranean Sea (*Rothwell et al., 2000*), or potentially off the East Coast of the United States (*Dugan and Flemings, 2000*). *Maslin et al. (2004)* find that 70% of the landslides in the North Atlantic over the last 45,000 years (45 kyr) occurred within the time windows of the two meltwater peaks, 15-13 and 11-8 kyr ago. These could have been driven by deglacial sediment loading or warming of the water column triggering hydrate melting.

Warming temperatures or sea-level changes may trigger the melting of hydrate deposits, provoking landslides (*Kvenvolden, 1999; Driscoll et al., 2000; Vogt and Jung, 2002*). Paull et al. (1991) calculate that landslides can release up to about 1-2 GtC as methane; 1 Gt is enough to alter the radiative forcing by about 0.25 watts per square meter (W m^{-2}). The origin of these estimates is discussed in the section on the Storegga Landslide.

4.1.3. Impact on structural-type hydrate deposits

In stratigraphic-type hydrate deposits, hydrate concentration is highest near the base of the stability zone, often hundreds of meters below the sea floor. In shallower waters, where the stability zone is thinner, models predict smaller inventories of hydrate. Therefore, most of the hydrates in stratigraphic-type deposits tend to be deep. In contrast with this, in a few parts of the world, transport of presumably gaseous methane through faults or permeable channels results in hydrate deposits that are abundant at shallow depths in the sediment column, closer to the sea floor. These "structural-type" deposits could be vulnerable to temperature-change-driven melting on a faster time scale than the stratigraphic deposits are expected to be.

The Gulf of Mexico contains structural-type deposits and is basically a leaky oil field (MacDonald et al., 1994, 2002, 2004; Sassen and MacDonald, 1994; Milkov and Sassen, 2000, 2001, 2003; Sassen et al., 2001a; Sassen et al., 2003). Natural oil seeps leave slicks on the sea surface that can be seen from space. Large chunks of methane hydrate have been found on the sea floor in contact with seawater (MacDonald et al., 1994). One of the three chunks MacDonald et al. saw had vanished when they returned a year later; presumably it had detached and floated away.

Collett and Kuuskraa (1998) estimate that 500 GtC might reside as hydrates in the Gulf sediments, but *Milkov (2004)* estimates only 5 GtC. The equilibrium temperature change in the deep ocean to a large, 5,000-GtC fossil fuel release could be 3°C (*Archer et al., 2004*). *Milkov and Sassen (2003)* subjected a two-dimensional model of the hydrate deposits in the Gulf to a 4°C temperature increase and predicted that 2 GtC from hydrate would melt. However, there are no observations to suggest that methane emission rates are currently accelerating, and temperature changes in Gulf of Mexico deep waters in the next 100 years are likely to be smaller than 3-4°C. *Sassen et al. (2001b)* find no molecular fractionation of gases in near-surface hydrate deposits that would be indicative of partial dissolution, and suggest that the reservoir may in fact be growing.

Other examples of structural deposits include the summit of Hydrate Ridge, off the coast of Oregon, USA (*Torres et al., 2004; Trehu et al., 2004b*), and the Niger Delta (*Brooks et al., 2000*). The distribution of hydrate at Hydrate Ridge indicates up-dip flow along sand layers (*Weinberger et al., 2005*). Gas is forced into sandy layers where it accumulates until the gas pressure forces it to vent to the surface (*Trehu et al., 2004a*). *Trehu et al. (2004b)* estimate that 30-40% of pore space is occupied by hydrate, while gas fractions are 2-4%. Methane emerges to the sea floor with bubble vents and subsurface flows of $1\ m\ s^{-1}$, and in regions with bacterial mats and vesicomyid clams (*Torres et al., 2002*). Further examples of structural deposits include the Peru Margin (*Pecher et al., 2001*) and Nankai Trough, Japan (*Nouze et al., 2004*).

Mud volcanoes are produced by focused-upward fluid flow into the ocean and are sometimes associated with hydrate and petroleum deposits. Mud volcanoes often trap methane in hydrate deposits that encircle the channels of fluid flow (*Milkov, 2000; Milkov et al., 2004*). The fluid flow channels associated with mud volcanoes are ringed with the seismic images of hydrate deposits, with authigenic carbonates, and with pockmarks (*Dim itrov and Woodside, 2003*) indicative of anoxic methane oxidation. *Milkov (2000)* estimates that mud volcanoes contain at most 0.5 GtC of methane in hydrate, about 100 times his estimate of the annual supply.

4.1.4. Fate of methane released as bubbles

Methane released from sediments in the ocean may reach the atmosphere directly, or it may dissolve in the ocean. Bubbles are not generally a very efficient means of transporting methane through the ocean to the atmosphere. *Rehder et al. (2002)* compared the dissolution kinetics of methane and argon and found enhanced lifetime of methane bubbles below the saturation depth in the ocean, about 500 m, because a hydrate film on the surface of the methane bubbles inhibited gas exchange. Bubbles dissolve more slowly from petroleum seeps, where oily films on the surface of the bubble inhibit gas exchange, also changing the shapes of the bubbles (*Leifer and MacDonald, 2003*). On a larger scale, however, *Leifer et al.*

(2000) diagnosed that the rate of bubble dissolution is limited by turbulent transport of methane-rich water out of the bubble stream into the open water column. The magnitude of the surface dissolution inhibition seems small; in the *Rehder et al. (2002)* study, a 2-cm bubble dissolves within 30 m above the stability zone, and only 110 m below the stability zone. Acoustic imaging of the bubble plume from Hydrate Ridge showed bubbles surviving from 600-700 m water depth, where they were released to just above the stability zone at 400 m *(Heeschen et al., 2003)*. One could imagine hydrate-film dissolution inhibition as a mechanism to concentrate the release of methane into the upper water column, but not really as a mechanism to get methane through the ocean directly to the atmosphere.

Methane can reach the atmosphere if the methane bubbles are released in waters that are only a few tens of meters deep, as in the case of melting the ice complex in Siberia *(Shakhova et al., 2005; Washburn et al., 2005; Xu et al., 2001)*, or during periods of lower sea level *(Luyendyk et al., 2005)*. If the rate of methane release is large enough, the rising column of seawater in contact with the bubbles may saturate with methane, or the bubbles can be larger, potentially increasing the escape efficiency to the atmosphere.

4.1.5. Fate of methane hydrate in the water column

Pure methane hydrate is buoyant in seawater, so floating hydrate is another source of methane delivery from the sediment to the atmosphere *(Brewer et al., 2002)*. In sandy sediment, the hydrate tends to fill the existing pore structure of the sediment, potentially entraining sufficient sediment to prevent the hydrate/sediment mixture from floating, while in fine-grained sediments, bubbles and hydrate grow by fracturing the cohesion of the sediment, resulting in irregular blobs of bubbles *(Gardiner et al., 2003; Boudreau et al., 2005)* or pure hydrate. *Brewer et al. (2002)* and *Paull et al. (2003)* stirred surface sediments from Hydrate Ridge using the mechanical arm of a submersible remotely operated vehicle and found that hydrate did manage to shed its sediment load enough to float. Hydrate pieces of 0.1 m survived a 750-m ascent through the water column. *Paull et al. (2003)* described a scenario for a submarine landslide in which the hydrates would gradually make their way free of the turbidity current comprised of the sediment and seawater slurry.

4.1.6. Fate of dissolved methane in the water column

Methane is unstable to bacterial oxidation in oxic seawater. *Rehder et al. (1999)* inferred a methane oxidation lifetime in the high-latitude North Atlantic of 50 years. Methane oxidation is faster in the deep ocean near a particular methane source, where its concentration is higher (turnover time 1.5 years), than it is in the surface ocean (turnover time of decades) *(Valentine et al., 2001)*. Water-column concentration and isotopic measurements indicate complete water-column oxidation of the released methane at Hydrate Ridge *(Grant and Whiticar, 2002; Heeschen et al., 2005)*.

An oxidation lifetime of 50 years leaves plenty of time for transport of methane gas to the atmosphere. Typical gas-exchange time scales for gas evasion from the surface ocean would be about 3-5 m per day. A surface mixed layer 100 m deep would approach equilibrium (degas) in about a month. Even a 1,000-m-thick winter mixed layer would degas about 30% during a 3-month winter window. The ventilation time of subsurface waters depends on the depth and the fluid trajectories in the water *(Luyten et al., 1983)*, but 50 years is enough time that a significant fraction of the dissolved methane from bubbles might reach the atmosphere before it is oxidized.

4.2. Geologic Data Relevant to Past Hydrate Release

4.2.1. The Storegga landslide

One of the largest exposed submarine landslides in the ocean is the Storegga Landslide in the Norwegian continental margin (*Mienert et al., 2000, 2005; Bryn et al., 2005*). The slide excavated on average the top 250 m of sediment over a swath hundreds of kilometers wide, stretching halfway from Norway to Greenland (Figure 5.10). There have been comparable slides on the Norwegian margin every approximately 100 kyr, roughly synchronous with the glacial cycles (*Solheim et al., 2005*). The last one, Storegga proper, occurred about 8,150 years ago, after deglaciation. It generated a tsunami in what is now the United Kingdom (*D'Hondt et al., 2004; Smith et al., 2004*). The Storegga slide area contains methane hydrate deposits as indicated by a bottom simulating seismic reflector (BSR) (*Bunz and Mienert, 2004; Mienert et al., 2005; Zillmer et al., 2005a, b*) corresponding to the base of the hydrate stability zone (HSZ) at 200-300 m, and pockmarks (*Hovland et al., 2005*) indicating gas expulsion from the sediment.

The proximal cause of the slide may have been an earthquake, but the sediment column must have been destabilized by either or both of two mechanisms. One is the rapid accumulation of glacial sediment shed by the Fennoscandian ice sheet (*Bryn et al., 2005*). As explained above, rapid sediment loading traps porewater in the sediment column faster than it can be expelled by the increasing sediment load. At some point, the sediment column floats in its own porewater (*Dugan and Flemings, 2000*). This mechanism has the capacity to explain why the Norwegian continental margin, of all places in the world, might have landslides synchronous with climate change.

The other possibility is the dissociation of methane hydrate deposits by rising ocean temperatures. Rising sea level is also a player in this story, but a smaller one. Rising sea level tends to increase the thickness of the stability zone by increasing the pressure. A model of the stability zone shows this effect dominating deeper in the water column (*Vogt and Jung, 2002*); the stability zone is shown increasing by about 10 m for sediments in water depth below about 750 m. Shallower sediments are impacted more by long-term temperature changes, reconstructions of which show warming of 5-6°C over a thousand years or so, 11-12 kyr ago. The landslide occurred 2-3 kyr after the warming (*Mienert et al., 2005*). The slide started at a few hundred meters water depth, just off the continental slope, just where *Mienert et al. (2005)* calculate the maximum change in HSZ. *Sultan et al. (2004)* predict that warming in the near-surface sediment would provoke hydrate to dissolve by increasing the saturation methane concentration. This form of dissolution differs from heat-driven direct melting, however, in that it produces dissolved methane, rather than methane bubbles. *Sultan et al. (2004)* assert that melting to produce dissolved methane increases the volume, although laboratory analyses of volume changes upon this form of melting are equivocal. In any case, the volume changes are much smaller than for thermal melting that produces bubbles.

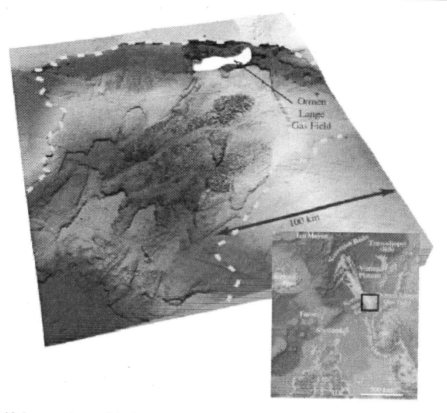

Figure 5.10. Image and map of the Storegga Landslide from *Masson et al. (2006)*. The slide excavated on average the top 250 m of sediment over a swath hundreds of kilometers wide. Colors indicate water depth, with yellow-orange indicating shallow water, and green-blue indicating deeper water

The amount of methane released by the slide can be estimated from the volume of the slide and the potential hydrate content. Hydrate just outside the slide area has been estimated by seismic methods to fill as much as 10% of the porewater volume, in a layer about 50 m thick near the bottom of the stability zone (*Bunz and Mienert, 2004*). If these results were typical of the entire. 10^4 km^2 area of the slide, the slide could have released 1- 2 GtC of methane in hydrate (*Paull et al., 1991*).

If 1 GtC CH4 reached the atmosphere all at once, it would raise the atmospheric concentration from today's value of ~1,700 ppb to ~2200 ppb, trapping about 0.25 additional W/m^2 of greenhouse heat, or more, considering indirect feedbacks. The methane radiative forcing would subside over a time scale of a decade or so, as the pulse of released methane was oxidized to CO2, and the atmospheric methane concentration relaxed toward the long-term steady-state value. The radiative impact of the Storegga Landslide would then be somewhat smaller in magnitude but opposite in sign to the eruption of a large volcano, such as theMount. Pinatubo eruption (-2 W/m^2), but it would last longer (10 years for methane and 2 years for a volcano).

It is tantalizing to wonder if there could be any connection between the Storegga Landslide and the 8.2-kyr climate event (*Alley and Agustsdottir, 2005*), which may have been been triggered by freshwater release to the North Atlantic. However, ice cores record a 75-ppb drop in methane concentration during the 8.2-kyr event (*Kobashi et al., 2007*), not a rise. A slowdown of convection in the North Atlantic would have cooled the overlying waters.

Maslin et al. (2004) suggested that an apparent correlation between the ages of submarine landslides in the North Atlantic region and methane variations during the deglaciation supported the hypothesis that clathrate release by this mechanism influenced atmospheric methane. The lack of response for Storegga, by far the largest landslide known, and a relatively weak association of other large slides with increased methane levels (Figure 5.11) suggest that it is unlikely that submarine landslides caused the atmospheric methane variations during this time period.

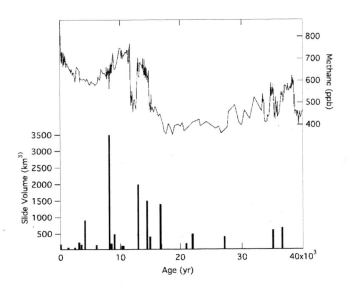

Figure 5.11. Timing of submarine landslides in the North Atlantic region and preindustrial ice core methane variations. Landslide data from *Maslin et al. (2004)*. Methane data from *Brook et al. (2000)* and *Kobashi et al. (2007)*. Abbreviations: km³, cubic kilometers; yr, year; ppb, parts per billion

Much of our knowledge of the Storegga Landslide is due to research sponsored by the Norwegian oil industry, which is interested in tapping the Ormen Lange gas field within the headlands of the Storegga slide but is concerned about the geophysical hazard of gas extraction (*Bryn et al., 2005*). Estimates of potential methane emission from the Storegga slide range from 1 to 5 GtC, which is significant but not apocalyptic. As far as can be determined, the Storegga Landslide had no impact on climate.

4.2.2. The Paleocene-Eocene Thermal Maximum

About 55 million years ago, the $\delta^{13}C$ signature of carbon in the ocean and on land decreased by 2.5-5 per mil (‰) on a time scale of less than 10 kyr, then recovered in parallel on a time scale of ~120-220 kyr (*Kennett and Stott, 1991; Zachos et al., 2001*).

Associated with this event, commonly called the Paleocene-Eocene Thermal Maximum (PETM), the $\delta^{18}O$ of CaCO3 from intermediate depths in the ocean decreased by 2-3‰, indicative of a warming of about 5°C (Figure 5.12). The timing of the spikes is to a large extent synchronous. Planktonic foraminifera and terrestrial carbon records show a $\delta^{13}C$ perturbation a bit earlier than benthic foraminifera do, suggesting that the carbon spike invaded the deep ocean from the atmosphere (*Thomas et al., 2002*). Similar events, also associated with transient warmings have been described from other times in geologic history

(*Hesselbo et al., 2000; Jenkyns, 2003*). The PETM is significant to the present day because it is an analog to the potential fossil fuel carbon release if we burn all the coal reserves.

The change in isotopic composition of the carbon in the ocean is attributed to the release of some amount of isotopically light carbon to the atmosphere. However, it is not clear where the carbon came from, or how much of it there was. The magnitude of the carbon shift depends on where it was recorded. The surface change recorded in CaCO3 in soils (*Koch et al., 1992*) and in some planktonic foraminifera (*Thomas et al., 2002*) is twice as large a change as is reported for the deep sea. Land records may be affected by changes in plant fractionation, driven by changing hydrological cycle (*Bowen et al., 2004*). Ocean records may be affected by $CaCO_3$ dissolution (*Zachos et al., 2005*) resulting in diagenetic imprints on the remaining $CaCO_3$, a necessity to use multiple species, or simple inability to find $CaCO_3$ at all.

We can estimate the change in the carbon inventory of the ocean by specifying an atmospheric partial pressure of CO_2 value (pCO_2), a mean ocean temperature, and insisting on equilibrium with $CaCO_3$ (*Zeebe and Westbroek, 2003*). The ocean was warmer, prior to the PETM event, than it is today. Atmospheric pCO_2 was probably at least 560 ppm at this time (*Huber et al., 2002*). The present-day inventory of CO2 in the ocean is about 40,000 GtC. According to simple thermodynamics, neglecting changes in the biological pump or circulation of the ocean, the geological steady-state inventory for late Paleocene, pre-PETM time could have been on the order of 50,000 GtC.

The lighter the isotopic value of the source, the smaller the amount of carbon that must be released to explain the isotopic shift (Figure 5.12, top). Candidate sources include methane, which can range in its $\delta^{13}C$ isotopic composition from –30 to –1 10%. If the ocean $\delta^{13}C$ value is taken at face value, and the source was methane at –60%, then 2,000 GtC would be required to explain the isotopic anomaly. If the source were thermogenic methane or organic carbon at $\delta^{13}C$ of about –25%, then 10,000 GtC would be required.

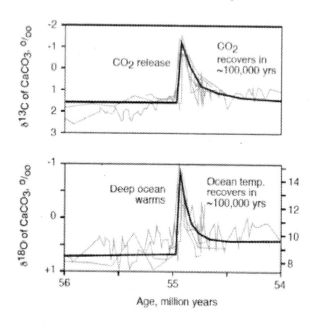

Figure 5.12. Carbon (top) and oxygen (bottom) isotope record for benthic foraminifera from sites in the south Atlantic and western Pacific Oceans for the Paleocene-Eocene Thermal Maximum (PETM), from *Zachos et al. (2001)*, modified by *Archer (2007)*. ‰, per mil

Buffett and Archer (2004) find that the steady-state hydrate reservoir size in the ocean is extremely sensitive to the temperature of the deep sea. At the temperature of Paleocene time but with everything else as in the present-day ocean, they predict less than a thousand GtC of methane in steady state. As the ocean temperature decreases, the stability zone gets thinner and covers less area. Their model was able to fit 6,000 GtC in the Arctic Ocean, however, using 6°C temperatures from CCSM (*Huber et al., 2002*) (which may be too cold) and assuming that the basin had been anoxic (*Sluijs et al., 2006*).

Marine organic matter has an isotopic composition of –20‰ and would require 6,000 GtC to explain the isotopic anomaly. *Svensen et al. (2004)* proposed that lava intrusions into organic-rich sediments could have caused the isotopic shift. They cite evidence that the isotopic composition of methane produced from magma intrusion should be –35 to – 50‰, requiring therefore 2,500-3,500 GtC to explain the isotope anomaly in the deep ocean. If CO_2 were also released, from metamorphism of $CaCO_3$, the average isotopic composition of the carbon spike would be lower, and the mass of carbon greater. *Storey et al. (2007)* showed that the opening of the North Atlantic Ocean and associated igneous intrusions and volcanism correspond in time with the PETM. However, the time scale of carbon release (<10 kyr), indicated by the isotopic shft, is likely more abrupt than one would expect from this kind of volcanic activity. Furthermore, the volcanic activity continued for hundreds of thousands of years, leaving still unexplained the reason for the fast (<10,000 years) carbon isotope excursion.

A comet impact might have played a role in the PETM, and while the isotopic composition of comets is not well constrained, carbon in cometary dust tends to be about – 45‰ (*Kent et al., 2003*). *Kent et al. (2003)* calculate that an 11 km comet containing 20-25% organic matter, a rather large icy tarball, could deliver 200 GtC, enough to decrease the $\delta^{13}C$ of the atmosphere and upper ocean by 0.4‰. It is unlikely that a comet could deliver thousands of GtC, however. An impact strike to a carbonate platform or an organic-rich sediment of some sort could release carbon, but it would take a very large crater to release thousands of gigatons of carbon.

Volcanic carbon has an isotopic composition of –7‰, requiring a huge carbon release of ~20,000 GtC to explain the PETM. Excess carbon emissions have been attributed to superplume cycles in the mantle and flood-basalt volcanic activity (*Larson, 1991*). *Schmitz et al. (2004)* and *Bralower et al. (1997)* find evidence of increased volcanic activity during the PETM interval but view the activity as rearranging ocean circulation, triggering methane release, rather than being a major primary source of carbon itself, presumably because the potential volcanic carbon source is too slow.

Acidification of the ocean by invasion of CO_2 drove a shoaling of the depth of $CaCO_3$ preservation in the Atlantic (*Zachos et al., 2005*) although, curiously, the signal is much smaller in the Pacific (*Zachos et al., 2003*). The magnitude of the carbonate compensation depth (CCD) shift in the Atlantic would suggest a large carbon addition, on the order of 5,000 GtC or more (*Archer et al., 1997*).

A large carbon release is also supported by the warming inferred from the $\delta^{18}O$ spike. The benthic $\delta^{18}O$ record is clearly interpretable as a temperature change, at a depth of several kilometers in the ocean, from about 8° to about 14°C, in a few thousand years. Warming is also implied by Mg/Ca ratios in $CaCO_3$ (*Zachos et al., 2003*) and other tracers (Sluijs et al., 2006; 2007). The temperature can be altered by both CH_4 and CO_2. *Schmidt and Shindell (2003)* calculated that the steady-state atmospheric CH4 concentration during the period of excess emission (ranging from 500-20,000 years) would be enough to explain the temperature change. However, the atmospheric-methane concentration anomaly would decay away a few decades after the excess emission ceased. At this point the temperature anomaly would die away also. Hence, as soon as the carbon isotopic composition stopped plunging negatively, the oxygen isotopic composition should recover as the ocean cools. The carbon isotopic composition meanwhile should remain light for hundreds of thousands of years (*Kump and Arthur, 1999*) until the carbon reservoir isotopic composition reapproached a steady-state value. The record shows instead that the oxygen and carbon isotopic anomalies recovered in parallel (Figure 5.12). This suggests that CO_2 is the more likely greenhouse warmer rather than CH_4. It could be that the time scale for the pCO_2 to reach steady state might be different than the time scale for the isotopes to equilibrate, analogous to the equilibration of the surface ocean by gas exchange: isotopes take longer. However, in the *Kump and Arthur (1999)* model results, pCO_2 seems to take longer to equilibrate than $\delta^{13}C$. The first-order result is that the CO_2 and $\delta^{13}C$ time scales are much more similar than the CH_4 and $\delta^{13}C$ time scales would be.

A warming of 5°C would require somewhere between one and two doublings of the atmospheric CO_2 concentration, if the climate sensitivity is in the range of IPCC predictions of 2.5 – 4.5°C. Beginning from 600 ppm, we would increase the pCO_2 of the atmosphere to somewhere in the range of 1,200 – 2,400 ppm. The amount of carbon required to achieve this value for hundreds of thousands of years (after equilibration with the ocean and with the $CaCO_3$ cycle) would be of order 20,000 GtC. This would imply a mean isotopic composition of the spike of mantle isotopic composition, not isotopically light methane. The amount of carbon required to explain the observed $\delta^{18}O$ would be higher if the initial atmospheric pCO_2 were higher than the assumed 600 ppm. The only way that a biogenic methane source could explain the warming is if the climate sensitivity were much higher in the Paleocene than it seems to be today, which seems unlikely because the ice albedo feedback amplifies the climate sensitivity today (*Pagani et al., 2006*).

The bottom line conclusion about the source of the carbon isotopic excursion is that it is still not clear. There is no clear evidence in favor of a small, very isotopically depleted source of carbon. Mechanistically, it is easier to explain a small release than a large one, and this is why methane has been a popular culprit for explaining the $\delta^{13}C$ shift. Radiative considerations argue for a larger carbon emission, corresponding to a less fractionated source than pure biogenic methane. Thermogenic methane might do, such as the release of somewhat more thermogenic methane than in Gulf of Mexico sediments, if there were a thermogenic deposit that large. Perhaps it was some combination of sources, an initial less-fractionated source such as marine organic matter or a comet, followed by hydrate release.

The PETM is significant to the present day because it is a close analog to the potential fossil fuel carbon release if we burn all the coal reserves. There are about 5,000 GtC in coal, while oil and traditional natural gas deposits are hundreds of Gt each (*Rogner, 1997*). The recovery time scale from the PETM (140 kyr) is comparable to the model predictions, based on the mechanism of the silicate weathering thermostat (400 kyr time scale, *Berner et al., 1983*).

The magnitude of the PETM warming presents an important and currently unanswered problem. A 5,000-GtC fossil fuel release will warm the deep ocean by perhaps 2-4°C, based on paleoclimate records and model results (*Martin et al., 2005*). The warming during the PETM was 5°C, and this was from an atmospheric CO_2 concentration higher than today (at least 600 ppm), so that a further spike of only 2,000 GtC (based on methane isotopic composition) would have only a tiny radiative impact, not enough to warm the Earth by 5°C. One possible explanation is that our estimates for the climate sensitivity are too low by a factor of 2 or more. However, as mentioned above, one might expect a decreased climate sensitivity for an ice-free world rather than for the ice-age climate of today.

Another possible explanation is that the carbon release was larger than 2,000 GtC. Perhaps the global average $\delta^{13}C$ shift was as large as recorded in soils (*Koch et al., 1992*) and some planktonic foraminifera (*Thomas et al., 2002*). The source could have been thermogenic methane, or maybe it was not methane at all but CO2, derived from some organic pool such as sedimentary organic carbon (*Svensen et al., 2004*). At present, the PETM serves as a cautionary tale about the long duration of a release of new CO2 to the atmosphere (*Archer, 2005*). However, our current understanding of the processes responsible for the $\delta^{13}C$ spike is not strong enough to provide any new constraint to the stability of the methane hydrate reservoir in the immediate future.

4.2.3. Santa barbara basin and the clathrate gun hypothesis

Nisbet (2002) and *Kennett et al. (2003)* argue that methane from hydrates is responsible for the deglacial rise in the Greenland methane record between 20,000 and 10,000 years ago, and for abrupt changes in methane at other times (Figure 5.6C). *Kennett et al.(2000)* found episodic negative $\delta^{13}C$ excursions in benthic foraminifera in the Santa Barbara basin, which they interpret as reflecting release of hydrate methane during warm climate intervals. Biomarkers for methanotrophy are found in greater abundance and indicate greater rates of reaction during warm intervals in the Santa Barbara basin (*Hinrichs et al., 2003*) and in the Japanese coastal margin (*Uchida et al., 2004*). *Cannariato and Stott (2004)*, however, argued that these results could have arisen from contamination or subsequent diagenetic overprints. *Hill et al. (2006)* measured the abundance of tar in Santa Barbara basin sediments, argued that tar abundance was proportional to methane emissions, and described increases in tar abundance and inferred destabilization of methane hydrates associated with warming during the last glacial-interglacial transition.

As discussed in Section 1, there are several arguments against the hypothesis of a clathrate role in controlling atmospheric methane during the last glacial period. Perhaps the most powerful so far is that the isotopic ratio of deuterium to hydrogen (D/H) in ice core methane for several abrupt transitions in methane concentration indicates a freshwater source, rather than a marine source, apparently ruling out much of a role for marine hydrate methane release (*Sowers, 2006*). However, the D/H ratio has not yet been measured for the entire ice core record. The timing of the deglacial methane rise was also more easily explained by wetland emissions than by catastrophic methane release (*Brook et al., 2000*). The interhemispheric gradient of methane tells us that the deglacial increase in atmospheric methane arose in part from high northern latitudes (*Dallenbach et al., 2000*), although more work is needed to verify this conclusion because constraining the gradient is analytically difficult. The deglacial methane rise could therefore be attributed at least in part to methanogenesis from decomposition of thawing organic matter from high-latitude wetlands. Regardless of the source of the methane, the climate forcing from the observed methane record (Figure 5.6C and D) is too weak to argue for a dominant role for methane in the glacial cycles (*Brook et al., 2000*).

4.3. Review of model results addressing past and future methane hydrate destabilization

4.3.1. Climate impact of potential release

Probably the most detailed analysis to date of the potential for methane release from hydrates on a century time scale is the study of *Harvey and Huang (1995)*. Their study calculated the inventory of hydrate and the potential change in that inventory with an ocean warming. They treated as a parameter the fraction of methane in bubbles that could escape the sediment column to reach the ocean, and evaluated the sensitivity of the potential methane release to that escaped fraction. Our picture of methane release mechanisms has been refined since 1995, although it remains difficult to predict the fate of methane from melted hydrates. *Harvey and Huang (1995)* did not treat the invasion of heat into the ocean or into the sediment column. Their conclusion was that the radiative impact from hydrate methane will be much smaller than that of CO_2, or even between different scenarios for CO_2 release. The calculation should be redone, but it is unlikely that an updated calculation would change the bottom-line conclusion.

Schmidt and Shindell (2003) showed that the chronic release of methane from a large hydrate reservoir over thousands of years can have a significant impact on global climate. The accumulating CO_2 from the oxidation of the methane also has a significant climate impact. New CO_2 from methane oxidation accumulates in the atmosphere/ocean/terrestrial biosphere carbon pool and persists to affect climate for hundreds of thousands of years (*Archer, 2005*). If a pool of methane is released over a time scale of thousands of years, the climate impact from the accumulating CO_2 concentration may exceed that from the steady-state increase in the methane concentration (*Harvey and Huang, 1995; Dickens, 2001a; Schmidt and Shindell, 2003; Archer and Buffett, 2005*). After the emission stops, methane drops quickly to a lower steady state, while the CO_2 persists.

If hydrates melt in the ocean, much of the methane would probably be oxidized in the ocean rather than reaching the atmosphere directly as methane. This reduces the century time scale climate impact of melting hydrate, but on time scales of millennia and longer the climate impact is the same regardless of where the methane is oxidized. Methane oxidized to CO_2 in the ocean will equilibrate with the atmosphere within a few hundred years, resulting in the same partitioning of the added CO_2 between the atmosphere and the ocean regardless of its origin. The rate and extent to which methane carbon can escape the sediment column in response to warming is very difficult to constrain at present. It depends on the stability of the sediment slope to sliding, and on the permeability of the sediment and the hydrate stability zone's cold trap to bubble methane fluxes.

4.4. Conclusions about potential for abrupt release of methane from marine hydrates

On the time scale of the coming century, it appears likely that most of the marine hydrate reservoir will be insulated from anthropogenic climate change. The exception is in shallow ocean sediments where methane gas is focused by subsurface migration. The most likely response of these deposits to anthropogenic climate change is an increased background rate of chronic methane release, rather than an abrupt release. Methane gas in the atmosphere is a transient species, its loss by oxidation continually replenished by ongoing release. An increase in the rate of methane emission to the atmosphere from melting hydrates would increase the steady-state methane concentration of the atmosphere. The potential rate of methane emission from hydrates is more speculative than the rate from other methane sources such as the decomposition of peat in thawing permafrost deposits, or anthropogenic emission from agricultural, livestock, and fossil fuel industries, but the potential rates appear to be comparable to these sources.

5. Terrestrial Methane Hydrates

There are two sources for methane in hydrates, biogenic production by microbes degrading organic matter in anaerobic environments, and thermogenic production at temperatures above 110°C, typically at depths greater than about 15 km. Terrestrial methane hydrates are primarily biogenic (*Archer, 2007*). They form and are stable under ice sheets (thicker than ~250 m) and within permafrost soils at depths of about 150 to 2,000 m below the surface (*Kvenvolden, 1993; Harvey and Huang, 1995*). Their presence is known or inferred from geophysical evidence (e.g., well logs) on Alaska's North Slope, the Mackenzie River delta (Northwest Territories) and Arctic islands of Canada, the Messoyakha Gas Field and two other regions of western Siberia, and two regions of northeastern Siberia (*Kvenvolden and Lorenson, 2001*). Samples of terrestrial methane hydrates have been recovered from 900 to 1,110 m depth in the Mallik core in the Mackenzie River delta (*Kvenvolden and Lorenson, 2001; Uchida et al., 2002*).

5.1. Terrestrial methane hydrate pool size and distribution

While most methane hydrates are marine, the size of the contemporary terrestrial methane hydrate pool, although unknown, may be large. Estimates range from less than 10 Gt CH4 (*Meyer, 1981*) to more than 18,000 Gt CH4 (*Dobrynin et al., 1981*) (both cited in *Harvey and Huang, 1995*). More recent estimates are 400 Gt CH4 (*MacDonald, 1990*), 800 Gt CH4 (*Harvey and Huang, 1995*), and 4.5-400 GtC; this is a small fraction of the ocean methane hydrate pool size (see Sec. 4).

Terrestrial methane hydrates are a potential fossil energy source. Recovery can come from destabilization of the hydrates by warming, reducing the pressure, or injecting a substance (e.g., methanol) that shifts the stability line (see Box 5.1). The Messoyakha Gas Field in western Siberia, at least some of which lies in the terrestrial methane hydrate stability zone, began producing gas in 1969, and some production is thought to have come from methane hydrates, though methanol injection made this production very expensive (*Kvenvolden, 1993; Krason, 2000*). A more recent review of the geological evidence for methane production from hydrates at Messoyakha by *Collett and Ginsburg (1998)* could not confirm unequivocally that hydrates contributed to the produced gas. Due to low costs of other available energy resources, there had not been significant international industrial interest in hydrate methane extraction from 1970 to 2000 (*Kvenvolden, 2000*), and the fraction of terrestrial methane hydrate that is or will be technically and economically recoverable is not well established. In the United States , the Methane Hydrate Research and Development Act of 2000 and its subsequent 2005 Amendment have fostered the National Methane Hydrates R&D Program, supporting a wide range of laboratory, engineering, and field projects with one focus being on developing the knowledge and technology base to allow commercial production of methane from domestic hydrate deposits by the year 2015, beginning with Alaska's North Slope. Estimates of technically and economically recoverable methane in hydrates are being developed (*Boswell, 2005, 2007*).

5.2. Mechanisms to destabilize terrestrial methane hydrates

Terrestrial methane hydrates in permafrost are destabilized if the permafrost warms sufficiently or if the permafrost hydrate is exposed through erosion (see Box 5.3). Destabilization of hydrates in permafrost by global warming is not expected to be significant over the next few centuries (*Nisbet, 2002*; see Sec. 5.4). *Nisbet (2002)* notes that although a warming pulse will take centuries to reach permafrost hydrates at depths of several hundred meters, once a warming pulse enters the soil/sediment, it continues to propagate downward and will eventually destabilize hydrates, even if the climate has subsequently cooled.

Terrestrial methane hydrates under an ice sheet are destabilized if the ice sheet thins or retreats. The only globally significant ice sheets now existing are on Greenland and Antarctica; maps of the global distribution of methane hydrates do not show any hydrates under either ice sheet (*Kvenvolden, 1993*). It is likely, however, that hydrates formed under Pleistocene continental ice sheets (e.g., *Weitemeyer and Buffett, 2006*; see Sec. 5.3.1).

Terrestrial methane hydrates can also be destabilized by thermokarst erosion (a melt-erosion process) of coastal-zone permafrost. Ice complexes in the soil melt where they are exposed to the ocean along the coast, the land collapses into the sea, and more ice is exposed (*Archer, 2007*). The Siberian coast is experiencing very high rates of coastal erosion (*Shakhova et al., 2005*). Methane hydrates associated with this permafrost become destabilized through this process, and methane is released into the coastal waters (*Shakhova et al., 2005*). Magnitudes of the emissions are discussed below.

De Batist et al. (2002) analyzed seismic reflection data from Lake Baikal sediments, the only freshwater nonpermafrost basin known to contain gas hydrates, and infer that hydrate destabilization is occurring in this tectonically active lacustrine basin via upward flow of hydrothermal fluids advecting heat to the base of the hydrate stability zone. If occurring, this means of destabilization is very unlikely to be important globally, as the necessary geological setting is rare.

Mining terrestrial hydrates for gas production will necessarily destabilize them, but presumably most of this methane will be captured, used, and the carbon emitted to the atmosphere as CO_2.

5.3. Evidence of past release of terrestrial hydrate methane

No direct evidence has been identified of past release of terrestrial hydrate methane in significant quantities. Analyses related to the PETM and clathrate gun hypothesis discussed in Sec. 4 have focused on methane emissions from the larger and more vulnerable marine hydrates. Emissions from terrestrial hydrates may have contributed to changes in methane observed in the ice core record, but there are so far no distinctive isotopic tracers of terrestrial hydrates, as is the case for marine hydrate (*Sowers, 2006*).

5.3.1. Quantity of methane released from terrestrial hydrates in the past

Weitemeyer and Buffett (2006) modeled the accumulation and release of biogenic methane from terrestrial hydrates below the Laurentide and Cordilleran ice sheets of North

America during the last glaciation. Methane was generated under the ice sheet from anaerobic decomposition of buried, near-surface soil organic matter, and hydrates formed if the ice sheet was greater than ~250 m thick. Hydrate destabilization arose from pressure decreases with ice sheet melting/thinning. They simulated total releases for North America of about 40-100 Tg CH_4, with most of the deglacial emissions occurring during periods of glacial retreat during a 500-year interval around 14 kyr before present (BP), and a 2,000-year interval centered on about 10 kyr BP. The highest simulated emission rates (~15-35 Tg CH_4 yr^{-1}) occurred during the dominant period of ice sheet melting around 11-9 kyr BP.

Shakova et al. (2005) measured supersaturated methane concentrations in northern Siberian coastal waters. This supersaturation is thought to arise from degradation of coastal shelf hydrate, hydrate that had formed in permafrost when the shelf was exposed during low sea level of the last glacial maximum. Methane concentrations in the Laptev and East Siberian Seas were supersaturated up to 800% in 2003 and 2500% in 2004. From this and an empirical model of gas flux between the atmosphere and the ocean, they estimated summertime (i.e., ice-free) fluxes of up to 0.4 Mg CH_4 km^{-2} y^{-1} (or 0.4 g CH_4 m^{-2} y^{-1}). They assume that the methane flux from the sea floor is of the same order of magnitude, and may reach 1-1.5 g CH_4 m^{-2} y^{-1}. These fluxes are low compared to wetland fluxes (typically ~1-100 g CH_4 m^{-2} y^{-1}; *Bartlett and Harriss, 1993*), but applied across the total area of shallow Arctic shelf, the total annual flux for this region may be as high as 1-5 Tg CH_4 y^{-1}, depending on degree of oxidation in the seawater. (See Table 5.1 above for global methane emissions by source.)

5.3.2. Climate impact of past methane release from terrestrial hydrates

Most studies of climate impacts from possible past methane hydrate releases have considered large releases from marine hydrates (see Sec. 4 above). It is generally not well known what fraction of the methane released from hydrate destabilization is either trapped in overlying sediments or oxidized to carbon dioxide before reaching the atmosphere *(Reeburgh, 2004)*, and the same considerations are relevant to release from terrestrial sources.

Weitemeyer and Buffett (2006) estimated intervals of 500-2,000 years when methane hydrate destabilization from retreat of the North American ice sheet caused increases of atmospheric methane of 10-200 ppb, with the largest perturbation at 11-9 kyr before present. Any effect of methane oxidation before reaching the atmosphere was ignored; this oxidation would have reduced the impact on the atmospheric methane burden. This atmospheric perturbation is equivalent to about 2-25% of pre-industrial Holocene atmospheric methane burdens, and roughly equivalent to a radiative forcing of 0.002 – 0.1 W m^{-2} (using contemporary values for methane radiative efficiency and indirect effects from *Ramaswamy et al., 2001*).

Thermokarst erosion on the Arctic coast of Siberia is thought to cause hydrate destabilization and emissions of methane that are at most 1% of total global methane emissions *(Shakhova et al. 2005)*, and so this process is very unlikely to be having a large climatic impact.

5.4. Estimates of future terrestrial hydrate release and climatic impact

Harvey and Huang (1995) modeled terrestrial methane hydrate release due to global warming (step function temperature increases of 5°C, 10°C, and 15°C, and the propagation of this heat into hydrate-bearing permafrost). Over the first few centuries the methane release is very small, and after 1,000 years, the cumulative methane release is <1%, 2%, and 5% of the total terrestrial methane hydrate pool size, respectively; by 5,000 years this cumulative release has increased to 3%, 15%, and 30%, respectively. Even 5,000 years after a step function increase in temperature of 15°C, the radiative forcing caused by terrestrial hydrate melting (direct effects of methane plus methane converted to carbon dioxide) was only ~0.3 W/m^2.

Methane release from hydrate destabilization due to decaying ice sheets is unlikely to be substantial unless there are significant hydrate pools under Greenland and/or Antarctica, which does not seem to be the case. Thermoskarst erosion release is the only known present terrestrial hydrate methane source. This process can be expected to continue into the future, and it is very likely that emissions will remain a small fraction of the global methane budget and therefore have a small impact on radiative forcing. However, most recent modeling analyses have focused on marine hydrates (e.g., *Dickens, 2001c; Archer and Buffett, 2005*), and more work on the terrestrial hydrate reservoir is clearly needed.

5.5. Conclusions

No mechanisms have been proposed for the abrupt release of significant quantities of methane from terrestrial hydrates (*Archer, 2007*). Slow and perhaps sustained release from permafrost regions may occur over decades to centuries from mining extraction of methane from terrestrial hydrates in the Arctic (*Boswell, 2007*), over decades to centuries from continued thermokarst erosion of coastal permafrost in Eurasia (*Shakhova et al., 2005*), and over centuries to millennia from the propagation of any warming 100-1,000 m down into permafrost hydrates (*Harvey and Huang, 1995*).

6. Changes in Methane Emissions from Natural Wetlands

6.1. Introduction

Natural wetlands are most extensive at high northern latitudes, where boreal and arctic wetlands contain substantial carbon in peat and are frequently associated with permafrost, and in the tropics, often associated with river and lake floodplains. Annual methane emissions from tropical wetlands are roughly twice that from boreal/arctic wetlands. Globally, wetlands are the largest single methane source to the atmosphere, with recent emission estimates ranging from 100 to 231 Tg CH4 yr^{-1} (*Denman et al., 2007*), constituting more than 75% of the total estimated natural emissions. Variations in wetland distribution and saturation, in response to long-term variations in climate, are therefore thought to have been main determinants for variation in the atmospheric CH4 concentration in the past (*Chappellaz et al., 1990; Chappellaz et al., 1993a,b; Brook et al., 1996, 2000; Delmotte et al., 2004*). Recent interannual variations in methane emissions have been dominated by fluctuations in wetland emissions (*Bousquet et al., 2006*), although biomass burning also plays a significant role.

Methane emissions from natural wetlands are sensitive to temperature and moisture (see below), and thus to climate variability and change. Emissions can also be influenced by anthropogenic activities that impact wetlands such as pollution loading (e.g., *Gauci et al., 2004*), land management (e.g., *Minkkinen et al., 1997*), and water management (e.g., *St. Louis et al., 2000*). While these anthropogenic impacts can be expected to change in the coming decades, they are unlikely to be a source of abrupt changes in methane emissions from natural wetlands, so this section will focus on climate change impacts.

Global climate-model projections suggest that the tropics, on average, and the northern high latitudes are likely to become warmer and wetter during the 21st century, with greater changes at high latitudes (*Chapman and Walsh, 2007; Meehl et al., 2007*). Temperatures in the tropics by 2100 are projected to increase by 2-4°C (*Meehl et al., 2007*). Precipitation in the tropics is expected to increase in East Africa and Southeast Asia, show little change in West Africa and Amazonia, and decrease in Central America and northern South America (*Meehl et al., 2007*).

Warming in the northern high latitudes in recent decades has been stronger than in the rest of the world (*Serreze and Francis, 2006*), and that trend is projected to continue, with multimodel projections indicating that arctic land areas could warm by between 3.5° and 8°C by 2100 (*Meehl et al., 2007*). The northern high latitudes are also expected to see an increase in precipitation by more than 20% in winter and by more than 10% in summer. Climate change of this magnitude is expected to have diverse impacts on the arctic climate system (*A CIA, 2004*), including the methane cycle. Principal among the projected impacts is that soil temperatures are expected to warm and permafrost, which is prevalent across much of the northern high latitudes, is expected to thaw and degrade. Permafrost thaw may alter the distribution of wetlands and lakes through soil subsidence and changes in local hydrological conditions. Since methane production responds positively to soil moisture and summer soil temperature, the projected strong warming and associated landscape changes expected in the northern high latitudes, coupled with the large carbon source (northern peatlands have ~250 GtC as peat within 1 to a few meters of the atmosphere; *Turunen et al., 2002*), will likely lead to an increase in methane emissions over the coming century.

6.2. Factors controlling methane emissions from natural wetlands

Methane is produced as a byproduct of microbial decomposition of organic matter under anaerobic conditions that are typical of saturated soils and wetlands. As this methane migrates from the saturated soil to the atmosphere (via molecular diffusion, ebullition (bubbling), or plant-mediated transport), it can be oxidized to carbon dioxide by microbial methanotrophs in oxygenated sediment or soil. In wetlands, a significant fraction of the methane produced is oxidized by methanotrophic bacteria before reaching the atmosphere (*Reeburgh, 2004*). If the rate of methanogenesis is greater than the rate of methanotrophy and pathways for methane to diffuse through the soil are available, then methane is emitted to the atmosphere. Dry systems, where methanotrophy exceeds methanogenesis, can act as weak sinks for atmospheric methane (see Table 5.1). Methane emissions are extremely variable in space and time, and therefore it is difficult to quantify regional-scale annual emissions (*Bartlett and Harriss, 1993; Melack et al., 2004*). Recent reports of a large source (62-236 Tg $CH4$ yr^{-1}) of methane from an aerobic process in plants (*Keppler et al., 2006*) appear to be in overstated (*Dueck et al., 2007; Wang et al., 2008*).

There are relatively few field studies of methane fluxes from tropical wetlands around the world, but work in the Amazon and Orinoco Basins of South America has shown that methane emissions appear to be most strongly controlled in aquatic habitats by inundation depth and vegetation cover (e.g., flooded forest, floating macrophytes, open water) (*Devol et al., 1990; Bartlett and Harriss, 1993; Smith et al., 2000; Melack et al., 2004*). Wet season (high water) fluxes are generally higher than dry season (low water) fluxes (*Bartlett and Harriss, 1993*).

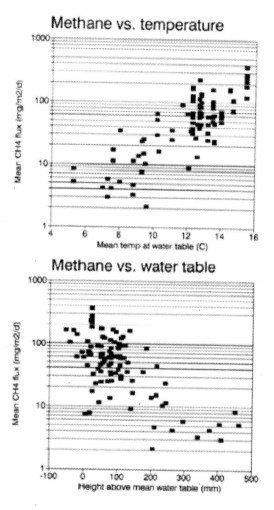

Figure 5.13. Relationships between water table height, temperature, and methane emissions for northern wetlands from *Bubier et al. (1995)*. Abbreviations: mg/m^2/d, milligrams per square meter per day; mm, millimeters; C, degrees Celsius

At high latitudes, the most important factors influencing methane fluxes are water table depth, soil or peat temperature, substrate type and availability, and vegetation type (Figure 5.13). Water table depth determines both the fraction of the wetland soil or peat that is anaerobic and the distance from this zone of methane production to the atmosphere (i.e., the length of the oxidation zone) and is often the single most important factor controlling emissions (*Bubier et al., 1995; Waddington et al., 1996; MacDonald et al., 1998*). The strong sensitivity of CH4 emissions to water table position suggests that changing hydrology of northern wetlands under climate change could drive large shifts in associated methane emissions.

Vegetation type controls plant litter tissue quality/decomposability, methanogen substrate input by root exudation (e.g., *King and Reeburgh, 2002*), and the potential for plant- mediated transport of methane to the atmosphere (e.g., *King et al., 1998; Joabsson and Christensen, 2001*). Substrate type and quality, generally related to quantity of root exudation and to vegetation litter quality and degree of decomposition, can directly affect potential methane production. Vegetation productivity controls the amount of organic matter available for decomposition.

In wetland ecosystems, when the water table is near the surface and substantial methane emissions occur, the remaining controlling factors rise in relevance. *Christensen et al. (2003)* find that temperature and microbial substrate availability together explain almost 100% of the variations in mean annual CH_4 emissions across a range of sites across Greenland, Iceland, Scandinavia, and Siberia. *Bubier et al. (1995)* find a similarly strong dependence on soil temperature at a northern peatland complex in Canada. The observed strong relationship between CH_4 emissions and soil temperature reflects the exponential increase in microbial activity as soil temperatures warm. The strong warming expected across the northern high latitudes is likely to be a positive feedback on methane emissions.

The presence or absence of permafrost can also have a direct influence on CH_4 emissions. Across the northern high latitudes, permafrost features such as ice wedges, ice lenses, thermokarst, and ice heaving determine the surface microtopography. Small variations in surface topography have a strong bearing on plant community structure and evolution as well as soil hydrologic and nutritional conditions (*Jorgenson et al., 2001, 2006*), all of which are controlling factors for methane emission. Projections of future methane emission are hampered by the difficulty of modeling landscape and watershed hydrology well enough at large scales to realistically represent small changes in wetland water table depth.

6.3. Observed and projected changes in Natural Wetlands

6.3.1. Observed changes in Arctic Wetlands and lakes

Increased surface ponding and wetland formation have been observed in warming permafrost regions (*Jorgenson et al., 2001, 2006*). These increases are driven primarily by permafrost-thaw-induced slumping and collapsing terrain features (thermokarst) that subsequently fill with water. For the Tanana Flats region in central Alaska, large-scale degradation of permafrost over the period 1949-95 is associated with substantial losses of birch forest and expansion of wetland fens (*Jorgenson et al., 2001*).

In recent decades, lake area and the number of lakes in discontinuous permafrost regions have decreased in western Siberia (*Smith et al., 2005*) and Alaska (*Riordan et al., 2006*) but have increased in continuous permafrost regions in northwestern Siberia (*Smith et al., 2005*). The differing trends in discontinuous and continuous permafrost zones can be understood if one considers that initial permafrost warming leads to development of thermokarst and lake and wetland expansion as the unfrozen water remains trapped near the surface by the icy soil beneath it. As the permafrost degrades more completely, lake or wetland drainage follows, as water more readily drains through the more ice-free soil to the ground-water system.

A strength of the *Smith et al. (2005)* study is that lake abundance is determined via satellite, permitting the study of thousands of lakes and evaluation of the net change across a broad area, which can in turn be attributed to regional driving mechanisms such as climate and permafrost degradation. A similar analysis for wetlands would be useful but is presently intractable because wetlands are not easy to pinpoint from satellite, as inundation, particularly in forested regions, cannot be easily mapped, and wetland-rich landscapes are often very spatially heterogeneous. (*Frey and Smith, 2007*).

Present-generation global climate or large-scale hydrologic models do not represent the thermokarst processes that appear likely to dictate large-scale changes in wetland extent over the coming century. However, wetland area can also respond to trends in precipitation minus evaporation (P–E). A positive P–E trend could lead, in the absence of large increases in runoff, to an expansion of wetland area and more saturated soil conditions, thereby increasing the area from which methane emission can occur. Most climate models predict that both arctic precipitation and evapotranspiration will rise during the 21st century if greenhouse gas concentrations in the atmosphere continue to rise. In at least one model, the NCAR CCSM3, the P–E trend is positive throughout the 21st century (*Lawrence and Slater, 2005*).

6.3.2. Observed and Projected Changes in Permafrost Conditions

There is a considerable and growing body of evidence that soil temperatures are warming, active layer thickness (ALT) is increasing, and permafrost is degrading at unprecedented rates (e.g., *Osterkamp and Romanovsky, 1999; Romanovsky et al., 2002, Smith et al., 2005; Osterkamp and Jorgenson, 2006*). Continuous permafrost in Alaska, which has been stable over hundreds, or even thousands, of years, has suffered an abrupt increase in degradation since 1982 that "appears beyond normal rates of change in landscape evolution" (*Jorgenson et al., 2006*). Similarly, discontinuous permafrost in Canada has shown a 200-300% increase in the rate of thawing over the 1995-2002 period relative to that of 1941-91 (*Camill, 2005*). *Payette et al. (2004)* present evidence of accelerated thawing of subarctic peatland permafrost over the last 50 years. An example of permafrost degradation and transition to wetlands in the Tanana Flats region of central Alaska is shown in Figure 5.14.

Model projections of soil temperature warming and permafrost degradation in response to the strong anticipated high-latitude warming vary considerably, although virtually all of them indicate that a significant amount of permafrost degradation will occur if the Arctic continues to warm (*Anisimov and Nelson, 1997; Stendel and Christensen, 2002; Zhang et al., 2003; Sazonova et al., 2004*). *Buteau et al. (2004)* find downward thawing rates of up to 13 cm yr^{-1} in ice-rich permafrost for a 5°C warming over 100 years. A collection of process-based models, both global and regional, all with varying degrees of completeness in terms of their representation of permafrost, indicates widespread large-scale degradation of permafrost (and by extension increased thermokarst development), sharply increasing ALTs, and a contraction of the area where permafrost can be found near the Earth's surface during the 21st century (*Lawrence and Slater, 2005; Euskirchen et al., 2006; Lawrence et al., 2008; Saito et al., 2007; Zhang et al., 2007*).

BOX 5.3. HIGH-LATITUDE TERRESTRIAL FEEDBACKS

In recent decades, the Arctic has witnessed startling environmental change. The changes span many facets of the arctic system including rapidly decreasing sea ice extent, melting glaciers, warming and degrading permafrost, increasing runoff to the Arctic Ocean, expanding shrub cover, and important changes to the carbon balance (*Serreze et al., 2000; A CIA, 2004; Hinzman et al., 2005*). The observed environmental trends are driven largely by temperatures that are increasing across the Arctic at roughly twice the rate of the rest of the world (*Serreze and Francis, 2006*). If the arctic warming continues and accelerates, as is predicted by all global climate models (*Chapman and Walsh, 2007*), it may invoke a number of feedbacks that have the potential to alter and possibly accelerate arctic *and* global climate change. If the feedbacks operate constructively, even relatively small changes in the Arctic could conspire to amplify global climate change. Continued environmental change, especially if it occurs rapidly, is likely to have adverse consequences for highly vulnerable arctic and global ecosystems and negative impacts on human activities, including costly damage to infrastructure, particularly in the Arctic..

The Arctic can influence global climate through both positive and negative feedbacks (Figure 5.15). For example, sea-ice retreat reduces surface albedo, enhances absorption of solar radiation, and ultimately leads to greater pan-Arctic warming. Large-scale thawing of permafrost alters soil structural (thermokarst) and hydrologic properties (*Jorgenson et al., 2001*) with additional effects on the spatial extent of lakes and wetlands (*Smith et al., 2005; Riordan et al., 2006*), runoff to the Arctic Ocean, ecosystem functioning (*Jorgenson et al., 2001; Payette et al., 2004*), and the surface energy balance. Warming is also expected to enhance decomposition of soil organic matter, releasing carbon to the atmosphere (a positive feedback) (*Zimov et al., 2006*) and also releasing nitrogen which, in nutrient-limited arctic ecosystems, may prompt shrub growth (a negative feedback due to carbon sequestration) (*Sturm et al., 2001*). This greening-of-the-Arctic negative feedback may itself be offset by a positive radiative feedback related to lower summer and especially winter albedos of shrubs and trees relative to tundra (*Chapin et al., 2005*), which promotes an earlier spring snowmelt that among other things affects soil temperature and permafrost (*Sturm et al., 2001*).

The future of the Arctic as a net sink or source of carbon to the atmosphere depends on the delicate balance between carbon losses through enhanced soil decomposition and carbon gains to the ecosystem related to the greening of the Arctic (*McGuire et al., 2006*). Irrespective of the carbon balance, anticipated increases in methane emissions mean that the Arctic is likely to be an effective greenhouse gas source (*Friborg et al., 2003; McGuire et al., 2006*).

The Arctic is a complex and interwoven system. On the basis of recent evidence of change, it appears that many of these processes are already operating. Whether or not the positive or negative feedbacks will dominate is a critical question facing climate science. In a recent paper reviewing the integrated regional changes in arctic climate feedbacks, *McGuire et al. (2006)* conclude that the balance of evidence indicates that the positive feedbacks to global warming will likely dominate over the next century, but their relationship to global climate change remains difficult to predict, especially since much of the research to date has considered these feedbacks in isolation.

6.4. Observed and modeled sensitivity of wetland methane emissions to climate change

Field studies indicate that methane emissions do indeed increase in response to soil warming and permafrost thaw. *Christensen et al. (2003)* note that a steady rise in soil temperature will enhance methane production from existing regions of methanogenesis that are characterized by water tables at or near the surface. While this aspect is important, changes in landscape-scale hydrology can cause significant change in methane emissions. For example, at a mire in subarctic Sweden, permafrost thaw and associated vegetation changes drove a 22-66% increase in CH_4 emissions over the period 1970 to 2000 (*Christensen et al., 2004*). *Bubier et al. (2005)* estimated that in a Canadian boreal landscape with discontinuous permafrost and ~30% wetland coverage, methane fluxes increased by ~60% from a dry year to a wet year, due to changes in wetland water table depth, particularly at the beginning and end of the summer. *Nykänen et al. (2003)* also found higher methane fluxes during a wetter year at a sub-Arctic mire in northern Finland. *Walter et al. (2006)* found that thawing permafrost along the margins of thaw lakes in eastern Siberia accounts for most of the methane released from the lakes. This emission, which occurs primarily through ebullition, is an order of magnitude larger where there has been recent permafrost thaw and thermokarst compared to where there has not. These hotspots have extremely high emission rates but account for only a small fraction of the total lake area. Methane released from these hotspots appears to be Pleistocene age, indicating that climate warming may be releasing old carbon stocks previously stored in permafrost (*Walter et al., 2006*). At smaller scales, there is strong evidence that thermokarst development substantially increases CH_4 emissions from high-latitude ecosystems. Mean CH_4 emission rate increases between permafrost peatlands and collapse wetlands of 13-fold (*Wickland et al., 2006*), 30-fold (*Turetsky et al., 2002*), and up to 19-fold (*Bubier et al., 1995*) have been reported.

A number of groups have attempted to predict changes in natural wetland methane emissions on a global scale. These studies broadly suggest that natural methane emissions from wetlands will rise as the world warms. *Shindell et al. (2004)* incorporate a linear parameterization for methane emissions, based on a detailed process model, into a global climate model and find that overall wetland methane emissions increased by 121 Tg CH_4 y^{-1}, 78% higher than their baseline estimate. They project a tripling of northern high-latitude methane emissions, and a 60% increase in tropical wetland methane emissions in a doubled CO_2 simulation. The increase is attributed to a rise in soil temperature in combination with wetland expansion driven by a positive P-E trend predicted by the model. About 80% of the increase was due to enhanced flux rates, and 20% due to expanded wetland area or duration of inundation. The predicted increase in the atmospheric methane burden was 1,000 Tg, ~20% of the current total, equivalent to an increase of ~430 ppb, assuming a methane lifetime of 8.9 years. Utilizing a similar approach but with different climate and emission models, *Gedney et al. (2004)* project that global wetland emissions (including rice paddies) will roughly double, despite a slight reduction in wetland area. The northern wetland methane emissions, in particular, increase by 100% (44 to 84 Tg CH4 yr^{-1}) in response to increasing soil temperatures and in spite of a simulated 10% reduction in northern wetland areal extent. Using a more process-based ecosystem model, which includes parameterizations for methane production and emission, *Zhuang et al. (2007)* model a doubling of methane emissions over the 21st century in Alaska, once again primarily in response to the soil temperature influence on methanogenesis, and secondarily to an increase in net primary productivity of Alaskan ecosystems. These factors outweigh a negative contribution to methane emissions related to a simulated drop in the water table. It is important to note that these models simulate only the direct impacts of climate change (altered temperature and moisture regimes, and in one case enhanced vegetation productivity) but not indirect impacts, such as changing landscape hydrology with permafrost degradation and changing vegetation distribution. At this time, it is not known whether direct or indirect effects will have a stronger impact on net methane emissions. These models all predict fairly smooth increases in annual wetland emissions, with no abrupt shifts in flux.

Figure 5.14. Transition from tundra (left, 1978) to wetlands (right, 1998) due to permafrost degradation over a period of 20 years (*Jorgensen et al., 2001*). Photographs, taken from the same location in Tanana Flats in central Alaska, courtesy of NOAA (obtained from http://www.arctic.noaa.gov/detect/land-tundra.shtml).

6.5. Conclusion about potential for abrupt release of methane from wetlands

Tropical wetlands are a stronger methane source than boreal and arctic wetlands and will likely continue to be over the next century, during which fluxes from both regions are expected to increase. However, four factors differentiate northern wetlands from tropical wetlands and make them more likely to experience a larger increase in fluxes: (1) high-latitude amplification of climatic warming will lead to a stronger temperature impact, (2) for regions with permafrost, warming-induced permafrost degradation could make more organic matter available for decomposition and substantially change the system hydrology, (3) the sensitivity of microbial respiration to temperature generally decreases with increasing temperatures (e.g., *Davidson and Janssens, 2006*), and (4) most northern wetlands have substantial carbon as peat. On the other hand, two characteristics of northern peatlands counter this: (1) northern peatlands are complex, adaptive ecosystems, with internal feedbacks and self-organizing structure (*Belyea and Baird, 2007*) that allow them to persist in a relatively stable state for millennia and that may reduce their sensitivity to hydrological change, and (2) much of the organic matter in peat is well-decomposed (e.g., *Frolking et al. 2001*) and may not be good substrate for methanogens.

The balance of evidence suggests that anticipated changes to northern wetlands in response to large-scale permafrost degradation, thermokarst development, a positive P-E trend in combination with substantial soil warming, enhanced vegetation productivity, and an abundant source of organic matter will likely conspire to drive a chronic increase in CH_4 emissions from the northern latitudes during the 21st century. Due to the strong interrelationships between temperature, moisture, permafrost, and nutrient and vegetation change, and the fact that negative feedbacks such as the draining and drying of wetlands are also possible, it is difficult to establish how large the increase will be over the coming century. Current models suggest that a doubling of CH_4 emissions from northern wetlands could be realized fairly easily. However, since these models do not realistically represent all the processes thought to be relevant to future northern high-latitude CH_4 emissions, much larger (or smaller) increases cannot be discounted.

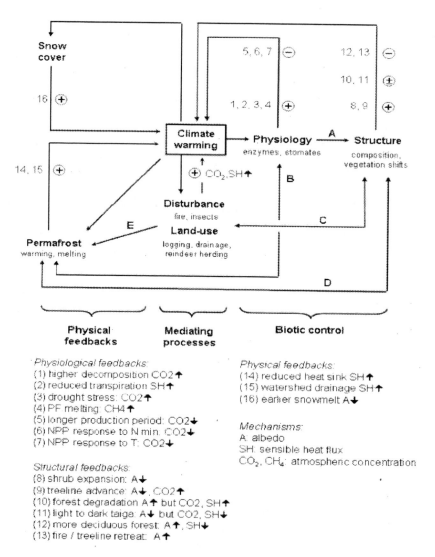

Figure 5.15. Terrestrial responses to warming in the Arctic that influence the climate system. Responses of permafrost on the left are coupled with functional (physiological) and structural biotic responses on the right either directly (arrows B and D) or through mediating processes of disturbance and land use (arrows C and E). Functional and structural biotic responses are also coupled (arrow A). Physical responses will generally result in positive feedbacks. In general, functional responses of terrestrial ecosystems act as either positive or negative feedbacks to the climate system. In contrast, most of the structural responses to warming are ambiguous because they result in both positive and negative feedbacks to the climate system. Abbreviation: NPP, net primary production. Figure adapted from *McGuire et al. (2006)*

It is worth noting that our understanding of the northern high-latitude methane cycle continues to evolve. For example, a recent field study suggests that prior estimates of methane emissions from northern landscapes may be biased low due to an underestimation of the contribution of ebullition from thermokarst hot spots in Siberian thaw lakes (*Walter et al., 2006*). Another recently discovered phenomenon is the cold adaptation of some methanogenic microorganisms that have been found in permafrost deposits in the Lena River basin (*Wagner et al., 2007*). These microbes can produce methane even in the very cold conditions of permafrost, often drawing on old soil organic matter. The activity levels of these cold-adapted methanogens are sensitive to temperature, and even a modest soil warming can lead to an accumulation of methane deposits which, under scenarios where permafrost degradation leads to thermokarst or coastal erosion, could be quickly released to the atmosphere.

These recent studies highlight the fact that key uncertainties remain in our understanding of natural methane emissions from wetlands and their susceptibility to climate change as well as in our ability to predict future emissions. Among the most important uncertainties in our understanding and required improvements to process-based models are (1) the contribution of ebullition and changes in ebullition to total methane emissions; (2) the rate of change in permafrost distribution and active layer thickness and associated changes in distribution of wetlands and lakes as well as, more generally, terrestrial ecosystems; (3) model representation of soil thermal and hydrologic processes and their response to climate change; (4) the contribution that shifts in vegetation and changes in peatland functioning will have on the methane cycle; and (5) representation of the highly variable and regionally specific methane production and emission characteristics. Even with resolution of these issues, all predictions of future methane emissions are based on the accurate simulation and prediction of high-latitude climate. Improvements of many aspects of modeling the high-latitude climate system are required, including improvements to the treatment of snow, polar clouds, subsoil processes, sub-polar oceans, and sea ice in global climate models.

7. Final Perspectives

Although the prospect of a catastrophic release of methane to the atmosphere as a result of anthropogenic climate change over the next century appears very unlikely based on current knowledge, many of the processes involved are still poorly understood, and developing a better predictive capability requires further work. On a longer time scale, methane release from hydrate reservoir is likely to be a major influence in global warming over the next 1,000 to 100,000 years. Changes in climate, including warmer temperatures and more precipitation in some regions, will likely increase the chronic emissions of methane from both melting hydrates and natural wetlands over the next century. The magnitude of this effect cannot be predicted with great accuracy yet, but is likely to be equivalent to the current magnitude of many anthropogenic methane sources, which have already more than doubled the levels of methane in the atmosphere since the start of the Industrial Revolution.

Appendix A. Glossary and Acronyms

Glossary

Ablation Loss of snow and ice, primarily by melting and calving.

Abrupt Climate Change A large-scale change in the climate system that takes place over a few decades or less, persists (or is anticipated to persist) for at least a few decades, and causes substantial disruptions in human and natural systems.

Albedo The fraction of solar radiation reflected by a surface or object, often expressed as a percentage.

Anthropogenic Resulting from or produced by human beings.

Atlantic meridional overturning circulation (AMOC) A northward flow of warm, salty water in the upper layers of the Atlantic, and a southward flow of colder water in the deep Atlantic.

Clathrate A substance in which a chemical lattice or cage of one type of molecule traps another type of molecule.

Climate system The climate system is the highly complex system consisting of five major components: the atmosphere, the hydrosphere, the cryosphere, the land surface and the biosphere, and the interactions between them. The climate system evolves in time under the influence of its own internal dynamics and because of external forcings such as volcanic eruptions, solar variations and anthropogenic forcings such as the changing composition of the atmosphere and land use change.

Climate feedback An interaction mechanism between processes in the climate system is called a climate feedback when the result of an initial process triggers changes in a second process that in turn influences the initial one. A positive feedback intensifies the original process, and a negative feedback reduces it.

Climate model A numerical representation of the climate system based on the physical, chemical and biological properties of its components, their interactions and feedback processes, and accounting for all or some of its known properties.

Climate variability Climate variability refers to variations in the mean state and other statistics (such as standard deviations, the occurrence of extremes, etc.) of the climate on all spatial and temporal scales beyond that of individual weather events. Variability may be due to natural internal processes within the climate system (internal variability), or to variations in natural or anthropogenic external forcing (external variability).

Cryosphere The component of the climate system consisting of all snow, ice and frozen ground (including permafrost) on and beneath the surface of the Earth and ocean.

Downscaling A method that derives local- to regional-scale (10 to 100 km) information from larger scale models or data analyses.

El Niño Southern Oscillation (ENSO) The term El Niño was initially used to describe a warm-water current that periodically flows along the coast of Ecuador and Perú, disrupting the local fishery. It has since become identified with a basin-wide warming of the tropical Pacific Ocean east of the dateline. This oceanic event is associated with a fluctuation of a global-scale tropical and subtropical surface pressure pattern called the Southern Oscillation. This coupled atmosphere-ocean phenomenon, with preferred time scales of two to about seven years, is collectively known as the El Niño-Southern Oscillation (ENSO). It is often measured by the surface pressure anomaly difference between Darwin and Tahiti and the sea surface temperatures in the central and eastern equatorial Pacific. During an ENSO event, the prevailing trade winds weaken, reducing upwelling and altering ocean currents such that the sea surface temperatures warm, further weakening the trade winds. This event has a great impact on the wind, sea surface temperature and precipitation patterns in the tropical Pacific. It has climatic effects throughout the Pacific region and in many other parts of the world, through global teleconnections. The cold phase of ENSO is called La Niña.

Forcing Any mechanism that causes the climate system to change or respond.

Greenhouse gases (GHG) Greenhouse gases are those gaseous constituents of the atmosphere, both natural and anthropogenic, that absorb and emit radiation at specific wavelengths within the spectrum of thermal infrared radiation emitted by the Earth's surface, the atmosphere itself, and by clouds. This property causes the greenhouse effect. Water vapor (H_2O), carbon dioxide (CO_2), nitrous oxide (N_2O), methane (CH_4) and ozone (O_3) are the primary greenhouse gases in the Earth's atmosphere. Moreover, there are a number of entirely human-made greenhouse gases in the atmosphere, such as the halocarbons and other chlorine- and bromine-containing substances, dealt with under the Montreal Protocol. Beside CO_2, N_2O and CH_4, the Kyoto Protocol deals with the greenhouse gases sulphur hexafluoride (SF6), hydrofluorocarbons (HFCs) and perfluorocarbons (PFCs).

Holocene Epoch The geological epoch extending back approximately 11,500 years from the present.

Ice sheet Glaciers of near-continental extent and of which there are at present two, the Antarctic Ice Sheet and the Greenland Ice Sheet.

La Niña The cold phase of the El Niño Southern Oscillation (ENSO).

Mass Balance The net glacier and ice-sheet annual gain or loss of ice/snow.

Medieval Warm Period An interval between AD 900 and 1300 in which some Northern Hemisphere regions were warmer than during the Little Ice Age that followed.

Megadrought Prolonged (multi-decadal) droughts such as those documented for the Medieval Period.

Methane Methane (CH_4) is the second most important greenhouse gas that humans directly influence, carbon dioxide (CO_2) being first.

Methane hydrate A solid in which methane molecules are trapped in a lattice of water molecules. On Earth, methane hydrate forms under high pressure – low temperature conditions in the presence of sufficient methane.

Paleoclimate Climate during periods prior to the development of measuring instruments, including historic and geologic time, for which only proxy climate records are available.

Permafrost Ground (soil or rock and included ice and organic material) that remains at or below 0°C for at least two consecutive years.

Proxy A local record (e.g., pollen, tree rings) that is interpreted, using physical and biophysical principles, to represent some combination of climate-related variations back in time. Climate-related data derived in this way are referred to as proxy data. Examples of proxies include pollen analysis, tree ring records, characteristics of corals and various data derived from ice cores.

Radiative forcing A change in the net radiation at the top of the troposphere caused by a change in the solar radiation, the infrared radiation, or other changes that affect the radiation energy absorbed by the surface (e.g., changes in surface reflection properties), resulting in a radiation imbalance. A positive radiative forcing tends to warm the surface on average, whereas a negative radiative forcing tends to cool it. Changes in GHG concentrations represent a radiative forcing through their absorption and emission of infrared radiation.

Sea level change Sea level can change, both globally and locally, due to (i) changes in the shape of the ocean basins, (ii) changes in the total mass of water and (iii) changes in water density.

Sea surface temperature (SST) The temperature in the top few meters of the ocean, measured by ships, buoys and drifters.

Sink Any process, activity or mechanism that removes a greenhouse gas, an aerosol or a precursor of a greenhouse gas or aerosol from the atmosphere.

Thermohaline circulation (THC) Currents driven by fluxes of heat and fresh water across the sea surface and subsequent interior mixing of heat and salt. The terms Atlantic Meridional Overturning Circulation (AMOC) and Thermohaline Circulation are often used interchangeably but have distinctly different meanings. The AMOC, by itself, does not include any information on what drives the circulation (See AMOC definition above). In contrast, THC implies a specific driving mechanism related to creation and destruction of buoyancy.

Tropopause That area of the atmosphere between the troposphere and the stratosphere.

Acronyms

ABW, AABW	Antarctic Bottom Water
ACC	Antarctic Circumpolar Current
AGCM	Atmospheric General Circulation Model
ALT	active layer thickness
AMO	Atlantic Multidecadal Oscillation
AMOC	Atlantic Meridional Overturning Circulation
AOGCM	Atmosphere-Ocean General Circulation Model
AOVGCM	Atmosphere-Ocean-Vegetation General Circulation Model
AR4	Fourth Assessment Report, IPCC
ATM	Airborne laser altimetry
AVGCM	Atmosphere-Vegetation General Circulation Model
BSR	bottom-simulating reflector
CCD	carbonate compensation depth
CCSM	Community Climate System Model
CCSP	Climate Change Science Program
CLIVAR	Climate Variability and Predictability
COGA	Climatological Ocean Global Atmosphere
COHMAP	Cooperative Holocene Mapping Project
CZCS	Coastal Zone Color Scanner
D/H	Isotopic ratio of deuterium to hydrogen
D-O	Dansgaard-Oeschger
DWF	deep water formation
EDGAR	Emission Database for Global Atmospheric Research
EDMN	EPICA Droning Maud Land
EGVM	Equilibrium Global Vegetation Model
ELMO	Eocene layer of mysterious origin
EMIC	Earth System Model of Intermediate Complexity
ENSO	El Niño/Southern Oscillation
EOF	empirical orthogonal function
EPICA	European Project for Ice Coring in Antarctica
ESRL	Earth System Research Laboratory
GCM	General Circulation Model
GFDL	Geophysical Fluid Dynamics Laboratory
GHCN	Global Historical Climatology Network
GHG	greenhouse gases
GHSZ	Gas Hydrate Stability Zone
GIA	glacial-isostatic adjustment
GIN	Greenland-Iceland-Norwegian
GISP2	Greenland Ice Sheet Project 2
GLIMS	Global Land Ice Measurements from Space
GLSDB	Global Lake Surface Database

GNAIW	Glacial North Atlantic Intermediate Water
GOGA	Global Ocean Global Atmosphere
GRACE	Gravity Recovery and Climate Experiment
GRIP	Greenland Ice Core Project
GSOP	Global Synthesis and Observations Panel
HSZ	Hydrate stability zone
ICEsat	Ice, Cloud, and land Elevation Satellite
InSAR	Interferometric Synthetic Aperture Radar
IPCC	Intergovernmental Panel on Climate Change
IR	infrared
IRD	ice-rafted debris
ISOMIP	Ice Shelf–Ocean Model Intercomparison Project
ITCZ	Intertropical Convergence Zone
LGM	Last Glacial Maximum
LIG	last interglaciation period
LIS	Laurentide Ice Sheet
LSW	Labrador Sea water
mascon	mass concentration
MCA	Medieval Climate Anomaly
MDR	main development region
MIS	Marine Isotope Stage
ML	mixed layer
MOC	Midocean current
MWP	Medieval Warm Period; meltwater pulse
NADA	North American Drought Atlas
NADW	North Atlantic Deep Water
NAM	Northern Annular Mode
NAO	North Atlantic Oscillation, also Northern Annular Mode
NCAR CCM3	National Center for Atmospheric Research Community Climate System Model 3
NOAA	National Oceanic and Atmospheric Administration
NGRIP	North Greenland Ice core Project
NRC	National Research Council
PDB	Pee Dee Belemnite
PDO	Pacific Decadal Oscillation
PDSI	Palmer Drought Severity Index
PETM	Paleocene-Eocene Thermal Maximum
PMIP	Paleoclimate Modeling Intercomparison Project
POGA	Pacific Ocean Global Atmosphere
POGA-ML	Pacific Ocean Global Atmosphere Mixed Layer Ocean
P-E	Precipitation minus evapotranspiration
RCM	Regional Climate Model
RF	radiative forcing
RSL	relative sea level
SAP	Synthesis and Assessment Product
SICI	Small Ice Cap Instability

SLE	sea level equivalent
SLP	sea level pressure
SLR	sea level rise
SLT	sea level equivalent
SMOW	Standard Mean Ocean Water
SRALT	satellite radar altimetry
SST	sea surface temperature
TAGA	Tropical Atlantic Global Atmosphere
THC	Thermohaline Circulation
TNA	Tropical North Atlantic
UNFCCC	United Nations Framework Convention on Climate Change
USGS	U.S. Geological Survey
VOC	Volatile Organic Carbon
WAIS	West Antarctic Ice Sheet
WDCGG	World Data Centre for Greenhouse Gases
WGMS	World Glacier Monitoring Service
WMO	World Meteorological Organization
WOCE	World Ocean Circulation Experiment
20C3M	20th Century Climate in Coupled Models

Abbreviations

a	year
BP	before present
dS/dt	surface elevation change with time
g	gram
G	giga
Gt	gigatonne
GtC	gigatonnes of carbon
hPa	hectoPascal
ka	thousand years (ago)
kyr	thousand years (ago)
kg	kilogram
km	kilometer
m	meter
Mg	megagram
mm	millimeters
Pa	Pascal
pCO_2	atmospheric partial pressure of CO_2
ppb	parts per billion
ppmV	parts per million as measured in volume
ppm	parts per million
PW	petawatt
s	second

s.d.	standard deviation
Sv	sverdrup
T	tera
Tg	teragram
W	watt
yr	year
μm	micrometer
‰	per mil

End Notes

[1] The term "forcing" is used throughout this chapter to indicate any mechanism that causes the climate system to change, or respond. Examples of forcings discussed in this chapter include freshwater forcing of ocean circulation, and changes in sea-surface temperatures and radiative forcing as a forcing of drought. As defined by the IPCC Third Assessment Report (Church et al., 2001), **radiative forcing** refers to a change in the net radiation at the top of the troposphere caused by a change in the solar radiation, the infrared radiation, or other changes that affect the radiation energy absorbed by the surface (e.g., changes in surface reflection properties), resulting in a radiation imbalance. A positive radiative forcing tends to warm the surface on average, whereas a negative radiative forcing tends to cool it. Changes in GHG concentrations represent a radiative forcing through their absorption and emission of infrared radiation.

[2] Refreezing at depth of percolating meltwater in spring and summer, and of retained capillary water during winter. Inability to measure these gains leads to a potentially significant systematic error in the net mass balance.

[3] Note that these values differ from those in the Krabill et al. publications primarily because they take account of possible surface lowering by accelerated snow densification as air temperatures rise; moreover, they probably underestimate total losses because the ATM surveys undersample thinning coastal glaciers.

REFERENCES

[1] Alley, R. B., et al. (2003). Abrupt climate change. *Science*, *299*, 2005-2010.

[2] Archer, D. (2007). Methane hydrate stability and anthropogenic climate change, *J. Geophys. Res. Biogeosciences*, in review.

[3] Berger, A. & Loutre, M. F. (2002). An exceptionally long Interglacial ahead? *Science*, *297*, 1287-1288.

[4] Bindoff, N. L. et al., (2007). Observations: Oceanic climate change and sea level. In: *Climate Change 2007: The Physical Science Basis. Contribution of Working Group I to the Fourth Assessment Report of the Intergovernmental Panel on Climate Change*

[5] Solomon, S., Qin, D., Manning, M., Chen, Z., Marquis, M., Averyt, K. B., Tignor, M. & Miller, H. L. (eds.)]. Cambridge University Press, Cambridge, United Kingdom, *3*, 85-432.

[6] Blunier, T. & Brook, E. J. (2001). Timing of millennial-scale climate change in Antarctica and Greenland during the last glacial period. *Science*, *291*, 109-112.

[7] Boswell, R. (2007). Resource potential of methane hydrate coming into focus. *J. Petrol. Sci. Engin.*, *56*, 9-13.

[8] Broecker, W. S. (1997). Thermohaline circulation, the Achilles heel of our climate system: Will man-made CO2 upset the current balance? *Science*, 278, 1582-1588.

[9] Brook, E. J., Sowers, T. & Orchardo, J. (1996). Rapid variations in atmospheric methane concentration during the past 110 ka. *Science*, *273*, 1087-1091.

[10] Changnon, S. A., Pielke, R. A., Jr., Changnon, D., Sylves, R. T. & Pulwarty, R. (2000). Human factors explain the increased losses from weather and climate extremes. *Bulletin of the American Meteorological Society*, *81*, 437-442.

[11] Church, J. A., et al., (2001). Changes in sea level, in Houghton, J. T., et al., eds., *Climate Change 2001: The Scientific Basis*. Working Group I, Third Assessment Report, Intergovernmental Panel on Climate Change, Cambridge University Press, 639-693

[12] Clark, P. U., Pisias, N. G., Stocker, T. S. & Weaver, A. J. (2002). The role of the thermohaline circulation in abrupt climate change: *Nature*, *415*, 863-869.

[13] Clarke, G. K. C., Leverington, D. W., Teller, J. T. & Dyke, A. S. (2004). Paleohydraulics of the last outburst flood from glacial Lake Agassiz and the 8200 BP cold event. *Quat. Sci. Rev.*, *23*, 3 89-407.

[14] Cook, E. R., Woodhouse, C., Eakin, C. M., Meko, D. M. & Stahle, D. W. (2004). Long-term aridity changes in the western United States. *Science*, *306*, 1015-1018.

[15] Dlugokencky, E. J., Masarie, K. A., Lang, P. M. & Tans, P. P. (1998). Continuing decline inthe growth rate of the atmospheric methane burden. *Nature*, *393*, 447-450.

[16] Emile-Geay, J., Cane, M. A., Seager, R., Kaplan, A. & Almasi, P. (2007). ENSO as a mediator for the solar influence on climate. *Paleoceanography*, *22*, doi:10.1029/ 2006 PA001304.

[17] Etheridge, D. M., Steele, L. P., Francey, R. J. & Langenfelds, R. L. (1998). Atmospheric methane between 1000 A.D. and present: Evidence of anthropogenic emissions and climatic variability. *J. Geophys. Res.*, *103*, 15979-15993.

[18] Fairbanks, R. G. (1989). A 17,000-year glacio-eustatic sea level record: influence of glacial melting dates on the Younger Dryas event and deep ocean circulation. *Nature*, *342*, 637-642.

[19] Gregory, J. M., et al., (2005). A model intercomparison of changes in the Atlantic thermohaline circulation in response to increasing atmospheric CO_2 concentration. *Geophys. Res. Lett.*, *32*, L12703, doi: 10.1029/2005GL023209.

[20] Grootes, P. M., Stuiver, M., White, J. W. C., Johnsen, S. J. & Jouzel, J. (1993). Comparison of oxygen isotope records from the GISP2 and GRIP Greenland ice cores. *Nature*, *366*, 552-554.

[21] Hansen, J. & Sato, M. (2001). Trends of measured climate forcing agents. *Proc. Natl. Acad. Sci.*, 98, 14778-14783, doi:10.1073/pnas.261553698.

[22] Harvey, L. D. D. & Huang, Z. (1995). Evaluation of the potential impact of methane clathrate destabilization on future global warming. *J. Geophysical Res.*, *100*, 2905-2926.

[23] Herweijer, C., Seager, R., Cook, E. R., & J. Emile-Geay, E. R. (2007). North American droughts of the last millennium from a gridded network of tree-ring data. *Journal of Climate*, *20*, 1353-1376.

[24] Hewitt, C. D., Broccoli, A. J., Crucifix, M., Gregory, J. M., Mitchell, J. F. B. & Stouffer, R. J. (2006). The effect of a large freshwater perturbation on the glacial North Atlantic ocean using a coupled general circulation model. *Journal of Climate*, *19*, 4436-4447.

[25] Hodell, D. A., Curtis, J. H. & Brenner, M. (1995). Possible role of climate in the collapse of Classic Maya civilization. *Nature*, *375*, 391-394.

[26] Holland, M. M., Bitz, C. M. & Tremblay, B. (2006). Future abrupt reductions in the summer Arctic sea ice. *Geophys. Res. Lett.*, 33, L23 503, doi: 10.1 029/2006GL028024.

[27] Huber, C., Leuenberger, M., Spahni, R., Fluckiger, J., Schwander, J., Stocker, T. F., Johnsen, S., Landais, A. & Jouzel, J. (2006). Isotope calibrated Greenland temperature record over marine isotope stage 3 and its relation to CH4. *Earth and Planetary Science Letters*, *243*, 504-519.

[28] IPCC, (2007). *Climate change 2007: The physical science basis.* Contribution of Working Group I to the Fourth Assessment Report of the Intergovernmental Panel on Climate Change: Solomon, S., Qin, D., Manning, M., Chen, Z., Marquis, M., Averyt, K. B., Tignor, M. & Miller, H. L. eds., Cambridge University Press, Cambridge, UK, *996*.

[29] Jorgenson, M. T., Racine, C. H., Walters, J. C. & Osterkamp, T. E. (2001). Permafrost degradation and ecological changes associated with a warming climate in central Alaska. *Clim. Change*, *48*, 55 1-579.

[30] Jorgenson, M. T., Shur, Y. L. & Pullman, E. R. (2006). Abrupt increase in permafrost degradation in Arctic Alaska. *Geophys. Res. Lett.*, *33*, L02503, doi: 10.1 029/2005GL024960

[31] Kuhlbrodt, T., Griesel, A., Montoya, M., Levermann, A., Hofmann, M. & Rahmstorf, S. (2007). On the driving processes of the Atlantic meridional overturning circulation. *Rev. Geophys.*, *45*, RG200 1, doi: 10.1 029/2004RG000 166.

[32] Kvenvolden, K. A. (1993). Gas hydrates-geological perspective and global change. *Reviews of Geophysics*, *31*, 173-187.

[33] MacFarling Meure, C., Etheridge, D., Trudinger, C., Steele, P. S., Langenfelds, R., van Ommen, T., Smith, A. & Elkins, J. (2006). Law Dome CO2, CH4, and N2O records extended to 2000 years BP. *Geophysical Research Letters*, *33*, L14810, doi: 10.1 029/2006GL0261 52.

[34] Mann, M. E., Cane, M. A., Zebiak, S. E. & Clement, A. (2005). Volcanic and solar forcing of the tropical Pacific over the past 1000 years. *J. Climate*, *18*, 447-456.

[35] Mao, W. L., Koh, C. A. & Sloan, E. D. (2007). Clathrate hydrates under pressure. *Physics Today*, October, 42-47.

[36] Meehl, G. A., et al., (2007). Global Climate Projections. In: *Climate Change The Physical Science Basis. Contribution of Working Group 1 to the Fourth Assessment Report of the Intergovernmental Panel on Climate Change* In: S., Solomon, D., Qin, M.,Manning, Z., Chen, M., Marquis, K. B., Averyt, M. Tignor, & H. L. Miller, (eds.). Cambridge University Press, Cambridge, United Kingdom, 748- 845.

[37] Meissner, K. J. & Clark, P. U. (2006). Impact of floods versus routing events on the thermohaline circulation. *Geophysical Research Letters*, 33, L26705. Mercer, J., 1978: West Antarctic ice sheet and CO2 greenhouse effect: A threat of disaster. *Nature*, *271*, 321-325.

[38] Newman, M., Compo, G. P. & Alexander, M. A. (2003). ENSO-forced variability of the Pacific decadal oscillation. *J. Climate*, *16*, 3853-3857.

[39] Nicholls, R. J. et al., (2007). Coastal systems and low-lying areas. In: *Climate Change Impacts, Adaptation and Vulnerability. Contribution of Working Group II to the Fourth Assessment Report of the Intergovernmental Panel on Climate Change* In: M. L., Parry, O. F., Canziani, J. P., Palutikof, P. J. van der Linden, & C. E. Hanson (eds.). Cambridge University Press, Cambridge, United Kingdom, 315-356.

[40] NRC, (2002). *Abrupt climate change: Inevitable surprises.* National Academy Press, *230.*

[41] Overpeck, J. T. & Cole, J. E. (2006). Abrupt change in Earth's climate system. *Annu. Rev. Environ. Resour., 31,* 1-3 1.

[42] Peltier, W. R., Vettoretti, G. & Stastna, M. (2006). Atlantic meridional overturning and climate response to Arctic Ocean freshening. *Geophysical Research Letters, 33,* 10.1029/2005GL025251.

[43] Pielke, R. A. & Landsea, C. W. (1998). Normalized hurricane damages in the United States, 1925-1995. *Weather and Forecasting,* September, 621-631.Rahmstorf, S., A semi-empirical approach to projecting future sea-level rise. *Science, 315,* 368-370.

[44] Ridley, J. K., Huybrechts, P., Gregory, J. M. & Lowe, J. A. (2005). Elimination of the Greenland ice sheet in a high CO2 climate. *Journal of Climate, 18,* 3409-3427.

[45] Ross, T. & Lott, N. (2003). A climatology of 1980-2003 extreme weather and climate events. National Climatic Data Center Technical Report No. 2003-01, 14.

[46] Rowley, R. J., Kostelnick, J. C., Braaten, D., Li, X. & Meisel, J. (2007). Risk of rising sea level to population and land area. Eos, *88,* 105-116.

[47] Sachs, J. P. & Anderson, R. F. (2005). Increased productivity in the subantarctic ocean during Heinrich events. *Nature, 434,* 1118-1121.

[48] Schulz, H., von Rad, U. & Erlenkeuser, H. (1998). Correlation between Arabian Sea and Greenland climate oscillations of the past 110,000 years. *Nature, 393,* 54-57.

[49] Seager, R., Ting, M., Held, I., Kushnir, Y., Lu, J., Vecchi, G., Huang, H. P., Harnik, N., Leetmaa, A., Lau, N. C., Li, C., Velez, J. & Naik, N. (2007). Model projections of an imminent transition to a more arid climate in southwestern North America. *Science, 316,* 1181-1184.

[50] Shakhova, N., Semiletov, I. & Panteleev, G. (2005). The distribution of methane on the Siberian Arctic shelves: Implications for the marine methane cycle. *Geophysical Research Letters, 32.*

[51] Shin, S. I., Sardeshmukh, P. D., Webb, R. S., Oglesby, R. J. & Barsugli, J. J. (2006). Understanding the mid-Holocene climate. *Journal of Climate, 19,* 2801-2818.

[52] Spence, J. P., Eby, M. & Weaver, A. J. (2008). The sensitivity of the Atlantic meridional overturning circulation to freshwater forcing at eddy-permitting resolutions. *Journal of Climate,* in press.

[53] Stahle, D. W., Fye, F. K., Cook, E. R. & Griffin, R. D. (2007). Tree-ring reconstructed megadroughts over North America since AD 1300. *Climatic Change,* doi 10.1007/s1 0584-006-9171 -x (in press).

[54] Stouffer, R. J. & others, (2006). Investigating the causes of the response of the thermohaline circulation to past and future climate changes: *Journal of Climate, 19,* 1365-1387.

[55] Stroeve, J., Holland, M. M., Meier, W., Scambos, T. & Serreze, M. (2007). Arctic sea ice decline: Faster than forecast, Geophys. Res. Lett., 34, L09501, doi: 10.1 029/2007GL029703.

[56] Tang, Y. M. & Roberts, M. J. (2005). The impact of a bottom boundary layer scheme on the North Atlantic Ocean in a global coupled climate model. *Journal of Physical Oceanography, 35,* 202-2 17.

[57] Thorpe, R. B., Wood, R. A. & Mitchell, J. F. B. (2004). The sensitivity of the thermohaline circulation response to preindustrial and anthropogenic greenhouse gas

forcing to the parameterization of mixing across the Greenland-Scotland ridge. *Ocean Modelling*, 7, 259-268.

[58] Vellinga, M. A. & Wood, R. A. (2007). Impacts of thermohaline circulation shutdown in the twenty-first century. *Clim. Change*, doi: 10.1007/s10584-006-9146-y. In: A. J. Weaver, & C. Hillaire-Marcel, (2004). Global warming and the next ice age. *Science*, 304, 400-402.

[59] Webb, T., III, Bartlein, P. J., Harrison, S. P. & Anderson, K. H. (1993). Vegetation, lake levels, and climate in eastern North America for the past 18,000 years. In: *Global Climates since the Last Glacial Maximum* In: H. E., Wright, Jr., J. E., Kutzbach, T., Webb III, W. F., Ruddiman, F. A. Street-Perrott, & P. J. Bartlein (eds.). University of Minnesota Press, Minneapolis.

[60] Weiss, H. & Bradley, R. S. (2001). What drives societal collapse? *Science*, 291, 609-610.

[61] Willebrand, J., Barnier, B., Boning, C., Dieterich, C., Killworth, P. D., Le Provost, C., Jia, Y., Molines, J. M. & New, A. L. (2001). Circulation characteristics in three eddy-permitting models of the North Atlantic. *Progress in Oceanography*, 48, 123-161.

[62] Williams, J. W., Shuman, B. N., Webb, T. III, Bartlein, P. J. & Leduc, P. L. (2004). Late- Quaternary vegetation dynamics in North America: Scaling from taxa to biomes. *Ecological Monographs*, 74, 309-334.

[63] Yancheva, G., Nowaczyk, N. R., Mingram, J., Dulski, P., Schettler, G., Negendank, J. F. W., Liu, J., Seligman, D. M., Peterson, L. C. & Haug, G. H. (2007). Influence of the intertropical convergence zone on the East Asian monsoon. *Nature*, 445, 74-77.

[64] Yokoyama, Y., Lambeck, K., De Deckker, P., Johnson, P. & Fifield, K. (2000). Timing for the maximum of the last glacial constrained by lowest sea-level observations. *Nature*, 406, 713–716.

[65] Zwally, H. J., Abdalati, W., Herring, T., Larson, K., Saba, J. & Steffen, K. (2002). Surface melt-induced acceleration of Greenland ice-sheet flow. *Science*, 297, 218-222. SAP 3.4: Abrupt Climate Change

[66] Abdalati, W., Krabill, W., Frederick, E., Manizade, S., Martin, C., Sonntag, J., Swift, R., Thomas, R., Wright, W. & Yungel, J. (2001). Outlet glacier and margin elevation changes: Near-coastal thinning of the Greenland ice sheet. *Journal of Geophysical Research*, 106(D24), 33,729-33,741.

[67] Ackert, R. P., Mukhopadhyay, S., Parizek, B. R. & Borns, H. W. (2007). Ice elevation near the West Antarctic Ice Sheet divide during the last glaciation. *Geophysical Research Letters*, 34, doi: 10.1 029/2007GL03 1412.

[68] Alley, R.B., S. Anandakrishnan, T.K.S. Dupont, Parizek, B.R. & D. Pollard, D., 2007: Effect of sedimentation on ice-sheet grounding-line stability. *Science*, 315, 1838- 1841.

[69] Anandakrishnan, S., Catania, G. A., Alley, R. B., Clark, R. B., Huybrechts, P. U. P. & Joughin, H. J. I. (2005). Ice-sheet and sea-level changes. *Science*, 310, 456-460.

[70] Arendt, A. A., Echelmeyer, K .A. W.D. Harrison, W. D., Lingle, C. S. & Valentine, V. B. (2002). Rapid wastage of Alaska glaciers and their contribution to rising sea level. *Science*, 297, 382-3 86.

[71] Arnold, N. & Sharp, M. (2002). Flow variability in the Scandinavian Ice Sheet: Modeling the coupling between ice sheet flow and hydrology. *Quaternary Science Reviews*, 21, 485-502.

[72] Arthern, R., Winebrenner, D. & Vaughan, D. (2006). Antarctic snow accumulation mapped using polarization of 4.3-cm wavelength microwave emission. *Journal of Geophysical Research, 111*, D06107.

[73] Arthern, R. & Wingham, D. (1998). The natural fluctuations of firn densification and their effect on the geodetic determination of ice sheet mass balance. *Climate Change, 40*, 605-624.

[74] Bales, R., McConnell, J., Mosley-Thompson, E. & Csatho, B. (2001). Accumulation over the Greenland ice sheet from historical and recent records. *Journal of Geophysical Research, 106*, 33813-33825.

[75] Bamber, J. L., Alley, R. B. & Joughin, I. (2007). Rapid response of modern day ice sheets to external forcing. *Earth and Planetary Science Letters, 257*, 1-13.

[76] Bard, E., et al., (1996). Deglacial sea level record from Tahiti corals and the timing of global meltwater discharge. *Nature, 382*, 241–244.

[77] Barletta, V. R., Sabadini, R. & Bordoni, A. (2008). Isolating the PGR signal in the GRACE data: Impact on mass balance estimates in Antarctica and Greenland. *Geophys. J. Int., 172*, 18-30.

[78] Bassett, S. E., Milne, G. A., Bentley, M. J. & Huybrechts, P. (2007). Modelling Antarctic sea-level data to explore the possibility of a dominant Antarctic contribution to meltwater pulse IA. *Quaternary Science Reviews* (in press).

[79] Bassett, S. E., Milne, G. A., Mitrovica, J. X. & Clark, P. U. (2005). Ice sheet and solid earth influences on far-field sea-level histories. *Science, 309*, 925-928.

[80] Bindschadler, R. A. (1983). The importance of pressurised subglacial water in separation and sliding at the glacier bed. *Journal of Glaciology, 29*, 3-19.

[81] Blunier, T. & Brook, E. J. (2001). Timing of millennial-scale climate change in Antarctica and Greenland during the last glacial period. *Science, 291*, 109-112.

[82] Boon, S. & Sharp, M. J. (2003). The role of hydrologically-driven ice fracture in drainage system evolution on an Arctic glacier. *Geophysical Research Letters, 30(18)*, 1916, doi: 10.1029/2003GL01 8034.

[83] Box, J. E., Bromwich, D. H., Veenhuis, B. A., Bai, L. S., Stroeve, J. C., Rogers, J. C., Steffen, K., Haran, T. & Wang, S. H. (2006). Greenland ice-sheet surface mass balance variability (1988-2004) from calibrated Polar MM5 output. *Journal of Climate, 19(12)*, 2783–2800.

[84] Bromwich, D. H., Guo, Z., Bai, L. & Chen, Q. S. (2004). Modeled Antarctic precipitation. Part I: Spatial and temporal variability. *Journal of Climate, 17(3)*, 427-447. C. S., Brown, M. F., Meier, & Post, A. (1983). *Calving speed of Alaska tidewater glaciers, with application to Columbia Glacier.* U.S. Geological Survey Professional Paper 1258-C.

[85] Chappell, J. (2002). Sea level changes forced ice breakouts in the Last Glacial cycle: New results from coral terraces. *Quaternary Science Reviews, 21*, 1-8.

[86] Chen J. L., Wilson, C. R., Blankenship, D. D. & Tapley, B. D. (2006). Antarctic mass rates from GRACE. *Geophysical Research Letters*, 33, L1 1502, doi: 10.1 029/2006GL026369.

[87] Church, J. A., Gregory, J. M. et al., (2001). Changes in sea level. *In: Climate Change The Scientific Basis. Contribution of Working Group I to the Third Assessment Report of the Intergovernmental Panel on Climate Change* [Houghton, J.T., et al. (eds.)].

Cambridge University Press, Cambridge, United Kingdom, *63*, 9-693. Clark, P.U., A.M. McCabe,

[88] Mix, A. C. & Weaver, A. J. (2004). The 19-kyr B. P. meltwater pulse and its global implications. *Science, 304*, 1141-1144.

[89] Clark, P. U., Mitrovica, J. X., Milne, G. A. & Tamisiea, M. (2002). Sea-level fingerprinting as a direct test for the source of global meltwater pulse IA. *Science, 295*, 2438-2441.

[90] Clark, P. U. & Mix, A. C. (2002). Ice sheets and sea level of the last glacial maximum. *Quaternary Science Reviews, 21*, 1229-1240.

[91] Clark, P. U., Alley, R. B., Keigwin, L. D., Licciardi, J. M., Johnsen, S. J. & Wang, H. (1996). Origin of the first global meltwater pulse following the last glacial maximum. *Paleoceanography, 11*, 563-577.

[92] Clarke, G. K. C. (1987). Fast glacier flow: ice streams, surging, and tidewater glaciers. *Journal of Geophysical Research, 92*, 8835-8841.

[93] Conway, H., Hall, B. L., Denton, G. H., Gades, A. M. & Waddington, E. D. (1999). Past and future grounding line retreat of the west Antarctic Ice Sheet. *Science, 286*, 280-283.

[94] Copland, L., Sharp, M .J. & Nienow, P. (2003). Links between short-term velocity variations and the subglacial hydrology of a predominantly cold polythermal glacier. *Journal of Glaciology, 49*, 337-348.

[95] Cuffey, K. M. & Marshall, S. J. (2000). Substantial contribution to sea-level rise during the last interglacial from the Greenland ice sheet. *Nature, 404*, 591-594.

[96] Cutler, K. B., et al., (2003). Rapid sea-level fall and deep-ocean temperature change since the last interglacial period. *Earth and Planetary Science Letters, 206*, 253-271.

[97] Davis, H. C., Li, Y., McConnell, J., Frey, M. M. & Edward Hanna, (2005). Snowfall-driven growth in east Antarctic ice sheet mitigates recent sea-level rise. *Science, 308(5730)*, 1898–1901.

[98] DeAngelis, H. & Skvarca, P. (2003). Glacier surge after ice shelf collapse. *Science, 299*, 1560-1562.

[99] DeConto, R. & Pollard, D. (2003). Rapid Cenozoic glaciation of Antarctica induced by declining atmospheric CO_2. *Nature, 421*, 245-249.

[100] Denton, G. H., Alley, R. B., Comer, G. C. & Broecker, W. S. (2005). The role of seasonality in abrupt climate change. *Quaternary Science Reviews, 24*, 1159-1182.

[101] Dyke, A. S. (2004). An outline of North American deglaciation with emphasis on central and northern Canada. In: *Quaternary Glaciations-Extant and Chronology* In: J. Ehlers, & P. L. Gibbard, (eds.). Elsevier Science and Technology Books, Amsterdam, Part II, 2b, 373-424.

[102] Edwards, R. L., et al., (1993). A large drop in atmospheric $^{14}C/^{12}C$ and reduced melting in the Younger Dryas, documented with ^{230}Th ages of corals. *Science, 260*, 962-968.

[103] Fahnestock, M. A., Abdalati, W. & Shuman, C. (2002). Long melt seasons on ice shelves of the Antarctic Peninsula: An analysis using satellite based microwave emission measurements. *Annals of Glaciology, 34*, 127-133.

[104] Fairbanks, R. G., (1989). A 17,000-year glacio-eustatic sea level record: Influence of glacial melting dates on the Younger Dryas event and deep ocean circulation. *Nature, 342*, 637-642.

[105] Fairbanks, R. G., Charles, C. D. & Wright, J. D. (1992). Origin of global meltwater pulses, *in Radiocarbon after four decades* [Taylor, R.E., et al. (eds.)]. New York, Springer-Verlag, 473–500.

[106] Flowers, G. & Clarke, G. (2002). A multi-component coupled model of glacier hydrology. *Journal of Geophysical Research,* 107-B11, 2287.

[107] Fricker, H. A., Young, N. W., Coleman, R., Bassis, J. N. & Minster, J. B. (2005). Multi-year monitoring of rift propagation on the Amery Ice Shelf, East Antarctica. *Geophysical Research Letters, 32,* L02502, doi: 10. 1029/2004GL02 1036.

[108] Giovinetto, M. B. & Zwally, J. (2000). Spatial distribution of net surface accumulation on the Antarctic ice sheet. *Annals of Glaciology, 31,* 171-178.

[109] Gregory, J. M., Huybrechts, P. & Raper, S. C. B. (2004). Threatened loss of the Greenland ice sheet. *Nature, 428,* 616.

[110] Grosfeld, K. & Sandhager, H. (2004). The evolution of a coupled ice shelf–ocean system under different climate states. *Global and Planetary Change, 42,* 107-132.

[111] Haeberli, W., Zemp, M., Hoelzle, M., Frauenfelder, R., Hoelzle, M. & Kääb, A. (2005). *Fluctuations of glaciers,* 1995-2000 (vol. *VIII*). International Commission on Snow and Ice of International Association of Hydrological Sciences/UNESCO, Paris. http://www.geo.unizh.ch/wgms.

[112] Hanebuth, T., Stattegger, K. & Grootes, P. M. (2000). Rapid flooding of the Sunda Shelf: A late-glacial sea-level record. *Science, 288,* 1033-1035.

[113] Hanna, E., Huybrechts, P., Janssens, I., Cappelen, J., Steffen, K. & Stephens, A. (2005). Runoff and mass balance of the Greenland ice sheet: 1958-2003. *Journal of Geophysical Research, 110,* D13 108, doi: 10. 1029/2004JD005641.

[114] Heroy, D. C. & Anderson, J. B. (2007). Radiocarbon constraints on Antarctic Peninsula Ice Sheet retreat following the last glacial maximum (LGM). *Quaternary Science Reviews, 26,* 1235-1247.

[115] Higgins, A. K. (1991). North Greenland glacier velocities and calf ice production. *Polarforschung, 60,* 1-23.

[116] Hock, R. (2003). Temperature index melt modeling in mountain areas. *Journal of Hydrology, 282,* 104-115, doi:10.1016/S0022-1694(03)00257-9.

[117] Holland, D. M., deYoung, B., Bachmayer, R. & Thomas, R. (2007a). *Ocean observations at Jakobshavn, abstract. XVI* Annual Meeting of the West Antarctic Ice Sheet Initiative.

[118] Holland, D. M., Jacobs, S. S. & Jenkins, A. (2003). Modeling Ross Sea ice shelf - ocean interaction. *Antarctic Science, 15,* 13-23.

[119] Holland, P. R., Jenkins, A. & Holland, D. M. (2007b). The nonlinear response of ice-shelf basal melting to variation in ocean temperature. (accepted, *Journal of Climate*).

[120] Horton, R., Herweijer, C., Rosenzweig, C., Liu, J., Gomitz, V. & Ruane, A. C. (2008). Sea level rise projections for current generation GCMs based on semi-empirical method: *Geophysical Research Letters, 35,* L027 15, doi: 10.1 029/2007GL032486.

[121] Horwath, M. & Dietrich, R. (2006). Errors of regional mass variations inferred from GRACE monthly solutions. *Geophysical Research Letters, 33,* L07502, doi: 10.1 029/2005GL0255550.

[122] Howat, I. M., Joughin, I., Tulaczyk, S. & Gogineni, S. (2005). Rapid retreat and acceleration of Helheim Glacier, east Greenland. *Geophysical Research Letters, 32*, (L22502).

[123] Hughes, T. J. (1973). Is the West Antarctic ice-sheet disintegrating. *Journal of Geophysical Research, 78*, 7884-7910.

[124] Huybrechts P. (2002). Sea-level changes at the LGM from ice-dynamic reconstructions of the Greenland and Antarctic ice sheets during the glacial cycles. *Quaternary Science Reviews, 21(1-3)*, 203-231.

[125] Huybrechts P., Gregory, J., Janssens, I. & Wild, M. (2004). Modelling Antarctic and Greenland volume changes during the 20th and 21st centuries forced by GCM time slice integrations. *Global and Planetary Change, 42(1-4)*, 83-105.

[126] Iken, A. & Bindschadler, R. A. (1986). Combined measurements of subglacial water pressure and surface velocity of Findelengletscher, Switzerland: Conclusions about drainage system and sliding mechanism. *Journal of Glaciology, 32*, 101- 119.

[127] ISOMIP Group, (2007). Ice Shelf-Ocean Model Intercomparison Project website. http://efdl.cims.nyu.edu/project_oisi/isomip/

[128] Ivins, E. R. & James, T. R. (2005). Antarctic glacial isostatic adjustment: A new assessment: *Antarctic Science, 17*, 537-549.

[129] Jacka, T., et al., (2004). Recommendations for the collection and synthesis of Antarctic Ice Sheet mass balance data. *Global and Planetary Change, 42(1-4)*, 1-15.

[130] Jacobs, S. S. & Giulivi, C. (1998). Interannual ocean and sea ice variability in the Ross Sea. *Antarctic Research Series, 75*, 135-150.

[131] Jacobs, S. S., Hellmer, H. H. & Jenkins, A. (1996). Antarctic ice sheet melting in the Southeast Pacific. *Geophysical Research Letters, 23(9)*, 957-960, doi: 10.1029/96GL00723.

[132] Jenkins, A. & Doake, C. S. M. (1991). Ice-ocean interaction on Ronne Ice Shelf, Antarctica. *Journal of Geophysical Research, 96(d1)*, 791-813, doi: 10.1 029/90JC01 952.

[133] Jezek, K. C., Gogineni, P. & Shanableh, M. (1994). Radar measurements of melt zones on the Greeland Ice Sheet. *Geophysical Research Letters, 21(1)*, 33-36, doi: 10.1029/ 93GL03377.

[134] Johannessen, O., Khvorostovsky, K., Miles, M. & Bobylev, L. (2005). Recent ice-sheet growth in the interior of Greenland. *Science, 310*, 1013-1016.

[135] Joughin, I., Abdalati, W. & Fahnestock, M. (2004). Large fluctuations in speed on Greenland's Jakobshavn Isbræ glacier. *Nature, 432*, 608-6 10.

[136] Joughin, I. & Bamber, J. (2005). Thickening of the Ice Stream Catchments Feeding the Filchner-Ronne Ice Shelf, Antarctica. *Geophysical Research Letters, 32*, L1 7503, doi: 10.1 029/2005GL023 844.

[137] Joughin, I., et al., (2003). Timing of recent accelerations of Pine Island Glacier, Antarctica. *Geophysical Research Letters, 30(13)*, 1706, 39-1–39-4.

[138] Joughin I., Tulaczyk, S., Bindschadler, R. & Price, S. F. (2002). Changes in west Antarctic ice stream velocities: Observation and analysis. *Journal of Geophysical Research*, 107(B11), 2289.

[139] Joughin I. & Tulaczyk, S. (2002). Positive mass balance of the Ross Ice Streams, West Antarctica. *Science, 295(5554)*, 476-480.

[140] Kamb, B. (1991). Rheological nonlinearity and flow instability in the deforming bed mechanism of ice stream motion. *Journal of Geophysical Research, 96*, 16585-16595.

[141] Kaser, G., Cogley, J. G., Dyurgerov, M. B., Meier, M. F. & Ohmura, A. (2006). Mass balance of glaciers and ice caps: consensus estimates for 1961-2004. *Geophysical Research Letters, 33*, L19501. doi:10.1029/2006GL02751 1.

[142] Kawamura, K., et al., (2007). Northern Hemisphere forcing of climatic cycles in Antarctica over the past 360,000 years. *Nature, 448*, 912-917.

[143] Koerner, R. M., (1989). Ice-core evidence for extensive melting of the Greenland Ice Sheet in the last interglacial. *Science, 244*, 964-968.

[144] Krabill, W., et al., (2000). Greenland Ice Sheet: High-elevation balance and peripheral thinning. *Science, 289*, 428-430.

[145] Krabill, W., et al., (2002). Aircraft laser altimetry measurement of elevation changes of the Greenland Ice Sheet: Technique and accuracy assessment. *Journal of Geodynamics, 34*, 357-376.

[146] Krabill, W., et al., (2004). Greenland Ice Sheet: increased coastal thinning. *Geophysical Research Letters, 31*, L24402, doi: 10.1 029/2004GL02 1533.

[147] Larsen, C. F., Motyka, R. J., Arendt, A. A., Echelmeyer, K. A. & Geissler, P. E. (2007). Glacier changes in southeast Alaska and northwest British Columbia and contribution to sea level rise. *Journal of Geophysical Research, 112*, F01007, doi: 10.1 029/2006JF000586.

[148] Lemke, P., Ren, J., Alley, R. B., Allison, I., Carrasco, J., Flato, G., Fujii, Y., Kaser, G., Mote, P., Thomas, R. H. & Zhang, T. (2007). Observations: changes in snow, ice and frozen ground. In: *Climate Change 2007: The Physical Science Basis. Contribution of Working Group I to the Fourth Assessment Report of the Intergovernmental Panel on Climate Change.* In: S. Solomon, et al. (eds.). Cambridge University Press, Cambridge, United Kingdom.

[149] Levitus, S., Antonov, J. I., Boyer, T. P. & Stephens, C. (2000). Warming of the World Ocean. *Science, 287(5461)*, 2225–2229, doi: 10.11 26/science.287.546 1.2225. In: E. L. Lewis, & R. G. Perkins, (1986). Ice pumps and their rates. *Journal of Geophysical Research, 91(1986)*, 11756–11762.

[150] Li, J. & Zwally, J. (2004). Modeling the density variation in shallow firn layer. *Annals of Glaciology, 38*, 309-3 13.

[151] Luthcke, S. B., Zwally, H. J., Abdalati, W., Rowlands, D. D., Ray, R. D., Nerem, R. S., Lemoine, F. G., McCarthy, J. J. & Chinn, D. S. (2006). Recent Greenland ice mass loss by drainage system from satellite gravity observations. *Science, 314(5803)*, 1286- 1289.

[152] MacAyeal, D. R. (1989). Large-scale flow over a viscous basal sediment: Theory and application to Ice Stream B, Antarctica. *Journal of Geophysical Research, 94*, 4071-4087.

[153] Mackintosh, A., White, D., Fink, D., Gore, D. B., Pickard, J. & Fanning, P. C. (2007). Exposure ages from mountain dipsticks in Mac. Robertson Land, East Antarctica, indicate little change in ice-sheet thickness since the last glacial maximum: *Geology*, 35, 55 1-554.

[154] Mair, D., Nienow, P., Willis, I. & Sharp, M. (2001). Spatial patterns of glacier motion during a high-velocity event: Haut Glacier d'Arolla, Switzerland. *Journal of Glaciology., 47*, 9-20.

[155] Meier, M. F., Dyurgerov, M. B., Rick, U. K., O'Neel, S., Pfeffer, W. T., Anderson, R. S., Anderson, S. P. & Glazovskiy, A. F. (2007). Glaciers dominate eustatic sea-level rise in the 21st century. *Science, 317,* 1064-1067.

[156] Meehl, G. A., Stocker, T. F., Collins, W. D., Friedlingstein, P., Gaye, A. T., Gregory, J. M., Kitoh, A., Knutti, R., Murphy, J. M., Noda, A., Raper, S. C. B., Watterson, I. G., Weaver, A. J. & Zhao, Z. C. (2007). Global Climate Projections. In: *Climate Change 2007: The Physical Science Basis. Contribution of Working Group 1 to the Fourth Assessment Report of the Intergovernmental Panel on Climate Change.* [Solomon, S., Qin, D., Manning, M., Chen, Z., Marquis, M., Averyt, K. B., Tignor, M. & Miller, H. L. (eds.). Cambridge University Press, Cambridge, United Kingdom and New York, NY, 748-845.

[157] Mercer, J. (1978). West Antarctic ice sheet and CO_2 greenhouse effect: A threat of disaster. *Nature, 271(5643),* 321-325.

[158] Monaghan, A. J., Bromwich, D. H., Fogt, R. L., Wang, S. H., Mayewski, P. A., Dixon, D. A., Ekaykin, A., Frezzotti, M., Goodwin, I. D., Isaksson, E., Kaspari, S. D., Morgan, V. I., Oerter, H., van Ommen, T., van der Veen, C. J. & Wen, J. S. (2006). Insignificant change in Antarctic snowfall since the International Geophysical Year. *Science, 313,* 827-830.

[159] Morris, E. M. & Vaughan, D. G. (1994). Snow surface temperatures in West Antarctica. *Antarctic Science, 6,* 529-535.

[160] Muhs, D. R., Simmons, K. R. & Steinke, B. (2002). Timing and warmth of the Last Interglacial period: new U-series evidence from Hawaii and Bermuda and a new fossil compilation for North America. *Quaternary Science Reviews, 21,* 1355-1383.

[161] Nakicenovic, N., et al., (2000). IPCC special report on emissions scenarios. Cambridge University Press, Cambridge, United Kingdom, *599.*

[162] O'Neel, S., Pfeffer, W. T., Krimmel, R. M. & Meier, M. F. (2005). Evolving force balance at Columbia Glacier, during its rapid retreat. *Journal of Geophysical Research, 110,* F03012, doi: 10.1029/2005JF000292.

[163] O'Neel, S., Echelmeyer, K. & Motyka, R. (2001). Short-term dynamics of a retreating tidewater glacier: LeConte Glacier, Alaska, USA. *Journal of Glaciology, 47,* 567- 578.

[164] Oerlemans, J., Dyurgerov, M. & van de Wal, R. S. W. (2007). Reconstructing the glacier contribution to sea-level rise back to 1850. *The Cryosphere, 1(1),* 59-65.

[165] Otto-Bliesner, S. J., Marshall, J. T., Overpeck, G. H. Miller, & Hu, A. X. (2006). Simulating arctic climate warmth and icefield retreat in the last interglaciation. *Science, 311,* 175 1-1753.

[166] Pagani, M., Zachos, J. C., Freeman, K. H., Tipple, B. & Bohaty, S. (2005). Marked decline in atmospheric carbon dioxide concentrations during the Paleogene. *Science, 309,* 600-603.

[167] Parizek, B. R. & Alley, R. B. (2004). Implications of increased Greenland surface melt under global-warming scenarios: Ice-sheet simulations. *Quaternary Science Reviews, 23,* 1013-1027.

[168] Paterson, W. S. B. (1994). *The Physics of Glaciers,* 3d ed. Elsevier Science Ltd., New York.

[169] Pattyn, F. (2002). Transient glacier response with a higher-order numerical ice-flow model. *Journal of Glaciology, 48(162),* 467-477.

[170] Pattyn, F., Huyghe, A., De Brabander, S. & De Smedt, B. (2006). The role of transition zones in marine ice sheet dynamics. *Journal of Geophysical Research*, 111(F02004), doi: 10.1029/2005JF000394.

[171] Payne, A. J., Vieli, A., Shepherd, A., Wingham, D. J. & Rignot, E. (2004). Recent dramatic thinning of largest West-Antarctic ice stream triggered by oceans. *Geophysical Research Letters*, *31*, (L23401).

[172] Peltier, W. R. (2005). On the hemispheric origins of meltwater pulse 1a: *Quaternary Science Reviews*, *24*, 1655-1671.

[173] Peltier, W. (2004). Global glacial isostatic adjustment and the surface of the ice-age Earth: the ICE-5G(VM2) model and GRACE. *Annual Review, Earth and Planetary Science Letters*, *32*, 111-149.

[174] Peltier, W. R., (1998). "Implicit ice" in the global theory of glacial isostatic adjustment. *Geophysical Research Letters*, *25*, 3955-3958.

[175] Peltier, W. R. & Fairbanks, R. G. (2006). Global glacial ice volume and last glacial maximum duration from an extended Barbados sea level record. *Quaternary Science Reviews*, *25*, 3322-3337.

[176] Petit, J. R., et al., (1999). Climate and atmospheric history of the past 420,000 years from the Vostok ice core. *Nature*, *399*, 429-43 6.

[177] Price, S. F., Conway, H. & Waddington, E. D. (2007). Evidence for late Pleistocene thinning of Siple Dome, West Antarctica. *Journal of Geophysical Research*, 112(F03021), doi: 10.1029/2006JF000725.

[178] Pritchard, H. D. & Vaughan, D. G. (2007). Widespread acceleration of tidewater glaciers on the Antarctic Peninsula. *Journal of Geophysical Research*, 112(F03S29), doi: 10.1 029/2006JF000597.

[179] Rahmstorf, S. (2007). A semi-empirical approach to projecting future sea-level rise. *Science*, *315*, 368-370.

[180] Ramillien, G., Lombard, A., Cazenave, A., Ivins, E. R., Llubes, M., Remy, F. & Biancale, R. (2006). Interannual variations of the mass balance of the Antarctica and Greenland ice sheets from GRACE. *Global and Planetary Change*, *53*,198-208.

[181] Raper, S. C. B. & Braithwaite, R. J. (2006). Low sea level rise projections from mountain glaciers and ice caps under global warming. *Nature*, *439*, 311-313.

[182] Raynaud, D., Chappellaz, J., Ritz, C. & Martinerie, P. (1997). Air content along the Greenland Ice Core Project core: A record of surface climatic parameters and elevation in central Greenland. *Journal of Geophysical Research*, *102*, 26607- 26613.

[183] Reeh, N., Mayer, C., Miller, H., Thomson, H. H. & Weidick, A. (1999). Present and past climate control on fjord glaciations in Greenland: Implications for IRD-deposition in the sea. *Geophysical Research Letters*, *26*, 1039-1042.

[184] Remy, F., Testut, L. & Legresy, B. (2002). Random fluctuations of snow accumulation over Antarctica and their relation to sea level change. *Climate Dynamics*, *19*, 267-276.

[185] Ridley, J. K., Huybrechts, P., Gregory, J. M. & Lowe, J. A. (2005). Elimination of the Greenland ice sheet in a high CO_2 climate: *Journal of Climate*, *18*, 3409-3427.

[186] Rignot, E. (2006). Changes in ice flow dynamics and ice mass balance of the Antarctic Ice Sheet. *Philosophical Transactions of the Royal Society of London, Series A*, *364(1844)*, 1637-1655.

[187] Rignot, E., Casassa, G., Gogineni, P., Krabill, W., Rivera, A. & Thomas, R. (2004a). Accelerated ice discharge from the Antarctic Peninsula following the collapse of Larsen B ice shelf. *Geophysical Research Letters*, 31(L18401), doi: 10.1 029/2004GL020697.

[188] Rignot, E. & Jacobs, S. (2002). Rapid bottom melting widespread near Antarctic Ice Sheet grounding lines. *Science, 296*, 2020-2023.

[189] Rignot, E. & Kanagaratnam, P. (2006). Changes in the velocity structure of the Greenland Ice Sheet. *Science, 311*, 986-990.

[190] Rignot, E., et al., (2004b). Improved estimation of the mass balance of the glaciers draining into the Amundsen Sea sector of West Antarctica from the CECS/NASA 2002 campaign. *Annals of Glaciology, 39.*

[191] Rignot, E., et al., (2005). Mass imbalance of Fleming and other glaciers, West Antarctic Peninsula. *Geophysical Research Letters*, 32(L07502).

[192] Rignot, E. & Thomas, R. H. (2002). Mass balance of polar ice sheets. *Science, 297(5586)*, 1502-1506.

[193] Rignot, E. J., Vaughan, D. G., Schmeltz, M., Dupont, T. & MacAyeal, D. R. (2002). Acceleration of Pine Island and Thwaites Glaciers, West Antarctica. *Annals of Glaciology, 34*, 189-194.

[194] Rignot, E., Bamber, J. L., van den Broeke, M. R., Davis, C., Li, Y., Jan van de Berg, W. & van Meijgaard, E. (2008). Recent Antarctic ice mass loss from radar interferometry and regional climate modelling, *Nature Geoscience, 1*, 106-110.

[195] Rinterknecht, V. R., Clark, P. U., Raisbeck, G. M., Yiou, F., Bitinas, A., Brook, E. J., Marks, L., Zelčs, V., Lunkka, J. P., Pavlovskaya, I. E., Piotrowski, J. A. & Raukas, A. (2006). The last deglaciation of the southeastern sector of the Scandinavian Ice Sheet. *Science, 311*, 1449-1452.

[196] Rohling, E. J., Grant, K., Hemleben, C. H., Siddall, M., Hoogakker, B. A. A. , Bolshaw, M. & Kucera, M. (2008). High rates of sea-level rise during the last interglacial period. Nature, *1*, 3 8-42.

[197] Röthlisberger, H. (1972). Water pressure in intra- and subglacial channels. *Journal of Glaciology, 11(62)*, 177-204.

[198] Rott, H., Rack, W., Skvarca, P. & de Angelis, H. (2002). Northern Larsen Ice Shelf, Antarctica: Further retreat after collapse. *Annals of Glaciology, 34*, 277-282.

[199] Scambos, T., Bohlander, J., Shuman, C. & Skvarca, P. (2004). Glacier acceleration and thinning after ice shelf collapse in the Larsen B embayment, Antarctica. *Geophysical Research Letters*, 31 (L18401), doi: 10,1 029/2004GL020670.

[200] Scambos, T., Hulbe, C. & Fahnestock, M. (2003). Climate-induced ice shelf disintegration in the Antarctic Peninsula. *Antarctic Research Series, 79*, 79-92.

[201] Scambos, T., Hulbe, C., Fahnestock, M. & Bohlander, J. (2000). The link between climate warming and break-up of ice shelves in the Antarctic Peninsula. *Journal of Glaciology, 46*, 5 16-530.

[202] Schoof, Ch., (2007). Marin ice-sheet dynamics. Part 1. The case of rapid sliding. *Journal of Fluid Mechanics, 573*, 27-55.

[203] Shackleton, N. J. (2000). The 100,000-year ice-age cycle identified and found to lag temperature, carbon dioxide, and orbital eccentricity. *Science, 289*, 1897-1902.

[204] Shepherd, A., Wingham, D. J. & Mansley, J. A. D. (2002). Inland thinning of the Amundsen Sea sector, West Antarctica. *Geophysical Research Letters, 29(10)*, 1364.

[205] Shepherd, A., Wingham, D., Payne, T. & Skvarca, P. (2003). Larsen Ice Shelf has progressively thinned. *Science, 302*, 856-859.

[206] Shreve, R. L. (1972). Movement of water in glaciers. *Journal of Glaciology, 11(62)*, 205- 214.

[207] Simms, A. R., Lambeck, K., Purcell, A., Anderson, J. B. & Rodriguez, A. B. (2007). Sea- level history of the Gulf of Mexico since the last glacial maximum with implications for the melting history of the Laurentide Ice Sheet. *Quaternary Science Reviews, 26*, 920-940.

[208] Steffen, K. & Box, J. E. (2001). Surface climatology of the Greenland ice sheet: Greenland climate network 1995–1999. *Journal of Geophysical Research, 106*, 33951- 33964.

[209] Steffen, K., Huff, R., Cullen, N., Rignot, E. & Bauder, A. (2004a). Sub-glacier ocean properties and mass balance estimates of Petermann Gletscher's floating tongue in Northwestern Greenland. American Geophysical Union, Fall Meeting, abstract #C31B- 0313.

[210] Steffen, K., Nghiem, S. V., Huff, R. & Neumann, G. (2004b). The melt anomaly of 2002 on the Greenland Ice Sheet from active and passive microwave satellite observations. *Geophysical Research Letters*, 31(L20402), doi: 10.1 029/ 2004 GL 020444.

[211] Stirling, C. H., Esat, T. M., Lambeck, K. & McCulloch, M. T. (1998). Timing and duration of the last interglacial: evidence for a restricted interval of widespread coral reef growth. *Earth and Planetary Science Letters, 160*, 745-762.

[212] Stirling, C. H., Esat, T. M., McCulloch, M. T. & Lambeck, K. (1995). High-precision U- series dating of corals from Western Australia and implications for the timing and duration of the Last Interglacial. *Earth and Planetary Science Letters, 135(1995)*, 115- 130.

[213] Stone, J. O., Balco, G. A., Sugden, D. E., Caffee, M. W., III Sass, L. C., Cowdery, S. G. & Siddoway, C. (2003). Holocene deglaciation of Marie Byrd Land, West Antarctica. *Science, 299*, 99-102.

[214] Stouffer, R. J., et al., (2006). Investigating the causes of the response of the thermohaline circulation to past and future climate changes. *Journal of Climate, 19*, 1365-13 87.

[215] Thomas, R., Abdalati, W., Frederick, E., Krabill, W., Manizade, S. & Steffen, K. (2003). Investigation of surface melting and dynamic thinning on Jakobshavn Isbræ, Greenland. *Journal of Glaciology, 49*, 23 1-239.

[216] Thomas, R., Davis, C., Frederick, E., Krabill, W., Li, Y., Manizade, S. & Martin, C. (2008). A comparison of Greenland ice-sheet volume changes derived from altimetry measurements. *Journal of Glaciology, 54(185)*, 203-212.

[217] Thomas, R., Frederick, E., Krabill, W., Manizade, S. & Martin, C. (2006). Progressive increase in ice loss from Greenland. *Geophysical Research Letters*, 2006GL026075R.

[218] Thomas, R., et al., (2001). Mass balance of higher-elevation parts of the Greenland ice sheet. *Journal of Geophysical Research, 106*, 33707-33716.

[219] Thomas, R. G., Rignot, E., Kanagaratnam, P., Krabill, W. & Casassa, G. (2005). Force-perturbation analysis of Pine Island Glacier, Antarctica, suggests caused for recent acceleration. *Annals of Glaciology, 39*, 133-138.

[220] Thomas, R. H. (2004). Force-perturbation analysis of recent thinning and acceleration of Jakobshavn Isbræ, Greenland. *Journal of Glaciology, 50(168)*, 57-66. In: R. H., Thomas, & C. R. Bentley, (1978). A model for Holocene retreat of the West Antarctic Ice Sheet. *Quaternary Research, 10*, 150-170.

[221] Toggweiler, J. R. & Samuels, B. (1995). Effect of Drake Passage on the global thermohaline circulation. *Deep-Sea Research, 42*, 477.

[222] Trenberth, K. E., Jones, P. D., Ambenje, P., Bojariu, R., Easterling, D., Tank, A. K., Parker, D., Rahimzadeh, F., Renwick, J. A., Rusticucci, M., Soden, B. & Zhai, P. (2007). Observations: Surface and atmospheric climate change. *In: Climate Change The Physical Science Basis. Contribution of Working Group I to the Fourth Assessment Report of the Intergovernmental Panel on Climate Change.* [Solomon, S., et al. (eds.)]. Cambridge University Press, Cambridge, United Kingdom.

[223] Tulaczyk, S. M., Kamb, B. & Engelhardt, H. F. (2001). Basal mechanics of ice stream B, West Antarctica I: Till mechanics. *Journal of Geophysical Research, 105(B1)*, 463-481. van de

[224] Berg, W. J., van den Broeke, M. R., Reijmer, C. H. & van Meijgaard, E. (2006). Reassessment of the Antarctic surface mass balance using calibrated output of a regional atmospheric climate model. *Journal of Geophysical Research, 111(D11104)*, doi: 10. 1029/2005JD006495.

[225] Van den Broeke, M. R., van de Berg, W. J. & van Meijgaard, E. (2006). Snowfall in coastal West Antarctica much greater than previously assumed. *Geophysical Research Letters*, 33(L02505), doi: 10.1 029/2005GL02523 9.

[226] van der Veen, C. J. (1993). Interpretation of short-term ice sheet elevation changes inferred from satellite altimetry. *Climate Change, 23*, 3 83-405.

[227] Vaughan, D, et al., (2003). Recent rapid regional climate warming on the Antarctic Peninsula. *Climate Change, 60*, 243-274.

[228] Vaughan, D. G., Bamber, J. L., Giovinetto, M., Russell, J. & Cooper, A. P. R. (1999). Reassessment of net surface mass balance in Antarctica. *Journal of Climate, 12(4)*, 933-946.

[229] Vaughan, D. G. & Doake, C. S. M. (1996). Recent atmospheric warming and retreat of ice shelves on the Antarctic Peninsula. *Nature, 379(6563)*, 328-331.

[230] Velicogna, I. & Wahr, J. (2005). Greenland mass balance from GRACE. *Geophys. Res. Lett.*, 32, L18505, doi:10.1029/2005GL023955.

[231] Velicogna, I. & Wahr, J. (2006a). Acceleration of Greenland ice mass loss in Spring 2004. *Nature, 443*, 329-33 1.

[232] Velicogna, I. & Wahr, J. (2006b). Measurements of time-variable gravity show mass loss in Antarctica. *Science, 311*, 1754-1756.

[233] Vieli, A. & Payne, A. J. (2005). Assessing the ability of numerical ice sheet models to simulate grounding line migration. *Journal of Geophysical Research*, 1 10(F01003).

[234] Waelbroeck, C., Labeyrie, L., Michel, E., Duplessy, J. C., McManus, J. F., Lambeck, K., Balbon, E. & Labracherie, M. (2002). Sea-level and deep water temperature changes derived from benthic foraminifera isotopic records. *Quaternary Science Reviews, 21*, 295-305.

[235] Walker, R. & Holland, D. M. (2007). A two-dimensional coupled model for ice shelf-ocean interaction. *Ocean Modelling, 17*, 123-139.

[236] Weaver, A. J., Saenko, O. A., Clark, P. U. & Mitrovica, J. X. (2003). Meltwater pulse 1A from Antarctica as a trigger of the Bølling-Allerød warm period. *Science, 299,* 1709-1713.

[237] Weertman, J. (1974). Stability of the junction between an ice sheet and an ice shelf. *Journal of Glaciology, 13,* 3-11.

[238] Wingham, D., Shepherd, A., Muir, A. & Marshall, G. (2006). Mass balance of the Antarctic ice sheet. *Philosophical Transactions of the Royal Society, A, 364,* 1627-1635.

[239] Yokoyama, Y., Lambeck, K., De Deckker, P., Johnson, P. & Fifield, K. (2000). Timing for the maximum of the last glacial constrained by lowest sea-level observations. *Nature, 406,* 713-716.

[240] Zwally, H. J., Abdalati, W., Herring, T., Larson, K., Saba, J. & Steffen, K. (2002a). Surface melt-induced acceleration of Greenland ice-sheet flow. *Science, 297(5579),* 218-222.

[241] Zwally, J., et al., (2002). ICESat's laser measurements of polar ice, atmosphere, ocean, and land. *Journal of Geodynamics, 34,* 405-445.

[242] Zwally, H. J., et al., (2005). Mass changes of the Greenland and Antarctic ice sheets and shelves and contributions to sea-level rise: 1992–2002. *Journal of Glaciology, 51(175),* 509-527.

[243] Adams, J. B., Mann, M. E. & Ammann, C. M. (2003). Proxy evidence for an El Niño-like response to volcanic forcing. *Nature, 426,* 274-27 8.

[244] Anderson, C. J., Arritt, R. W., Takle, E. S., Pan, Z. T., Gutowski, W. J., Otieno, F. O., da Silva, R., Caya, D., Christensen, J. H., Luthi, D., Gaertner, M. A., Gallardo, C., Giorgi, F., Hong, S. Y., Jones, C., Juang, H. M. H., Katzfey, J. J., Lapenta, W. M., Laprise, R., Larson, J. W., Liston, G. E., McGregor, J. L., Pielke, R. A., Roads, J. O. & Taylor, J. A. (2003). Hydrological processes in regional climate model simulations of the central United States flood of June-July 1993. *Journal of Hydrometeorology, 4(3),* 584-598.

[245] Anderson, L., Abbott, M. B., Finney, B. P. & Burns, S. J. (2005). Regional atmospheric circulation change in the North Pacific during the Holocene inferred from lacustrine carbonate oxygen isotopes, Yukon Territory, Canada. *Quaternary Research, 64(1),* 21-35.

[246] Asmerom, Y., Polyak, V., Burns, S. & Rassmussen, J. (2007). Solar forcing of Holocene climate: New insights from a speoleothem record, southwestern United States. *Geology, 35,* 1-4.

[247] Austin, P., Mackay, A. W., Palagushkina, O. & Leng, M. J. (2007). A high-resolution diatom-inferred palaeoconductivity and lake level record of the Aral Sea for the past 1600 yr. *Quaternary Research, 67,* 3 83-393.

[248] Backlund, P., Janetos, A., Schimel, D. S., Hatfield, J., Ryan, M., Archer, S. & Lettenmaier, D. (2008). Executive Summary. In: *The effects of climate change on agriculture, land resources, water resources, and biodiversity.* A Report by the U.S. Climate Change Science Program and the Subcommittee on Global Change Research. Washington, DC, USA, 362.

[249] Baker, V. R., Kochel, R. C. & Patton, P. C. (1988). *Flood Geomorphology.* New York, Wiley.

[250] Barber, D. C., Dyke, A., Hillaire-Marcel, C., Jennings, A. E., Andrews, J. T., Kerwin, M. W., Bilodeau, G., McNeely, R., Southon, J., Morehead, M. D. & Gagnon, J. M. (1999). Forcing of the cold event of 8,200 years ago by catastrophic drainage of Laurentide lakes. *Nature, 400(6742)*, 344-348.

[251] Barron, J. A., Bukry, D. & Dean, W. E. (2005). Paleoceanographic history of the Guaymas Basin, Gulf of California, during the past 15,000 years based on diatoms, silicoflagellates, and biogenic sediments. *Marine Micropaleontology, 56(3-4)*, 81-102.

[252] Bartlein, P. J. & Hostetler, S. W. (2004). Modeling paleoclimates. In: *The Quaternary Period in the United States.* In: A. R., Gillespie, S. C. Porter, & B. F. Atwater (eds.). Elsevier, Amsterdam.

[253] Bartlein, P. J. & Whitlock, C. (1993). Paleoclimatic interpretation of the Elk Lake pollen record. In: *Elk Lake, Minnesota: Evidence for Rapid Climate Change in the North-Central United States.* In: J. P. Bradbury, & W. E. Dean (eds.). Geological Society of America, Boulder, CO.

[254] Bartlein, P. J., Anderson, K. H., Anderson, P. M., Edwards, M. E., Mock, C. J., Thompson, R. S., Webb, R. S., Webb, T. III, & Whitlock, C. (1998). Paleoclimate simulations for North America over the past 21,000 years: Features of the simulated climate and comparisons with paleoenvironmental data. *Quaternary Science Reviews, 17(6-7)*, 549-5 85.

[255] Bell, G. D. & Janowiak, J. E. (1995). Atmospheric circulation associated with the Midwest floods of 1993. *Bulletin of the American Meteorological Society, 76(5)*, 68, 1-695.

[256] Bengtsson, L., Hodges, K. I. & Roeckner, E. (2006). Storm tracks and climate change. *Journal of Climate, 19*, 3518-3543.

[257] Benito, G., Lang, M., Barriendos, M., Llasat, M. C., Frances, F., Ouarda, T., Thorndycraft, V. R., Enzel, Y., Bardossy, A., Coeur, D. & Bobee, B. (2004). Use of systematic, palaeoflood and historical data for the improvement of flood risk estimation: Review of scientific methods. *Natural Hazards, 31(3)*, 623-643.

[258] Benson, L., Kashgarian, M., Rye, R., Lund, S., Paillet, F., Smoot, J., Kester, C., Mensing, S., Meko, D. & Lindström, S. (2002). Holocene multidecadal and multicentennial droughts affecting Northern California and Nevada. *Quaternary Science Reviews, 21*, 659-682.

[259] Berger, A. (1978). Long-term variations of caloric insolation resulting from the Earth's orbital elements. *Quaternary Research, 9*, 139-167.

[260] Berger, A. & Loutre, M. F. (1991). Insolation values for the climate of the last 10 million years. *Quaternary Science Reviews, 10*, 297-3 17.

[261] Biasutti, M. & Giannini, A. (2006). Robust Sahel drying in response to late 20th century forcings. *Geophysical Research Letters, 33*, doi: 10.1 029/2006GL026067.

[262] Bonan, G. B. & Stillwell-Soller, L. M. (1998). Soil water and the persistence of floods and droughts in the Mississippi River Basin. *Water Resources Research, 34*, 2693-2701.

[263] Booth, R. K., Notaro, M., Jackson, S. T. & Kutzbach, J. E. (2006). Widespread drought episodes in the western Great Lakes region during the past 2000 years: Geographic extent and potential mechanisms. *Earth and Planetary Science Letters, 242*, 415-427.

[264] Bosilovich, M. G. & Sun, W. Y. (1999). Numerical simulation of the 1993 Midwestern flood: Local and remote sources of water. *Journal of Geophysical Research-Atmospheres, 104*(D16), 19415-19423.

[265] Braconnot, P., Otto-Bliesner, B., Harrison, S., Joussaume, S., Peterchmitt, J. Y., Abe-Ouchi, A., Crucifix, M., Driesschaert, E., Fichefet, T., Hewitt, C. D., Kageyama, M., Kitoh, A., Laine, A., Loutre, M. F., Marti, O., Merkel, U., Ramstein, G., Valdes, P., Weber, S. L., Yu, Y. & Zhao, Y. (2007a). Results of PMIP2 coupled simulations of the Mid-Holocene and last glacial maximum—Part 1: Experiments and large-scale features. *Climate of the Past, 3(2)*, 261-277.

[266] Braconnot, P., Otto-Bliesner, B., Harrison, S., Joussaume, S., Peterchmitt, J. Y., Abe-Ouchi, A., Crucifix, M., Driesschaert, E., Fichefet, T., Hewitt, C. D., Kageyama, M., Kitoh, A., Loutre, M. F., Marti, O., Merkel, U., Ramstein, G., Valdes, P., Weber, L., Yu, Y. & Zhao, Y. (2007b). Results of PMIP2 coupled simulations of the Mid-Holocene and Last Glacial Maximum—Part 2: Feedbacks with emphasis on the location of the ITCZ and mid- and high latitudes heat budget. *Climate of the Past, 3(2)*, 279-296.

[267] Breshears, D. D., Cobb, N. S., Rich, P. M., Price, K. P., Allen, C. D., Balice, R. G., Romme, W. H., Kastens, J. H., Floyd, M. L., Belnap, J., Anderson, J. J., Myers, O. B. & Meyer, C. W. (2005). Regional vegetation die-off in response to global-change type drought. *Proceedings of the National Academy of Sciences, 102(42)*, 15 144-15148.

[268] Cane, M. A., Clement, A. C., Kaplan, A., Kushnir, Y., Pozdnyakov, D., Seager, R., Zebiak, S. E. & Murtugudde, R. (1997). Twentieth-century sea surface temperature trends. *Science, 275*, 957-960.

[269] Castro, C. L., McKee, T. B. & Pielke, R. A. Sr., (2001). The Relationship of the North American Monsoon to tropical and North Pacific sea surface temperatures as revealed by observational analyses. *Journal of Climate, 14*, 4449-4473.

[270] Castro, C. L., Pielke, R. A. Sr. & Adegoke, J. O. (2007a). Investigation of the summer climate of the contiguous U.S. and Mexico using the Regional Atmospheric Modeling System (RAMS). Part I: Model climatology (1950-2002). *Journal of Climate, 20*, 3866-3887.

[271] Castro, C. L., Pielke, R. A. Sr., Adegoke, J. O., Schubert, S. D. & Pegion, P. J. (2007b). Investigation of the summer climate of the contiguous U.S. and Mexico using the Regional Atmospheric Modeling System (RAMS). Part II: Model climate variability. *Journal of Climate, 20*, 3888-3901.

[272] Chiang, J. C. H. & Sobel, A. S. (2002). Tropical tropospheric temperature variations caused by ENSO and their influence on the remote tropical climate. *Journal of Climate, 15*, 2616-2631.

[273] Christensen, N. & Lettenmaier, D. P. (2006). A multimodel ensemble approach to assessment of climate change impacts on the hydrology and water resources of the Colorado River basin. *Hydrology and Earth System Sciences Discussion, 3*, 1-44.

[274] Claussen, M, (2001). Earth system models. In: Understanding the Earth System. [Ehlers, E. & T. Krafft (eds.)]. Springer, Berlin, 147-162.

[275] Claussen, M., Hoelzmann, P., Pachur, H. J., Kubatzki, C., Brovkin, V. & Ganopolski, A. (1999). Simulation of an abrupt change in Saharan vegetation in the mid-Holocene. *Geophysical Research Letters, 26(14)*, 2037-2040.

[276] Clement, A. C., Seager, R., Cane, M. A. & Zebiak, S. E. (1996). An ocean dynamical thermostat. *Journal of Climate, 9*, 2190-2196.

[277] Clement, R., Seager, R. & Cane, M. A. (2000). Suppression of El Niño during the mid-Holocene due to changes in the Earth's orbit. *Paleoceanography, 15*, 73 1-737.

[278] Climate Research Committtee, National Research Council. (1995). *Natural climate variability on decade-to-century time scales.* National Academy Press, Washington, DC, 644.

[279] Cobb, K., Charles, C. D., Cheng, H. & Edwards, R. L. (2003). El Niño/southern oscillation and tropical Pacific climate during the last millennium. *Nature, 424*, 27 1-276. COHMAP Members, 1988: Climatic changes of the last 18,000 years: Observations and model simulations. *Science, 241*, 1043-1052.

[280] Cole, J. E. & Cook, E. R. (1998). The changing relationship between ENSO variability and moisture balance in the continental United States. *Geophysical Research Letters, 25(24)*, 4529-4532.

[281] Cole, J. E., Overpeck, J. T. & Cook, E. R. (2002). Multiyear La Niña events and persistent drought in the contiguous United States. *Geophysical Research Letters, 29(13)*, 10.1029/2001GL013561.

[282] Committee on Radiative Forcing Effects on Climate, 2005: *Radiative forcing of climate change.* National Academy Press.

[283] Cook, E. R. & Jacoby, G. C. Jr., (1977). Tree-ring drought relationships in the Hudson Valley, New York. *Science, 198*, 399-401.

[284] Cook, B. I., Bonan, G. B. & Levis, S. (2006). Soil moisture feedbacks to precipitation in southern Africa. *Journal of Climate, 19*, 4198-4206.

[285] Cook, B. I., Miller, R. & Seager, R. (2008). Dust and sea surface temperature forcing of the 1930s Dust Bowl drought. *Geophys. Res. Lett.*, in press.

[286] Cook, E. R. & Krusic. P. J. (2004a). *The North American drought atlas.* Lamont-Doherty Earth Observatory and the National Science Foundation.

[287] Cook, E. R. & Krusic, P. J. (2004b). North American summer PDSI reconstructions. *IGBP PAGES/World Data Center for Paleoclimatology Data Contribution Series No. 2004-045.* NOAA/NGDC Paleoclimatology Program, Boulder, CO, 24. In: E. R., Cook, M. Kablack, & G. C. Jacoby, (1988). The 1986 drought in the southeastern United States: how rare an event was it? *Journal of Geophysical Research –Atmospheres, 93(D1)*, 14257-14260.

[288] Cook, E. R., Meko, D. M., Stahle, D. W. & Cleaveland. M. K. (1999). Drought reconstructions for the continental United States. *Journal of Climate, 12*, 1145-1162.

[289] Cook, E. R., Seager, R., Cane, M. A. & Stahle, D. W. (2007). North American drought: reconstructions, causes and consequences. *Earth Science Reviews, 81*, 93-134.

[290] Cook, E. R., Woodhouse, C., Eakin, C. M., Meko, D. M. & Stahle, D. W. (2004). Long-term aridity changes in the western United States. *Science, 306*, 1015-1018.

[291] Crucifix, M., Braconnot, P., Harrison, S. & Otto-Bliesner, B. (2005). Second phase of Paleoclimate Modelling Intercomparison Project. *Eos, Transactions of the American Geophysical Union, 86(28)*, 264.

[292] Dean, W. E., Forester, R. M. & Bradbury, J. P. (2002). Early Holocene change in atmospheric circulation in the Northern Great Plains: An upstream view of the 8.2 ka cold event. *Quaternary Science Reviews, 21(16-17)*, 1763-1775.

[293] Déry, S. J. & Brown, R. D. (2007). Recent Northern Hemisphere snow cover extent trends and implications for the snow-albedo feedback. *Geophysical Research Letters*, *34(22)*, doi: 1029/2007GL03 1474.

[294] Dyer, J. L. & Mote, T. L. (2006). Spatial variability and trends in obseved show depth over North America. *Geophysical Research Letters*, *33*, doi: 10.1 029/2006GRL027258.

[295] Dyke, A. S. (2004). An outline of North American deglaciation with emphasis on central and northern Canada. In: *Quaternary Glaciations—Extent and Chronology*, Part II: North America. In: J. Ehlers, & P. L. Gibbard (eds.). Elsevier, Amsterdam, 373-424.

[296] deMenocal, P., Ortiz, J., Guilderson, T., Adkins, J., Sarnthein, M., Baker, L. & Yarusinsky, M. (2000). Abrupt onset and termination of the African Humid Period: Rapid climate responses to gradual insolation forcing. *Quaternary Science Reviews*, *19(1-5)*, 347-361.

[297] Diffenbaugh, N. S. & Sloan, L. C. (2002). Global climate sensitivity to land surface change: The Mid Holocene revisited. *Geophysical Research Letters*, *29(10)*, 1476, doi: 10.1029/2002GL014880.

[298] Diffenbaugh, N. S., Ashfaq, M., Shuman, B., Williams, J. W. & Bartlein, P. J. (2006). Summer aridity in the United States: Response to mid-Holocene changes in insolation and sea surface temperature. *Geophysical Research Letters*, *33*, L22712, doi: 10.1029/2006GL028012.

[299] Douglass, A. E. (1929). The secret of the southwest solved with talkative tree rings. *National Geographic Magazine*, December, *73*, 6-770.

[300] Douglass, A. E. (1935). Dating Pueblo Bonito and other ruins of the southwest. National Geographic Society Contributed Technical Papers. *Pueblo Bonito Series*, *1*, 1-74.

[301] Dyke, A. S. (2004). An outline of North American deglaciation with emphasis on central and northern Canada. In: *Quaternary Glaciations—Extent and Chronology, Part II: North America.* In: J. Ehlers, & P. L. Gibbard (eds.). Elsevier, Amsterdam.

[302] Easterling, D. R., Evans, J. L., Groisman, P. Y., Karl, T. R., Kunkel, K. E. & Ambenje, P. (2000). Observed variability and trends in extreme climate events: A brief review. *Bulletin of the American Meteorological Society*, *81(3)*, 4 17-425.

[303] Egan, T. (2006). *The worst hard time*. Houghton Mifflin, 340.

[304] Ely, L. (1997). Response of extreme floods in the southwestern United States to climatic variations in the late Holocene. *Geomorphology*, *19*, 175-201.

[305] Ely, L. L., Enzel, Y., Baker, V. R. & Cayan, D. R. (1993). A 5000-year record of extreme floods and climate change in the southwestern United States. *Science, 262*, 410- 412.

[306] Emile-Geay, J., Cane, M. A., Seager, R., Kaplan, A. & Almasi, P. (2007). ENSO as a mediator for the solar influence on climate. *Paleoceanography*, *22*, doi: 10.1 029/2006PA001 304.

[307] Enfield, D. B., Mestas-Nunez, A. M. & Trimble, P. J. (2001). The Atlantic multidecadal oscillation and its relation to rainfall and river flows in the continental U.S. *Geophysical Research Letters*, *28(10)*, 2077-2080.

[308] Enzel, Y., Ely, L. L., House, P. K., Baker, V. R. & Webb, R. H. (1993). Paleoflood evidence for a natural upper bound to flood magnitudes in the Colorado River Basin. *Water Resources Research*, *29(7)*, 2287-2297.

[309] Esper, J., Cook, E. R. & Schweingruber, F. (2002). Low-frequency signals in long tree-ring chronologies for reconstructing past temperature variability. *Science, 295,* 2250-2253.

[310] Field, C. B., Mortsch, L. D., Brklacich, M., Forbes, D. L., Kovacs, P., Patz, J. A., Running, S. W. & Scott, M. J. (2007). North America. In: *Climate Change 2007: Impacts, Adaptation and Vulnerability. Contribution of Working Group II to the Fourth Assessment Report of the Intergovernmental Panel on Climate Change.* In: M. L., Parry, O. F., Canziani, J. P., Palutikof, P. J. van der Linden, & C. E. Hanson, (eds.). Cambridge University Press, Cambridge, United Kingdom, *6,* 17-652.

[311] Foley, J. A., Coe, M. T., Scheffer, M. & Wang, G. L. (2003). Regime shifts in the Sahara and Sahel: Interactions between ecological and climatic systems in northern Africa. *Ecosystems, 6(6),* 524-539.

[312] Forman, S. L., Oglesby, R. & Webb, R. S. (2001). Temporal and spatial patterns of Holocene dune activity on the Great Plains of North America: Megadroughts and climate links. *Global and Planetary Change, 29(1-2),* 1-29.

[313] Forster, P., Ramaswamy, V., Artaxo, P., Berntsen, T., Betts, R., Fahey, D. W., Haywood, J., Lean, J., Lowe, D. C., Myhre, G., Nganga, J., Prinn, R., Raga, G., Schulz, M. & Van Dorland, R. (2007). Changes in atmospheric constituents and in radiative forcing. In: *Climate Change 2007: The Physical Science Basis. Contribution of Working Group I to the Fourth Assessment Report of the Intergovernmental Panel on Climate Change.* In: S., Solomon, D., Qin, M., Manning, Z., Chen, M., Marquis, K. B. Averyt, M. Tignor, & H.L. Miller (eds.). Cambridge University Press, Cambridge, United Kingdom and New York.

[314] Fowler, H. J. & Kilsby, C. G. (2007). Using regional climate model data to simulate historical and future river flows in northwest England. *Climatic Change, 80,* 337- 367.

[315] Fritts, H. C. (1976). *Tree rings and climate.* Academic Press, London, 567.

[316] Fulp, T. (2005). How low can it go? *Southwest Hydrology,* March/April, 16-17, 28. In: F., Fye, D. W. Stahle, & E. R. Cook, (2003). Paleoclimatic analogs to 20th century moisture regimes across the USA. *Bulletin of the American Meteorological Society, 84(7),* 901-909.

[317] Fye, F. K., Stahle, D. W. & Cook, E. R. (2004). Twentieth century sea surface temperature patterns in the Pacific during decadal moisture regimes over the USA. *Earth Interactions, 8(22),* 1-22.

[318] Fye, F. K, Stahle, D. W., Cook, E. R. & Cleaveland, M. K. (2006). NAO influence on subdecadal moisture variability over central North America. *Geophysical Research Letters, 33,* L1 5707, doi: 10.1 029/2006GL026656.

[319] Gajewski, K., Viau, A., Sawada, M., Atkinson, D. & Wilson, S. (2001). Sphagnum peatland distribution in North America and Eurasia during the past 21,000 years. *Global Biogeochemical Cycles, 15(2),* 297-310.

[320] Gallimore, R., Jacob, R. & Kutzbach, J. (2005). Coupled atmosphere-ocean-vegetation simulations for modern and mid-Holocene climates: role of extratropical vegetation cover feedbacks. *Climate Dynamics, 25(7-8),* 755-776.

[321] Garcin, Y., Vincens, A., Williamson, D., Buchet, G. & Guiot, J. (2007). Abrupt resumption of the African Monsoon at the Younger Dryas-Holocene climatic transition. *Quaternary Science Reviews, 26(5-6),* 690-704.

[322] Gerten, D. & Adrian, R. (2000). Climate-driven changes in spring plankton dynamics and the sensitivity of shallow polymictic lakes to the North Atlantic oscillation. *Limnology and Oceanography*, *45*, 105 8-106.

[323] Giannini, A., Saravanan, R. & Chang, P. (2003). Oceanic forcing of Sahel rainfall on internanual to interdecadal timescales. *Science*, *302*, 1027-1030.

[324] Giorgi, F., Mearns, L. O., Shields, C. & Mayer, L. (1996). A regional model study of the importance of local versus remote controls of the 1988 drought and the 1993 flood over the central United States. *Journal of Climate*, *9(5)*, 1150-1162.

[325] Gleick, P. H. & Nash, L. (1991). *The societal and environmental costs of the continuing California drought*. Pacific Institute for Studies in Development, Environment, and Security, Oakland, CA.

[326] Goddard, L. & Graham, N. E. (1999). Importance of the Indian Ocean for simulating rainfall anomalies over eastern and southern Africa. *Journal of Geophysical Research*, *104*, 19099-19116.

[327] Graham, N. E., Hughes. M. K., Ammann, C. M., Cobb, K. M., Hoerling, M. P., Kennett , D. J., Kennett, J. P., Rein. B., Stott, L., Wigand, P. E. & Xu, T. (2007). Tropical Pacific- mid latitude teleconnections in medieval times. *Climatic Change*, *83*, 241-285.

[328] Griles, J. S. (2004). *Building on success, facing the challenges ahead*. Paper presented at Colorado River Water Users Association, Las Vegas NV, December 2004. http://www.usbr.gov/.

[329] Groisman, P. Y., Knight, R. W., Easterling, D. R., Karl, T. R., Hegerl, G. C. & Razuvaev, V. N. (2005). Trends in intense precipitation in the climate record. *Journal of Climate*, *18*, 1326-1350.

[330] Groisman, P. Y., Knight, R. W., Karl, T. R., Easterling, D. R., Sun, B. & Lawrimore, J. H. (2004). Contemporary changes of the hydrological cycle over the contiguous United States: Trends derived from in situ observations. *Journal of Hydrometeorology*, *5*, 64-85.

[331] Grootes, P. M., Stuiver, M., White, J. W. C., Johnsen, S. & Jouzel, J. (1993). Comparison of oxygen isotope records from the GISP2 and GRIP Greenland ice cores. *Nature*, *366(9)*, 552-554.

[332] Gutowski, W. J., Hegerl, G. C., Holland, G. J., Knutson, T. R., Mearns, L. O., Stouffer, R. J., Webster, P. J., Wehner, M. F. & Zwiers, F. W. (2008). Causes of Observed Changes in Extremes and Projections of Future Changes in *Weather and Climate Extremes in a Changing Climate. Regions of Focus: North America, Hawaii, Caribbean, and U.S. Pacific Islands*. In: T. R., Karl, G. A., Meehl, C. D. Miller, S. J., Hassol, A. M., Waple, & W. L. Murray, (eds.). A Report by the U.S. Climate Change Science Program and the Subcommittee on Global Change Research, Washington, DC.

[333] Hales, K., Neelin, J. D. & Zeng, N. (2006). Interaction of vegetation and atmospheric dynamical mechanisms in the mid-holocene African monsoon. *Journal of Climate*, *19(16)*, 4105-4120.

[334] Hall, R. I., Leavitt, P. R., Quinlan, R., Dixit, A. S. & Smol, J. P. (1999). Effects of agriculture, urbanization, and climate on water quality in the northern Great Plains. *Limnology and Oceanography*, *44*, 739-756.

[335] Hamlet, A. F., Mote, P. W., Clark, M. P. & Lettenmaier, D. P. (2007). Twentieth-century trends in runoff, evapotranspiration, and soil moisture in the western United States: *Journal of Climate*, *20*, 1468-1486.

[336] Hansen, Z. K. & Libecap, G. D. (2004). Small farms, externalities and the Dust Bowl of the 1930s. *Journal of Political Economy, 112,* 665-694.

[337] Harrison, S. P., Kutzbach, J. E., Liu, Z., Bartlein, P. J., Otto-Bliesner, B., Muhs, D., Prentice, I. C. & Thompson, R. S. (2003). Mid-Holocene climates of the Americas: A dynamical response to changed seasonality. *Climate Dynamics, 20(7-8),* 663-688.

[338] Haug, G. H., Hughen, K. A., Sigman, D. M., Peterson, L. C. & Rohl, U. (2001). Southward migration of the Intertropical Convergence Zone through the Holocene. *Science, 293,* 1304-1308.

[339] Hausmann, S. et al., (2002). Interactions of climate and land use documented in the varved sediments of Seebergsee in the Swiss Alps. *The Holocene, 12,* 279-289.

[340] Hegerl, G. C., Crowley, T. J., Allen, M., Hyde, W. T., Pollack, H. N., Smerdon, J. & Zorita, E. (2007). Detection of human influence on a new, validated 1500-year temperature reconstruction. *Journal of Climate, 20(4),* 650-666.

[341] Hegerl, G. C., Crowley, T. J., Baum, S. K., Kim, K. & Hyde, W. T. (2003). Detection of volcanic, solar and greenhouse gas signals in paleo-reconstructions of Northern Hemispheric temperature. *Geophysical Research Letters, 30(5),* 1242, doi: 10.1029/2002GL016635.

[342] Heim, R. R., Jr., (2002). A review of twentieth-century drought indices used in the United States. *Bulletin of the American Meteorological Society, 83(8),* 1149-1165.

[343] Held, I. M. & Soden, B. J. (2006). Robust responses of the hydrological cycle to global warming. *Journal of Climate, 19,* 5686-5699.

[344] Held, I. M., Delworth, T. L., Lu, J., Findell, K. L. & Knutson, T. R. (2005). Simulation of Sahel drought in the 20th and 21st centuries. *Proceedings of the National Academies of Science, 102,* 17891-17896.

[345] Herweijer, C., Seager, R. & Cook, E. R. (2006). North American droughts of the mid- to- late nineteenth century: A history, simulation and implication for medieval drought. *The Holocene, 16(2),* 159-171.

[346] Hendrix, C. S. & Glaser, S. M. (2007). Trends and triggers: Climate, climate change and civil conflict in sub-Saharan Africa. *Political Geography, 26,* 695-7, 15.

[347] Herweijer, C., Seager, R., Cook, E. R. & Emile-Geay, J. (2007). North American droughts of the last millennium from a gridded network of tree-ring data. *Journal of Climate, 20,* 1353-1376.

[348] Higgins, R. W., Yao, Y. & Wang, X. L. (1997). Influence of the North American monsoon system on the U.S. summer precipitation regime. *Journal of Climate, 10,* 2600-2622.

[349] Hirschboeck, K. K. (1989). Climate and floods. In: *Floods and Droughts: Hydrologic Perspectives on Water Issues.* U.S. Geologic Survey.

[350] Hirschboeck, K. K. (2003). Floods, paleofloods, and drought: Insights from the Upper Tails. CLIVAR/PAGES/IPCC Drought Workshop, November 18-21, Tucson, AZ, 6.

[351] Hoerling, M., Hurrell, J., Eischeid, J. & Phillips, A. (2006). Detection and attribution of twentieth century northern and southern African rainfall change. *Journal of Climate, 19,* 3989-4008.

[352] Hoerling, M. & Kumar, A. (2003). A perfect ocean for drought. *Science, 299,* 69 1-694.

[353] Hoerling, M. & Eischeid, J. (2007). Past peak water in the Southwest. *Southwest Hydrology,* 6, 18. In: M., Hoerling, G., Hegerl, D., Karoly, A. Kumar, & D. Rind, in prep. Attribution of the Causes of Climate Variations and Trends over North America

during the Modern Reanalysis Period in *Reanalysis of Historical Climate Data for Key Atmospheric Circulation Features: Implications for Attribution of Causes of Observed Change*. A Report by the U.S. Climate Change Science Program and the Subcommittee on Global Change Research, Washington, DC. [no editors listed in Public Review Draft].

[354] Hong, S. Y. & Pan, Z. T. (2000). Impact of soil moisture anomalies on seasonal, summertime circulation over North America in a regional climate model. *Journal of Geophysical Research, 105*, 29625–29634.

[355] House, P. K., Webb, R. H., Baker, V. R. & Levish, D. R. (2002). Ancient floods, modern hazards: Principles and applications of paleoflood hydrology. In: *Water Science and Application 5*. American Geophysical Union, Washington, DC.

[356] Hoyt, J. C. (1936). *Droughts of 1930-34. U.S. Geological Survey Water-Supply Paper, 680,* 106 p.

[357] Huang, H. P., Seager, R. & Kushnir, Y. (2005). The 1976/77 transition in precipitation over the Americas and the influence of tropical sea surface temperatures. *Climate Dynamics, 24*, 721-740.

[358] Hurrell, J. W. (1995). Decadal trends in the North Atlantic oscillation: Regional temperatures and precipitation. *Science, 269*, 676-679.

[359] IPCC, (2007). Climate Change The Physical Science Basis. Contribution of Working Group I to the Fourth Assessment Report of the Intergovernmental Panel on Climate Change In: S., Solomon, D., Qin, M., Manning, Z., Chen, M., Marquis, K. B., Averyt, M. Tignor, & H. L. Miller (eds.). Cambridge University Press, Cambridge, United Kingdom and New York, 996.

[360] Isenberg, A. C. (2000). *The destruction of the bison: An environmental history 1750-1920*. Cambridge University Press, Cambridge, United Kingdom, *206*.

[361] Jansen, E., Overpeck, J., Briffa, K. R., Duplessy, J. C., Joos, F., Masson-Delmotte, V., Olago, D., Otto-Bliesner, B., Peltier, W. R., Rahmstorf, S., Ramesh, R., Raynaud, D., Rind, D., Solomina, O., Villalba, R. & Zhang, D. (2007). Palaeoclimate. *In: Climate Change 2007. The Physical Science Basis. Contribution of Working Group I to the Fourth Assessment Report of the Intergovernmental Panel on Climate Change*. In: S., Solomon, D., Qin, M., Manning, Z., Chen, M., Marquis, K. B., Averyt, M. Tignor, & H. L. Miller (eds.). Cambridge University Press. Cambridge, United Kingdom, *43,* 3-497.

[362] Jolly, D., Harrison, S. P., Damnati, B. & Bonnefille, R. (1998). Simulated climate and biomes of Africa during the late Quaternary: Comparison with pollen and lake status data. *Quaternary Science Reviews, 17(6-7)*, 629-657.

[363] Jones, P. D. & Mann, M. E. (2004). Climate over past millennia. *Reviews of Geophysics*, 42, RG2002, doi: 10. 1029/2003RG000 143.

[364] Jones, R. N., Zhang, L., Chiew, F. H. S. & Boughton, W. C. (2006). Estimating the sensitivity of mean annual runoff to climate change using selected hydrological models. *Advances in Water Resources, 29(10)*, 1419-1429.

[365] Joos, F. & Spahni, R. (2008). Rates of change in natural and anthropogenic radiative forcing over the past 20,000 years: *Proceedings of the National Academy of Sciences of the United States of America, 105*, 1425-1430.

[366] Joussaume, S., Taylor, K. E., Braconnot, P., Mitchell, J. F. B., Kutzbach, J. E., Harrison, S. P., Prentice, I. C., Broccoli, A. J., Abe-Ouchi, A., Bartlein, P. J., Bonfils, C., Dong, B., Guiot, J., Herterich, K., Hewitt, C. D., Jolly, D., Kim, J. W., Kislov, A.,

Kitoh, A., Loutre, M. F., Masson, V., McAvaney, B., McFarlane, N., de Noblet, N., Peltier, W. R., Peterschmitt, J. Y., Pollard, D., Rind, D., Royer, J. F., Schlesinger, M. E., Syktus, J., Thompson, S., Valdes, P., Vettoretti, G., Webb, R. S. & Wyputta, U. (1999). Monsoon changes for 6000 years ago: Results of 18 simulations from the Paleoclimate Modeling Intercomparison Project (PMIP). *Geophysical Research Letters, 26(7)*, 859-862.

[367] Kalnay, E., Kanamitsu, M., Kistler, R., Collins, W., Deaven, D., Gandin, L., Iredell, M., Saha, S., White, G., Woollen, J., Zhu, Y., Leetmaa, A., Reynolds, B., Chelliah, M., Ebisuzaki, W., Higgins, W., Janowiak, J., Mo, K., Ropelewski, C., Wang, J., Jenne, R. & Joseph, D. (1996). The NCEP/NCAR 40-Year Reanalysis Project. *Bulletin of the American Meteorological Society, 77*, 437-471.

[368] Kaplan, A., Cane, M. A., Kushnir, Y., Clement, A. C., Blumenthal, M. B. & Rajagopalan, B. (1998). Analyses of global sea surface temperature 1856-1991. *Journal of Geophysical Research, 103*(C9), 18567-18589.

[369] Kaplan, J. O., Bigelow, N. H., Prentice, I. C., Harrison, S. P., Bartlein, P. J., Christensen, T. R., Cramer, W., Matveyeva, N. V., McGuire, A. D., Murray, D. F., Razzhivin, V. Y., Smith, B., Walker, D. A., Anderson, P. M., Andreev, A. A., Brubaker, L. B., Edwards, M. E. & Lozhkin, A. V. (2003). Climate change and Arctic ecosystems: 2. Modeling, paleodata-model comparisons, and future projections. *Journal of Geophysical Research, 108*(D19), 8171, doi: 10.1029/2002JD002559.

[370] Karl, T. R. & Young, P. J. (1987). The 1986 Southeast drought in historical perspective. *Bulletin of the American Meteorological Society, 68(7)*, 773-778.

[371] Karoly, D. J., Risbey, A., Reynolds, A. & Braganza, K. (2003). Global warming contributes to Australia's worst drought. *Australasian Science*, April, 14-17.

[372] Karspeck, A. R., Seager, R. & Cane, M. A. (2004). Predictability of tropical Pacific decadal variability in an intermediate model. *Journal of Climate, 17*, 1167-1180.

[373] King, A. W., Dilling, L., Zimmerman, G. P., Fairman, D. M., Houghton, R. A., Marland, G., Rose, A. Z. & Wilbanks, T. J. (2007). What Is the Carbon Cycle and Why Care? In: *The First State of the Carbon Cycle Report (SOCCR): The North American Carbon Budget and Implications for the Global Carbon Cycle.*. A Report by the U.S. Climate Change Science Program and the Subcommittee on Global Change

[374] Research King, A. W., Dilling, L., Zimmerman, G. P., Fairman, D. M., Houghton, R. A., Marland, G., Rose, A. Z. & Wilbanks, T. J. (eds.). National Oceanic and Atmospheric Administration, National Climatic Data Center, Asheville, NC, USA, pp. 15-20.

[375] Knight, J. R., Allan, R. J., Folland, C. K., Vellinga, M. & Mann, M. E. (2005). A signature of persistent natural thermohaline circulation cycles in observed climate. *Geophysical Research Letters, 32*, doi: 10.1 029/2005GL02423 3.

[376] Knight, D. B., Davis, R. E., Sheridan, S. C., Hondula, D. M., Sitka, L. J., Deaton, M., Lee, T. R., Gawtry, S. D., Stenger, P. J., Mazzei, F. & Kenny, B. P. (2008). Increasing frequencies of warm and humid air masses over the conterminous United States from 1948 to 2005. *Geophysical Research Letters, 35*, doi: 10.1029/2008GL033697.

[377] Knox, J. C. (1972). Valley alluviation in southwestern Wisconsin. *Annals of the Association of American Geographers, 62(3)*, 401-410.

[378] Knox, J. C. (1985). Responses of floods to Holocene climatic-change in the Upper Mississippi Valley. *Quaternary Research, 23(3)*, 287-300.

[379] Knox, J. C. (1993). Large increases in flood magnitude in response to modest changes in climate. *Nature, 361*, 430-432.

[380] Knox, J. C. (2000). Sensitivity of modern and Holocene floods to climate change. *Quaternary Science Reviews, 19*, 43, 9-457.

[381] Kochel, R. C. & Baker, V. R. (1982). Paleoflood. *Hydrology, 215(4531)*, 353-361.

[382] Kohfeld, K. E. & Harrison, S. P. (2000). How well can we simulate past climates? Evaluating the models using global palaeoenvironmental datasets. *Quaternary Science Reviews*, 19(1-5), 321-346.

[383] Koster, R., Dirmeyer, P. A., Guo, Z., Bonan, G., Chan, E., Cox, P., Gordon, C. T., Kanae, S., Kowalczyk, E. & Lawrence, E. et al., (2004). Regions of strong coupling between soil moisture and precipitation. *Science, 305*, 1138-1140.

[384] Kröpelin, S., Verschuren, D., Lezine, A. M., Eggermont, H., Cocquyt, C., Francus, P., Cazet, J. P., Fagot, M., Rumes, B., Russell, J. M., Darius, F., Conley, D. J., Schuster, M., von Suchodoletz, H. & Engstrom, D. R. (2008). Climate-driven ecosystem succession in the Sahara: the past 6000 years. *Science, 320*, 765-768.

[385] Kundzewicz, Z. W., Mata, L. J., Arnell, N. W., Döll, P., Kabat, P., Jiménez, B., Miller, K. A., Oki, T., Sen, Z. & Shiklomanov, I. A. (2007). Freshwater resources and their management. In: *Climate Change 2007. Impacts, Adaptation and Vulnerability. Contribution of Working Group II to the Fourth Assessment Report of the Intergovernmental Panel on Climate Change.* In: M. L., Parry, O. F. Canziani, J .P., Palutikof, P. J. van der Linden, & C. E. Hanson (eds.). Cambridge University Press, Cambridge, United Kingdom, 173-2 10.

[386] Kunkel, K. E. (2003). North American trends in extreme precipitation. *Natural Hazards, 29(2)*, 291-305.

[387] Kunkel, K. E., Changnon, S. A. & Angel, J. R. (1994). Climatic aspects of the 1993 Upper Mississippi River Basin flood. *Bulletin of the American Meteorological Society, 75(5)*, 811-822.

[388] Kunkel, K. E., Easterling, D. R., Redmond, K. & Hubbard, K. (2003). Temporal variations of extreme precipitation events in the United States: 1895-2000. *Geophysical Research Letters, 30(17)*.

[389] Kunkel, K. E., Bromirski, P. D., Brooks, H. E., Cavazos, T., Douglas, A. V., Easterling, D. R., Emanuel, K. A., Ya. P., Groisman, G. J., Holland, T. R., Knutson, J. P., Kossin, P. D., Komar, D. H. & Levinson, R. L. Smith, (2008). Observed Changes in Weather and Climate Extremes in *Weather and Climate Extremes in a Changing Climate. Regions of Focus: North America, Hawaii, Caribbean, and U.S. Pacific Islands.*

[390] Kushnir, Y. (1994). Interdecadal variations in North Atlantic sea surface temperature and associated atmospheric conditions. *Journal of Climate, 7*, 141-157.

[391] Kutzbach, J. E. & Liu, Z. (1997). Response of the African monsoon to orbital forcing and ocean feedbacks in the middle Holocene. *Science, 278*, 440-443.

[392] Kutzbach, J, E. & Ruddiman, W. F. (1993). Model description, external forcing, and surface boundary conditions. *In: Global Climates since the Last Glacial Maximum.* In: H. E., Wright, Jr., J. E., Kutzbach, T., Webb III, W. F., Ruddiman, F. A. Street-Perrott, & P. J. Bartlein (eds.). University of Minnesota Press, Minneapolis.

[393] Kutzbach, J. E. & Street-Perrott, F. A. (1985). Milankovitch forcing of flucuations in the level of tropical lakes from 18 to 0 kyr BP. *Nature, 317*, 130-134.

[394] Kutzbach, J. E. & Otto-Bliesner, B. L. (1982). The sensitivity of the African-Asian monsoonal climate to orbital parameter changes for 9000 yr B.P. in a low- resolution general circulation model. *Journal of the Atmospheric Sciences, 39(6),* 1177-1188.

[395] Laird, K. R., Fritz, S. C., Maasch, K. A. & Cumming, B. F. (1996). Greater drought intensity and frequency before A.D. 1200 in the northern Great Plains, U.S.A. *Nature, 384,* 552-554.

[396] LaMarche, V. C. Jr., (1974). Paleoclimatic inferences from long tree-ring records. *Science, 183,* 1043-1048.

[397] Lamb, H. H. (1965). The early medieval warm epoch and its sequel. *Palaeogeography, Palaeoclimatology, Palaeoecology, 1,* 13-37.

[398] Lau, N. C., Leetmaa, A. & Nath, M. J. (2006). Attribution of atmospheric variations in the 1997-2003 period to SST anomalies in the Pacific and Indian Ocean basins. *Journal of Climate, 19,* 3607-3628.

[399] Lebo, M. E., Reuter, J. H., Goldman, C. R. & Rhodes, C. L. (1994). Interannual variability of nitrogen limitation in a desert lake: influence of regional climate. *Canadian Journal of Fisheries and Aquatic Sciences, 51,* 862-872.

[400] Lettenmaier, D., Major, D., Poff, L. & Running, S. (2008). Water Resources. In: *The effects of climate change on agriculture, land resources, water resources, and biodiversity.* A Report by the U.S. Climate Change Science Program and the subcommittee on Global change Research. Washington, DC., USA, *362.*

[401] Levis, S., Bonan, G. B. & Bonfils, C. (2004). Soil feedback drives the mid-Holocene North African monsoon northward in fully coupled CCSM2 simulations with a dynamic vegetation model. *Climate Dynamics, 23(7-8),* 79 1-802.

[402] Libecap, G. D. & Hansen, Z. K. (2002). "Rain Folows the Plough" and dryfarming doctrine: The climate information problem and homestead failure in the upper Great Plains 1890-1925. *Journal of Economic History, 62,* 86-120.

[403] Li, W., Fu, R. & Dickinson, R. E. (2006). Rainfall and its seasonality over the Amazon in the 21st century as assessed by the coupled models for the IPCC AR4. *Journal of Geophysical Research, 111,* doi: 10. 1029/2005JD006355.

[404] Liu, A. Z., Ting, M. & Wang, H. (1998). Maintenance of circulation anomalies during the 1988 drought and 1993 floods over the United States. *Journal of the Atmospheric Sciences, 55(17),* 28 10-2832.

[405] Liu, Z. & Alexander, M. (2007). Atmospheric bridge, oceanic tunnel, and global climatic teleconnections. *Reviews of Geophysics, 45,* RG2005, doi: 10.1 029/ 2005 RG 0001 72.

[406] Liu, Z., Brady, E. & Lynch-Steiglitz, J. (2003). Global ocean response to orbital forcing in the Holocene. *Paleoceanography, 18(2),* 1054, doi: 10.1 029/2002PA000826.

[407] Liu, Z., Harrison, S. P., Kutzbach, J. & Otto-Bliesner, B. (2004). Global monsoons in the mid-Holocene and oceanic feedback. *Climate Dynamics,* 22(2-3), 157-182.

[408] Liu, Z., Wang, Y., Gallimore, R., Gasse, F., Johnson, T., deMenocal, P., Adkins, J., Notaro, M., Prentice, I. C., Kutzbach, J., Jacob, R., Behling, P., Wang, L. & Ong, E. (2007). Simulating the transient evolution and abrupt change of Northern Africa atmosphere-ocean-terrestrial ecosystem in the Holocene. *Quaternary Science Reviews, 26(13-14),* 1818-1837.

[409] Liu, Z. Y., Wang, Y., Gallimore, R., Notaro, M. & Prentice, I. C. (2006). On the cause of abrupt vegetation collapse in North Africa during the Holocene: Climate variability

vs. vegetation feedback. *Geophysical Research Letters*, *33*, L22709, doi: 10.1 029/2006GL02 8062.

[410] Lobell, D. B., Burke, M. B., Tebaldi, C., Mastrandrea, M. D., Falcon, W. P. & Naylor, R. L. (2008). Prioritizing climate change adaptation needs for food security in 2030. *Science*, *319*, 607-610.

[411] Lu, J., Vecchi, G. & Reichler, T. (2007). Expansion of the Hadley Cell under global warming. *Geophysical Research Letters*, *34*, doi: 10.1 02 9/2006GL02 8443.

[412] MacDonald, G. M., Beilman, D. W., Kremenetski, K. V., Sheng, Y. W., Smith, L. C. & Velichko, A. A. (2006). Rapid early development of circumarctic peatlands and atmospheric CH4 and CO2 variations. *Science*, *314(5797)*, 285-288.

[413] Maguire, R., (2005). The effects of drought in lower basin river operations. *Southwest Hydrology*, March/April, 22-23, 38.

[414] Mann, M. E. & Emanuel, K. A. (2005). Hurricane trends linked to climate change. *Eos*, *87*, 233.

[415] Mann, M. E., Cane, M. A., Zebiak, S. E. & Clement, A. (2005). Volcanic and solar forcing of the tropical Pacific over the past 1000 years. *Journal of Climate*, *18*, 447-456. Manuel, J., 2008: Drought in the Southeast: Lessons for water management. Environmental Health Perspective, *116*, A1 68-A1 71.

[416] Mason, J. A., Swinehart, J. B., Goble, R. J. & Loope, D. B. (2004). Late-Holocene dune activity linked to hydrological drought, Nebraska Sand Hills, USA. *Holocene*, *14(2)*, 209-2 17.

[417] Maruer, E. P. (2007). Uncertainty in hydrologic impacts of climate change in the Sierra Nevada, California, under two emissions scenarios. *Climatic Change*, *82*, 309- 325.

[418] Mayewski, P. A., Karlén, W., Maasch, K. A., Meeker, L. D., Meyerson, E. A., Gasse, F., van Kreveld, S., Holmgren, K., Lee-Thorp, J., Rosqvist, G., Rack, F., Staubwasser, M., Schneider, R. R., Steig, E. J., Rohling, E. E. & Stager, J. C. (2004). Holocene climate variability. *Quaternary Research*, *62(3)*, 243-255.

[419] McCabe, G. J., Palecki, M. A. & Betancourt, J. L. (2004). Pacific and Atlantic Ocean influences on multidecadal drought frequency in the United States. *Proceedings of the National Academies of Science*, *101(12)*, 4136-4141.

[420] Meehl, G. A., Stocker, T. F., Collins, W. D., Friedlingstein, P., Gaye, A. T., Gregory, J. M., Kitoh, A., Knutti, R., Murphy, J. M., Noda, A., Raper, S. C. B., Watterson, I. G., Weaver, A. J. & Zhao, Z. C. (2007). Global climate projections. In: *Climate Change 2007. The Physical Science Basis. Contribution of Working Group 1 to the Fourth Assessment Report of the Intergovernmental Panel on Climate Change*. In: S., Solomon, D., Qin, M., Manning, Z., Chen, M., Marquis, K. B., Averyt, M., Tignor, & H. L. Miller (eds.). Cambridge University Press, Cambridge, United Kingdom and New York.

[421] Meko, D. M., Woodhouse, C. A., Baisan, C. A., Knight, T., Lukas, J. J., Hughes, M. K. & Salzer, M. W. (2007). Medieval drought in the upper Colorado River Basin. *Geophysical Research Letters*, *34*, L10705.

[422] Miao, X. D., Mason, J. A., Swinehart, J. B., Loope, D. B., Hanson, P. R., Goble, R. J. & Liu, X. D. (2007). A 10,000 year record of dune activity, dust storms, and severe drought in the central Great Plains. *Geology*, *35(2)*, 119-122.

[423] Mock, C. J. & Brunelle-Daines, A. R. (1999). A modern analogue of western United States summer palaeoclimate at 6000 years before present. *The Holocene, 9(5)*, 541-545.

[424] Mosley, M. P. & McKerchar, A. I. (1993). Streamflow. In: Handbook of Hydrology. [Maidment, D.R. (ed.)]. McGraw-Hill, New York, *8*, 1-8.39

[425] Morrill, C., Overpeck, J. T. & Cole, J. E. (2003). A synthesis of abrupt changes in the Asian summer monsoon since the last deglaciation. *Holocene, 13(4)*, 465-476.

[426] Moore, J. N., Harper, J. T. & Greenwood, M. C. (2007). Significance of trends toward earlier snowmelt runoff, Columbia and Missouri Basin headwaters, western United States. *Geophysical Research Letters, 34(16)*, doi: 10.1 029/2007GL03 2022.

[427] Mote, P., Hamlet, A. & Salathé, E. (2008). Has spring snowpack declined in the Washinton Cascades. *Hydrology and Earth System Sciences, 12*, 193-206.

[428] Mote, P. W. (2006). Climate-driven variability and trends in mountain snowpack in western North America: *Journal of Climate, 19*, 6209-6220.

[429] Mote, P. W., Hamlet, A. F., Clark, M. P. & Lettenmaier, D. P. (2005). Declining mountain snowpack in western north America. *Bulletin of the American Meteorological Society, 86*, 39-49.

[430] Namias, J. (1966). Nature and possible causes of the northeastern United States drought during 1962-65. *Monthly Weather Review, 94(9)*, 543-554.

[431] Namias, J. (1991). Spring and summer 1988 drought over the Great Plains: Causes and predictions. *Journal of Climate, 4*, 54-65.

[432] National Assessment, (2000). Water: The potential consequences of climate variability and change for the water resources of the United States. Report of the Water Sector Assessment Team of the National Assessment of the Potential Consequences of Climate Variability and Change for the U.S. Global Change Research Program. [Gleick, P. (ed.)]. U.S. Global Change Research Program, 151.

[433] National Integrated Drought Information System Implementation Team, 2007: A pathway for national resilience. http://www.drought.

[434] Nemec, J. & Schaake, J. C. (1982). Sensitivity of water resource systems to climate variation. *Hydrologic Sciences Journal, 27*, 327-343.

[435] Newman, M., Compo, G. P. & Alexander, M. A. (2003). ENSO-forced variability of the Pacific Decadal Oscillation. *Journal of Climate, 16*, 3853-3 857.

[436] Nicholls, N. (2004). The changing nature of Australian droughts. *Climatic Change, 63*, 323-336.

[437] Nicholson, S. E., Some, B. & Kone, B. (2000). An analysis of recent rainfall conditions in West Africa, including the rainy seasons of the 1997 and 1998 La Niña years. *Journal of Climate, 13*, 2628-2640.

[438] Nigam, S. & Ruiz-Barradas, A. (2006). Great Plains hydroclimate variability: The view from North American regional reanalysis. *Journal of Climate, 19*, 3004-30 10.

[439] Nordas, R. & Gleditsch, N. P. (2007). Climate change and conflict. *Political Geography, 26*, 627-63 8.

[440] Notaro, M., Wang, Y., Liu, Z., Gallimore, R. & Levis, S. (2008). Combined statistical and dynamical assessment of simulated vegetation-rainfall interactions in North Africa during the mid-Holocene. *Global Change Biology, 14(2)*, 347-368, doi:10.1 11 1/j.1365-2486.2007.01495.x

[441] O'Connor, J. E. & Costa, J. E. (2004). The world's largest floods, past and present: Their causes and magnitudes. U.S. Geological Survey. http:// purl.access.gpo.gov/ GPO/LPS56009.

[442] O'Connor, J. E., Costa, J. E. & Geological Survey (U.S.), (2003). Large floods in the United States where they happen and why. U.S. Geological Survey. http://purl.access.gpo.gov/GPO/LPS37149.

[443] Oglesby, R. J. (1991). Springtime soil moisture, natural climatic variability, and North American drought as simulated by the NCAR community climate model 1. *Journal of Climate, 4(9)*, 890-897.

[444] Oglesby, R. J. & Erickson, D. J. III, (1989). Soil moisture and the persistence of North American drought. *Journal of Climate, 2(11)*, 1362-1380.

[445] Ostler, D. A, (2005). Upper Colorado River basin perspectives on the drought. *Southwest Hydrology*, March/April, *18*, 29.

[446] Otto-Bliesner, B. L., Brady, E. C., Clauzet, G., Tomas, R., Levis, S. & Kothavala, Z. (2006). Last Glacial Maximum and Holocene climate in CCSM3. *Journal of Climate, 19(11)*, 2526-2544.

[447] Overland, J. E., Percival, D. B. & Mofjeld, H. O. (2006). Regime shifts and red noise in the North Pacific. In: Deep-Sea Research Part I. *Oceanographic Research Papers, 53(4)*, 582-588.

[448] Pal, J. S., Giorgi, F. & Bi, X. Q. (2004). Consistency of recent European summer precipitation trends and extremes with future regional climate projections. *Geophysical Research Letters, 31(13)*.

[449] Pal, J. S. & Eltahir, E. A. B. (2002). Teleconnections of soil moisture and rainfall during the 1993 midwest summer flood. *Geophysical Research Letters, 29(18)*.

[450] Palmer, W. C. (1965). *Meteorological drought.* Weather Bureau Research Paper 45, U.S. Department of Commerce, Washington, DC, 58.

[451] Palmer, T. N. & Brankoviç, C. (1989). The 1988 US drought linked to anomalous sea surface temperature. *Nature, 338(6210)*, 54-57.

[452] Peteet, D. (2001). Late glacial climate variability and general circulation model (GCM) experiments: An overview. In: *Interhemispheric Climate Linkages.* [Markgraf, V. (ed.)]. Academic Press, San Diego.

[453] Pederson, D. C., Peteet, D. M., Kurdyla, D. & Guilderson, T. (2005). Medieval Warming, Little Ice Age, and European impact on the environment during the last millennium in the lower Hudson Valley, New York, USA. *Quaternary Research, 63*, 23 8-249.

[454] Peterson, T. C., Anderson, D. M., Cohen, S. J., Cortez-Vázquez, M., Murnane, R. J., Parmesan, C., Phillips, D., Pulwarty, R. S., Stone, J. M. R. (2008). Why Weather and Climate Extremes Matter in Weather and Climate Extremes in a Changing Climate. Regions of Focus: North America, Hawaii, Caribbean, and U.S. Pacific Islands. T. R., G. A. Karl, C. D., Meehl, S. J., Miller, A. M., Hassol, W. L. Waple, & Murray (eds.). A Report by the U.S. Climate Change Science Program and the Subcommittee on Global Change Research, Washington, DC.

[455] Pilgrim, D. H. & Cordery, I. (1993). Flood runoff. In: *Handbook of Hydrology.* [Maidment, D.R. (ed.)]. McGraw-Hill, New York, 9.1-9.42.

[456] Poore, R. Z., Pavich, M. J. & Grissino-Mayer, H. D. (2005). Record of the North American southwest monsoon from Gulf of Mexico sediment cores. *Geology, 33(3),* 209- 212.

[457] Raleigh, C. & Urdal, H. (2007). Climate change, environmental degradation and armed conflict. *Political Geography, 26,* 674-694.

[458] Rasmussen, J. B. T., Polyak, V. J. & Asmerom, Y. (2006). Evidence for Pacific-modulated precipitation variability during the late Holocene from the southwestern USA. *Geophysical Research Letters, 33(8).*

[459] Rayner, N. A., Parker, D. E., Horton, E. B., Folland, C. K., Alexander, L. V., Rowell, D. P. Kent, E. C. & Kaplan, A. (2003). Globally complete analyses of sea surface temperature, sea ice and night marine air temperature, 1871-2000. *Journal of Geophysical Research, 108,* 4407, doi: 10. 1029/2002JD002670.

[460] Renssen, H., Brovkin, V., Fichefet, T. & Goosse, H. (2006). Simulation of the Holocene climate evolution in Northern Africa: The termination of the African Humid Period. *Quaternary International, 150,* 95-102.

[461] Rosenzweig, C., Major, D. C., Demong, K., Stanton, C., Horton, R. & Stults, M. (2007). Managing climate change risks in New York City's water systems: Assessment and adaptation planning. *Mitigation and Adaptation Strategies for Global Change,* DOI 10.1007/s1 1027-006-9070-5.

[462] Ruddiman, W. F. (2006). Orbital changes and climate. *Quaternary Science Reviews,* 25(23- 24), 3092-3 112.

[463] Rudnick, D. L. & Davis, R. E. (2003). Red noise and regime shifts. *Deep_Sea Research I, 50,* 691-699.

[464] Ruiz-Barradas, A. & Nigam, S. (2005). Warm season rainfall over the U.S. Great Plains in observations, NCEP and ERA-40 Reanalyses, and NCAR and NASA atmospheric model simulations. *Journal of Climate, 18,* 1808-1830.

[465] Saltzman, B. (2002). *Dynamical paleoclimatology: Generalized theory of global climate change.* Academic Press, San Diego.

[466] Schindler, D. W., Bayley, S. E., Parker, B. R., Beaty, K. G. & Cruikshank, D. R. (1996a). The effects of climatic warming on the properties of boreal lakes and streams at the Experimental Lakes Area, northwestern Ontario. *Limnology and Oceanography, 41,* 1004-1017.

[467] Schindler, D. W., Curtis, P. J., Parker, B. R. & Stainton, M. P. (1996b). Consequences of climate warming and lake acidification for UV-B penetration in North American boreal lakes. *Nature, 379,* 705-708.

[468] Schneider, E. K. (1977). Axially symmetric steady-state models of the basic state for instability and climate studies. Part II: Non-linear calculations. *Journal of the Atmospheric Sciences, 34,* 280-296.

[469] Schneider, N. & Cornuelle, B. D. (2005). The forcing of the Pacific decadal oscillation. *Journal of Climate, 18(21),* 43, 55-4373.

[470] Schubert, S. D., Suarez, M. J., Region, P. J., Koster, R. D. & Bacmeister. J. T. (2004a). Causes of long-term drought in the United States Great Plains. *Journal of Climate, 17,* 485-503.

[471] Schubert, S. D., Suarez, M. J., Region, P. J., Koster, R. D. & Bacmeister, J. T. (2004b). On the cause of the 1930s Dust Bowl. *Science, 303,* 1855-1859.

[472] Seager, R. (2007). The turn-of-the-century drought across North America: Dynamics, global context and past analogues. *Journal of Climate, 20*, 5527-5552.

[473] Seager, R., Tzanova, A. & Nakamura, J. (2008). Drought in the Southeastern United States: Causes, variability over the last millennium and the potential for future hydroclimatic change. *Journal of Climate* (in review).

[474] Seager, R., Harnik, N., Kushnir, Y., Robinson, W. A. & Miller, J. (2003). Mechanisms of hemispherically symmetric climate variability. *Journal of Climate, 16*, 296-2978.

[475] Seager, R., Harnik, N., Robinson, W. A., Kushnir, Y., Ting, M., H. Huang, P. & Velez. J. (2005a). Mechanisms of ENSO-forcing of hemispherically symmetric precipitation variability. *Quarterly Journal of the Royal Meteorological Society, 131*, 1501- 1527.

[476] Seager, R., Kushnir, Y., Herweijer, C., Naik, N. & Velez. J. (2005b). Modeling of tropical forcing of persistent droughts and pluvials over western North America: 1856-2000. *Journal of Climate, 18*, 4068-4091.

[477] Seager, R., Burgman, R., Kushnir, Y., Clement, A., Naik, N. & Velez, J. (2007a). Tropical Pacific forcing of North American medieval megadroughts: Testing the concept with an atmosphere model forced by coral-reconstructed SSTs. *Journal of Climate*, submitted.

[478] Seager, R., Graham, N., Herweijer, C., Gordon, A. L., Kushnir, Y. & Cook. E. (2007b). Blueprints for Medieval hydroclimate. *Quaternary Science Reviews, 26*, 2322- 2336.

[479] Seager, R., Kushnir, Y., Ting, M., Cane, M., Naik, N. & Velez, J. (2007c). Would advance knowledge of 1930s SSTs have allowed prediction of the Dust Bowl drought? *Journal of Climate*, in press.

[480] Seager, R., Ting, M., Held, I., Kushnir, Y., Lu, J., Vecchi, G., Huang, H. P., Harnik, N., Leetmaa, A., Lau, N. C., Li, C., Velez, J. & Naik, N. (2007d). Model projections of an imminent transition to a more arid climate in southwestern North America. *Science, 316*, 1181-1184.

[481] Seidel, D. J., Fu, Q., Randel, W. J. & Reichler, T. J. (2008). Widening of the tropical belt in a changing climate. *Nature, Geoscience*, 1, 2 1-24. Published online: 2 December 2007 doi: 10.103 8/ngeo.2007.38.

[482] Shin, S. I., Sardeshmukh, P. D., Webb, R. S., Oglesby, R. J. & Barsugli, J. J. (2006). Understanding the mid-Holocene climate. *Journal of Climate, 19(12)*, 2801-2818, doi:10.1 175/JCLI3733.1.

[483] Shinker, J. J., Bartlein, P. J. & Shuman, B. (2006). Synoptic and dynamic climate controls of North American mid-continental aridity. *Quaternary Science Reviews, 25*, 1401-1417.

[484] Shuman, B., Bartlein, P., Logar, N., Newby, P. & Webb, T. III, (2002). Parallel climate and vegetation responses to the early Holocene collapse of the Laurentide Ice Sheet. *Quaternary Science Reviews, 21(16-17)*, 1793-1805.

[485] Shuman, B., Bartlein, P. J. & Webb, T. III, (2005). The magnitudes of millennial- and orbital-scale climatic change in eastern North America during the Late Quaternary. *Quaternary Science Reviews, 24(20-21)*, 2 194-2206.

[486] Shuman, B. & Finney, B. (2006). Late-Quaternary lake-level changes in North America. In: *Encyclopedia of Quaternary Science*. [Elias, S. (ed.)]. Elsevier, Amsterdam.

[487] Shuman, B., Bartlein, P. J. & Webb, T. III, (2007). Response to "Comments on: 'The magnitude of millennial- and orbital-scale climatic change in eastern North America

during the Late-Quaternary' by Shuman et al." *Quaternary Science Reviews, 26(1-2)*, 268-273.

[488] Shuman, B., Williams, J. W., Diffenbaugh, N. S., Ashfaq, M. & Bartlein, P. J. in review: The effects of insolation and soil-moisture behavior on moisture patterns and variability in the central United States during the mid-Holocene. *Quaternary Science Reviews*.

[489] Sinha, A., Cannariato, K., Stott, L. D., Chaing, H., Edwards, R. L., Yadava, M. G., Ramesh, R. & Singh, I. B. (2007). A 900-year (600 to 1500 A.D.) record of the Indian summer monsoon precipitation from the core monsoon zone of India. *Geosphysical Research Letters 34*, doi: 10.1 029/2007GL03043 1.

[490] Soden, B. J., Wetherald, R. T., Stenchikov, L. & Robock, A. (2002). Global cooling after the reuption of Mount Minatubo: A test of climate feedback by water vapor. *Science, 296*, 727-730.

[491] Stahle, D. W., Cleaveland, M. K. & Hehr, J. G. (1988). North Carolina climate change reconstructed from tree rings: A.D. 372 to 1985. *Science, 240*, 15 17-1520.

[492] Stahle, D. W., Cleaveland, M. K., Blanton, D. B., Therrell, M. D. & Gay, D. A. (1998). The lost colony and Jamestown droughts. *Science, 280(5363)*, 564-567.

[493] Stahle, D. W., Cook, E. R., Cleaveland, M. K., Therrell, M. D., Meko, D. M., Grissino-Mayer, H. D., Watson, E. & Luckman, B. H. (2000). Tree-ring data document 16th century megadrought over North America. *Eos, Transactions of the American Geophysical Union, 81(12)*, 121, 125.

[494] Stahle, D. W., Fye, F. K., Cook, E. R. & Griffin, R. D. (2007). Tree-ring reconstructed megadroughts over North America since AD 1300. *Climatic Change*, doi 10.1007/s1 0584-006-9171 -x (in press).

[495] Stine, S. (1994). Extreme and persistent drought in California and Patagonia during mediaeval time. *Nature, 369*, 546-549.

[496] Stockton, C. W. & Jacoby, G. C. (1976). Long-term surface-water supply and streamflow trends in the Upper Colorado River basin. *Lake Powell Research Project Bulletin No. 18*. National Science Foundation, 70 pp.

[497] Stone, J. R. & Fritz, S. C. (2006). Multidecadal drought and Holocene climate instability in the Rocky Mountains. *Geology, 34(5)*, 409-4, 12.

[498] Stuiver, M., Grootes, P. M. & Braziunas, T. F. (1995). The GISP2 & δ^{18}O climate record of the past 16 500 years and the role of the sun, ocean, and volcanoes. *Quaternary Research, 44(3)*, 341-354.

[499] Sutton, R. T. & Hodson, D. R. L. (2005). Atlantic Ocean forcing of North American and European summer climate. *Science, 309*, 115-118.

[500] Swetnam, T. W. & Betancourt, J. L. (1998). Mesoscale disturbance and ecological response to decadal climate variability in the American southwest. *Journal of Climate, 11*, 3128-3147.

[501] Texier, D., de Noblet, N., Harrison, S. P., Haxeltine, A., Jolly, D., Joussaume, S., Laarif, F., Prentice, I. C. & Tarasov, P. (1997). Quantifying the role of biosphere-atmosphere feedbacks in climate change: coupled model simulations for 6000 years BP and comparison with palaeodata for northern Eurasia and northern Africa. *Climate Dynamics*, 13, 865-882.

[502] Thompson, R. S., Whitlock, C., Bartlein, P. J., Harrison, S. P. & Spaulding, W. G. (1993). Climatic changes in western United States since 18,000 yr B.P. In: *Global*

Climates since the Last Glacial Maximum. In: H. E., Wright, Jr., J. E., Kutzbach, T., Webb III, W. F., Ruddiman, F. A. Street-Perrott, & P. J. Bartlein, (eds.). University of Minnesota Press, Minneapolis.

[503] Trenberth, K. E. & Guillemot, C. J. (1996). Physical processes involved in the 1988 drought and 1993 floods in North America. *Journal of Climate, 9(6),* 1288-1298.

[504] Trenberth, K. E., Branstator, G. W., Karoly, D., Kumar, A., Lau, N. C. & Ropelewski, C. (1998). Progress during TOGA in understanding and modeling global teleconnections associated with tropical sea surface temperature. *Journal of Geophysical Research, 103,* 1429 1-14324.

[505] Trenberth, K. E., Dai, A., Rasmussen, R. M. & Parsons, D. B. (2003). The changing character of precipitation. *Bulletin of the American Meteorological Society, 84,* 1205-12 17.

[506] Trenberth, K. E., Jones, P. D., Ambenje, P., Bojariu, R., Easterling, D., Klein Tank, A., Parker, D., Rahimzadeh, F., Renwick, J. A., Rusticucci, M., Soden, B. & Zhai, P. (2007). Observations: Surface and atmospheric climate change. In: *Climate Change The Physical Science Basis. Contribution of Working Group 1 to the Fourth Assessment Report of the Intergovernmental Panel on Climate Change.* In: S., Solomon, D., Qin, M., Manning, Z., Chen, M., Marquis, K. B., Averyt, M., Tignor, & H. L. Miller, (eds.). Cambridge University Press, Cambridge, United Kingdom, and New York. USBR (U.S. Department of the Interior, Bureau of Reclamation), 2005: Water 2025:

[507] Preventing crises and conflict in the west. http://www.doi.gov/water2025/.

[508] USBR (U.S. Department of the Interior, Bureau of Reclamation), 2007: Draft environmental impact statement, Colorado River interim guidelines for lower basin shortages and coordinated operations for Lake Powell and Lake Mead.

[509] Vecchi, G. A., Soden, B. J., Wittenberg, A. T., Held, I. M., Leetmaa, A. & Harrison, M. J. (2006). Weakening of tropical Pacific atmospheric circulation due to anthropogenic forcing. *Nature, 441,* 73-76.

[510] Vera, C., Higgins, W., Amador, J., Ambrizzi, T., Garreaud, R., Gochis, D., Gutzler, D., Lettenmaier, D., Marengo, J., Mechoso, C. R., Nogues-Paegle, J., Dias, P. L. S. & Zhang, C. (2006). Toward a unified view of the American Monsoon Systems. *Journal of Climate, 19(20),* 4977-5000.

[511] Viau, A. E., Gajewski, K., Sawada, M. C. & Fines, P. (2006). Millennial-scale temperature variations in North America during the Holocene. *Journal of Geophysical Research, 111,* D09102, doi:10.1029/2005JD006031.

[512] Wang, C., Enfield, D. B., Lee, S. K. & Landsea, C. W. (2006). Influences of the Atlantic warm pool on western hemisphere summer rainfall and Atlantic hurricanes. *Journal of Climate, 19(12),* 3011-3028.

[513] Wang, Y. J., Cheng, H., Edwards, R. L., He, Y. Q., Kong, X. G., An, Z. S., Wu, J. Y., Kelly, M. J., Dykoski, C. A. & Li, X. D. (2005). The Holocene Asian monsoon: Links to solar changes and North Atlantic climate. *Science, 308(5723),* 854-857.

[514] Wang, Y., Notaro, M., Liu, Z., Gallimore, R., Levis, S. & Kutzbach, J. E. (2007). Detecting vegetation-precipitation feedbacks in mid-Holocene North Africa from two climate models. *Climate of the Past Discussions, 3,* 961-975.

[515] Webb, T., III, Anderson, K. H., Bartlein, P. J. & Webb, R. S. (1998). Late Quaternary climate change in eastern North America: a comparison of pollen-derived estimates with climate model results. *Quaternary Science Reviews, 17(6-7),* 587- 606.

[516] Webb, T., III, Bartlein, P. J., Harrison, S. P. & Anderson, K. H. (1993a). Vegetation, lake levels, and climate in eastern North America for the past 18,000 years. In: *Global Climates since the Last Glacial Maximum*. [Wright, H.E., Jr., J.E. Kutzbach, T. Webb III, W.F. Ruddiman, F.A. Street-Perrott, and P.J. Bartlein (eds.)]. University of Minnesota Press, Minneapolis.

[517] Webb, T., III, Cushing, E. J. & Wright, H. E. Jr., (1983). Holocene changes in the vegetation of the Midwest. In: *Late-Quaternary Environments of the United States, vol. 2*. [Wright, H.E., Jr. (ed.)]. University of Minnesota Press, Minneapolis.

[518] Webb, T., III, Ruddiman, W. F., Street-Perrott, F. A., Markgraf, V., Kutzbach, J. E., Bartlein, P. J., Wright, H. E., Jr. & Prell, W. L. (1993b). Climatic changes during the past 18,000 years; Regional syntheses, mechanisms, and causes. In: *Global Climates Since the Last Glacial Maximum*. In: H. E., Wright, Jr., J. E., Kutzbach, T., Webb III, W. F., Ruddiman, F. A., Street-Perrott, & P. J. Bartlein (eds.)]. University of Minnesota Press, Minneapolis.

[519] Weldeab, S., Lea, D. W., Schneider, R. R. & Andersen, N. (2007). 155,000 years of West African Monsoon and ocean thermal evolution. *Science, 316*, doi:10.1 126/science.1 140461.

[520] West, E. (1995). *The way to the west: Essays on the Great Plains*. New Mexico Press, 244.

[521] Western Governor's Association, 2004, Creating a drought warning system for the 21st century: The National Integrated Drought Information System. http:// www. westgov. org/wga/publicat/nidis.pdf

[522] Wilby, R. L. & Wigley, T. M. L. (2002). Future changes in the distribution of daily precipitation totals across North America. *Geophysical Research Letters, 29*, doi: 10.1029/2001GL013048.

[523] Williams, J. W. (2002). Variations in tree cover in North America since the last glacial maximum. *Global and Planetary Change, 35*, 1-23.

[524] Williams, J. W., Shuman, B. N., Webb, T. III, Bartlein, P. J. & Leduc, P. L. (2004). Late- quaternary vegetation dynamics in North America: Scaling from taxa to biomes. *Ecological Monographs, 74(2)*, 309-334.

[525] Wohl, E. E. (2000). *Inland flood hazards: Human, riparian and aquatic communities*. Cambridge University Press, Cambridge, United Kingdom, and New York.

[526] Wohlfahrt, J., Harrison, S. P. & Braconnot, P. (2004). Synergistic feedbacks between ocean and vegetation on mid- and high-latitude climates during the mid- Holocene. *Climate Dynamics, 22(2-3)*, 223-23 8.

[527] Woodhouse, C. A. & Overpeck, J. T. (1998). 2000 years of drought variability in the central United States. *Bulletin of the American Meteorological Society, 79*, 2693-2714.

[528] Woodhouse C. A., Gray, S. T. & Meko, D. M. (2006). Updated streamflow reconstructions for the Upper Colorado River basin. *Water Resources Research, 42*, W05415, doi: 10.1 029/2005WR004455.

[529] Woodhouse, C. W., Kunkel, K. E., Easterling, D. R. & Cook, E. R. (2005). The 20th century pluvial in the western United States. *Geophysical Research Letters, 32*, L07701, doi: 10.1 029/2005GL02241 3.

[530] Worster, D. (1979). *Dust Bowl: The Southern Plains in the 1930s*. Oxford University Press, New York, 277.

[531] Worster, D. (1985). *Rivers of empire: Water, aridity and the growth of the American West*. Oxford University Press, New York, *416*.

[532] Wright, H. E., Jr., Kutzbach, J. E., Webb, T., III, Ruddiman, W. F., Street-Perrott, F. A. & Bartlein, P. J. (1993). *Global climates since the last glacial maximum*. In: H. E., Wright, Jr., J. E., Kutzbach, T., Webb, III, W. F., Ruddiman, F. A., Street-Perrott, & P. J., Bartlein (eds.)]. University of Minnesota Press, Minneapolis.

[533] Wright, H. E., Jr., Hu, F. S., Stefanova, I., Tian, J. & Brown, T. A. (2004). A chronological framework for the Holocene vegetational history of central Minnesota: The Steel Lake pollen record. *Quaternary Science Reviews, 23(5-6)*, 611-626.

[534] Yin, J. (2005). A consistent poleward shift of the storm tracks in simulations of 21st century climate. *Geophysical Research Letters, 32*, L06105, doi: 10.1 029/2004GL022058.

[535] Zhang, X., Zwiers, F. W., Hegerl, G. C., Lambert, F. H., Gillett, N. P. & Solomon, S. (2007). Detection of human influence on 20th century precipitation trends. *Nature, 448*, 461-466, doi:10.1038.

[536] Zhang, Y., Wallace, J. M. & Battisti, D. S. (1997). ENSO-like decade-to-century scale variability: 1900-93. *Journal of Climate, 10*, 1004-1020.

[537] Zhao, Y., Braconnot, P., Marti, O., Harrison, S. P., Hewitt, C., Kitoh, A., Liu, Z., Mikolajewicz, U., Otto-Bliesner, B. & Weber, S. L. (2005). A multi-model analysis of the role of the ocean on the African and Indian monsoon during the mid- Holocene. *Climate Dynamics*, 25(7-8), 777-800.

[538] Zhao, Y., Braconnot, P., Harrison, S. P., Yiou, P. & Marti, O. (2007). Simulated changes in the relationship between tropical ocean temperatures and the western African monsoon during the mid-Holocene. *Climate Dynamics, 28(5)*, 533-551.

[539] Adcroft, A., Scott, J. R. & Marotzke, J. (2001). Impact of geothermal heating on the global ocean circulation. *J. Geophys. Res., 28(9)*, 1735-173 8.

[540] Adkins, J. F. & Boyle, E. (1997). Changing atmospheric $\Delta^{14}C$ and the record of deepwater paleoventilation ages. *Paleoceanography, 12*, 337-344.

[541] Adkins, J. F., Cheng, H., Boyle, E. A., Druffel, E. R. M. & Edwards, R. L. (1998). Deep-sea coral evidence for rapid change in ventilation of the deep North Atlantic 15,400 years ago. *Science, 280*, 725-728.

[542] Adkins, J. F., McIntyre, K. & Schrag, D. P. (2002). The salinity, temperature and $\delta^{18}O$ of the glacial deep ocean. *Science, 298*, 1769-1773.

[543] Alley, R. B. (2007). Wally was right: Predictive ability of the North Atlantic "conveyor belt" hypothesis for abrupt climate change. *Annual Reviews of Earth and Planetary Sciences, 35*, 241-272.

[544] Alley, R. B. & Agustdottir, A. M. (2005). The 8k event: Cause and consequences of a major Holocene abrupt climate change. *Quat. Sci. Rev., 24*, 1123-1149.

[545] Altabet, M. A., Higginson, M. J. & Murray, D. W. (2002). The effect of millennial-scale changes in Arabian ea denitrification on atmospheric CO2. *Nature, 414*, 159-162.

[546] Arneborg, L. (2002). Mixing efficiencies in patchy turbulence. *J. Phys. Oceanogr., 32*, 1496-1506.

[547] Arzel, O., Fichefet, T. & Goosse, H. (2006). Sea ice evolution over the 20th and 21st centuries as simulated by the current AOGCMs. *Ocean Modelling, 12*, 401-415.

[548] Bacon, M. P. & Anderson, R. F. (1982). Distribution of thorium isotopes between dissolved and particulate forms in the deep sea. *J. Geophys. Res., 87*, 2045-2056.

[549] Barnett, T. P., Dumenil, L., Schlese, U., Roeckner, E. & Latif, M. (1989). The effect of Eurasian snow cover on regional and global climate variations. *J. Atmos. Sci., 46*, 661-685.

[550] Behl, R. & Kennett, J. P. (1996). Brief interstadial events in the Santa Barbara basin, NE Pacific, during the past 60 kyr. *Nature, 379*, 243-246.

[551] Bender, M., Sowers, T., Dickson, M. L., Orchardo, J., Grootes, P., Mayewski, P. A. & Meese, D. A. (1994). Climate correlations between Greenland and Antarctica during the past 100,000 years, *Nature, 372*, 663-666.

[552] Bender, M. L., Malaize, B., Orchardo, J., Sowers, T. & Jouzel, J. (1999). High-precision correlations of Greenland and Antarctic ice core records over the last 100 kyr. In: *Mechanisms of global climate change at millennial time scales.* [Clark, P.U., R.S. Webb, and L.D. Keigwin (eds.)]. *Geophys. Monogr. Ser., 112*, 149-164.

[553] Benway, H. M., Mix, A. C., Haley, B. A. d Klinkhammer, G. P. (2006). Eastern Pacific Warm Pool paleosalinity and climate variability: 0-30 kyr. *Paleoceanography, 21*, PA3008, doi: 10.1029/2005PA001208.

[554] Berger, A. & Loutre, M. F. (2002). An exceptionally long Interglacial ahead? *Science, 297*, 1287-1288.

[555] Bickert, T. & Mackensen, A. (2004). Late Glacial to Holocene changes in South Atlantic deep water circulation. In: *The South Atlantic in the Late Quaternary: Reconstruction of Material Budget and Current Systems.* [Wefer, G., et al. (eds.)]. Springer-Verlag, Berlin, 67 1-693.

[556] Bindoff, N. L., Willebrand, J., Artale, V., Cazenave, A., Gregory, J., Gulev, S., Hanawa, K., Le Quéré, C., Levitus, S., Nojiri, Y., Shum, C. K., Talley, L. D. & Unnikrishnan, A. (2007). Observations: Oceanic climate change and sea level, In: *Climate change 2007: The physical science basis. Contribution of Working Group I to the Fourth Assessment Report of the Intergovernmental Panel on Climate Change* In: S., Solomon, D., Qin, M., Manning, Z., Chen, M., Marquis, K. B., Averyt, M. Tignor, & H. L. Miller (eds.). Cambridge University Press, Cambridge, United Kingdom, and New York, 996.

[557] Black D. E., Peterson, L. C., Overpeck, J. T., Kaplan, A., Evans, M. N. & Kashgarian, M. (1999). Eight centuries of North Atlantic Ocean atmosphere variability. *Science, 286*, 1709-1713.

[558] Blunier, T. & Brook, E. J. (2001). Timing of millennial-scale climate change in Antarctica and Greenland during the last glacial period. *Science, 291*, 109-112.

[559] Blunier, T., et al., (1998). Asynchrony of Antarctic and Greenland climate change during the last glacial period. *Nature, 394*, 73, 9-743.

[560] Boessenkool, K. P., Hall, I. R., Elderfield, H. & Yashayaev, I. (2007). North Atlantic climate and deep-ocean flow speed changes during the last 230 years. *Geophys. Res. Lett., 34*, L13614, doi:10.1029/2007GL030285.

[561] Bond, G. C., Broecker, W. S., Johnsen, S., McManus, J., Labeyrie, L., Jouzel, J. & Bonani, G. (1993). Correlations between climate records from North Atlantic sediments and Greenland ice. *Nature, 365*, 143-147.

[562] Bond, G. & Lotti, R. (1995). Iceberg discharges into the North Atlantic on millennial time scales during the last glaciation. *Science, 267*, 1005-1010.

[563] Bond, G. C., Showers, W., Cheseby, M., Lotti, R., Almasi, P., deMenocal, P., Priore, P., Cullen, H., Hajdas, I. & Bonani, G. (1997). A pervasive millennial-scale cycle in North Atlantic Holocene and glacial climates. *Science, 278,* 1257-1266.

[564] Boning, C. W., Scheinert, M., Dengg, J., Biastoch, A. & Funk, A. (2006). Decadal variability of subpolar gyre transport and its reverberation in the North Atlantic overturning. *Geophysical Research Letters, 33,* 5.

[565] Boyle, E. A., (1992). Cadmium and $\delta^{13}C$ paleochemical ocean distributions during the stage 2 glacial maximum. *Annual Review of Earth and Planetary Sciences, 20,* 245-287.

[566] Boyle, E. A. (2000). Is ocean thermohaline circulation linked to abrupt stadial/interstadial transitions? *Quat. Sci. Rev., 19,* 255-272.

[567] Boyle, E. A. & Keigwin, L. D. (1982). Deep circulation of the North Atlantic over the last 200,000 years: Geochemical evidence. *Science, 218,* 784-787.

[568] Boyle, E. A. & Keigwin, L. (1987). North-Atlantic thermohaline circulation during the past 20,000 years linked to high-latitude surface-temperature. *Nature, 330,* 35-40.

[569] Braconnot, P., et al., (2007). Results of PMIP2 coupled simulations of the Mid-Holocene and Last Glacial Maximum. Part I: Experiments and large-scale features. *Climate of the Past, 3,* 261-277.

[570] Broecker, W. S. (1994). Massive iceberg discharges as triggers for global climate change. *Nature, 372,* 421-424.

[571] Broecker, W. S. (1998). Paleocean circulation during the last deglaciation: A bipolar seesaw? *Paleoceanography, 13,* 119-121.

[572] Broecker, W. S., Peteet, D. M. & Rind, D. (1985). Does the ocean-atmosphere system have more than one stable mode of operation? *Nature, 315,* 21-26.

[573] Broecker, W. S., et al., (1989). Routing of meltwater from the Laurentide ice-sheet during the Younger Dryas cold episode. *Nature, 341,* 3 18-21.

[574] Bryan, F. O., Danabasoglua, G., Nakashikib, N., Yoshidab, Y., Kimb, D. H., Tsutsuib, J. & S. C. Doney, et al., (2006). Response of the North Atlantic thermohaline circulation and ventilation to increasing carbon dioxide in CCSM3. *Journal of Climate, 19,* 2382-2397.

[575] Bryan, F. O., Hecht, M. W. & Smith, R. D. (2007). Resolution convergence and sensitivity studies with North Atlantic circulation models. Part I: The western boundary current system. *Ocean Modelling, 16,* 141-159.

[576] Bryden, H. L., Longworth, H. R. & Cunningham, S. A. (2005). Slowing of the Atlantic meridional overturning circulation at 25 degrees N. *Nature, 438,* 655-657.

[577] Bryden, H. L., Roemmich, D. H. & Church, J. A. (1991). Ocean heat transport across 24 degrees N in the Pacific. *Deep-Sea Research Part A-Oceanographic Research Papers, 38,* 297-324.

[578] Carlson, A. E., Clark, P. U., Haley, B. A., Klinkhammer, G. P., Simmons, K., Brook, E. J. & Meissner, K. J. (2007). Geochemical proxies of North American freshwater routing during the Younger Dryas cold event. Proceed. Nat. Acad. Sci., 104, 6556-6561.

[579] Carton, J. A., Chepurin, G., Cao, X. H. & Giese, B. (2000). A simple ocean data assimilation analysis of the global upper ocean 1950-95. Part I: Methodology. *Journal of Physical Oceanography, 30,* 294-309.

[580] Charles, C. D., Lynch-Stieglitz, J., Ninneman, U. S. & Fairbanks, R. G. (1996). Climate connections between the hemispheres revealed by deep-sea sediment/ice core correlations. *Earth Planet. Sci. Let.*, *142*, 19-27.

[581] Cheng, W., Bitz, C. M. & Chiang, J. C. H. (2007). Adjustment of the global climate to an abrupt slowdown of the Atlantic meridional overturning circulation. In: Ocean Circulation: Mechanisms and Impacts. *Geophysical Monobraph Series*, *173*, American Geophysical Union, 10.1029/173GM 19.

[582] Chiang, J. C. H., Biasutti, M. & Battisti, D. S. (2003). Sensitivity of the Atlantic intertropical convergence zone to last glacial maximum boundary conditions. *Paleoceanography*, *18*, 1094, doi: 10.1 029/2003PA0009 16.

[583] Clark, P. U., Hostetler, S. W., Pisias, N. G., Schmittner, A. & Meissner, K. J. (2007). Mechanisms for a ~7-kyr climate and sea-level oscillation during marine isotope stage 3. In: *Ocean Circulation: Mechanisms and Impacts*. In: A., Schmittner, J. Chiang, & Hemming, S. (eds.). American Geophysical Union, Geophysical Monograph 173, Washington, D.C., 209-246.

[584] Clark, P. U., Pisias, N. G., Stocker, T. S. & Weaver, A. J. (2002a). The role of the thermohaline circulation in abrupt climate change. *Nature*, *415*, 863-869.

[585] Clark, P. U., Mitrovica, J. X., Milne, G. A. & Tamisiea, M. (2002b). Sea-level fingerprinting as a direct test for the source of global meltwater pulse IA. *Science*, *295*, 2438-2441.

[586] Clarke, G. K. C., Leverington, D. W., Teller, J. T. & Dyke, A. S. (2004). Paleohydraulics of the last outburst flood from glacial Lake Agassiz and the 8200 BP cold event. *Quat. Sci. Rev.*, *23*, 3 89-407.

[587] Climap, (1981). Seasonal reconstructions of the earth's surface at the last glacial maximum. CLIMAP, 18.

[588] Conkright M. E., Locarnini, R. A., Garcia, H. E., O'Brien, T. D., Boyer, T. P., Stephens, C. & Antonov, J. I. (2002). World ocean atlas 2001: Objective analyses, data statistics, and figures: CD-ROM documentation. National Oceanographic Data Center, Silver Spring, MD, *17*.

[589] Cottet-Puinel, M., Weaver, A. J., Hillaire-Marcel, C., de Vernal, A., Clark, P. U. & Eby, M. (2004). Variation of Labrador Sea water formation over the last glacial cycle in a climate model of intermediate complexity. *Quat. Sci. Rev.*, *23*, 449-465.

[590] Cuffey, K. M. & Clow, G. D. (1997). Temperature, accumulation, and ice sheet elevation in central Greenland through the last deglacial transition. *Jour. Geophys. Res.*, *102*, 26,383-26,396.

[591] Cunningham, S. A., Kanzow, T., Rayner, D., Baringer, M. O., Johns, W. E., Marotzke, J., Longworth, H., Grant, E., Hirschi, J., Beal, L., Meinen, C. S. & Bryden, H. (2007). Temporal variability of the Atlantic meridional overturning circulation at 25°N. *Science*, in press.

[592] Curry, R., Dickson, B. & Yashayaev, I. (2003). A change in the freshwater balance of the Atlantic Ocean over the past four decades. *Nature*, *426*, 826-829

[593] Curry W. B. & Lohmann, G. P. (1982). Carbon isotopic changes in benthic foraminifera from the western South Atlantic: Reconstructions of glacial abyssal circulation patterns. *Quat. Res.*, *18*, 218-235.

[594] Curry, W. B., Marchitto, T. M., McManus, J. F., Oppo, D. W. & Laarkamp, K. L. (1999). Millennial-scale changes in ventilation of the thermocline, intermediate, and

deep waters of the glacial North Atlantic. In: Mechanisms of global climate change at millennial time scales. In: P. U., Clark, R. S. Webb, & L. D. Keigwin, (eds.). American Geophysical Union, *Geophysical Monograph 112*, Washington, DC, 59-76.

[595] Curry, W. B. & Oppo, D. W. (2005). Glacial water mass geometry and the distribution of $\delta^{13}C$ of $\sum CO_2$ in the western Atlantic Ocean. *Paleoceanography, 20(1)*, PA1017.

[596] Dahl, K. A., Broccoli, A. J. & Stouffer, R. J. (2005). Assessing the role of North Atlantic freshwater forcing in millennial scale climate variability: A tropical Atlantic perspective. *Climate Dynamics, 24*, 325-346.

[597] Dansgaard, W., White, J. W. C. & Johnsen, S. J. (1989). The abrupt termination of the Younger Dryas climate event. *Nature, 339*, 532-534.

[598] de Humbolt, A., (1814). Voyage aux regions equinoxiales du nouveaux continent, fait en 1799-1804 par Al. de Humboldt et A. Bonpland. Part 1. Relation historique, 1, F. Schoell, Paris. [H.M. Williams, translator (3d ed.), 1822, Longman, Hurst, Rees, Orme and Brown, London, 1, 293 p.

[599] Delworth, T. L. & Mann, M. E. (2000). Observed and simulated multidecadal variability in the Northern Hemisphere. *Clim. Dyn., 16*, 661-676.

[600] Delworth, T. L., Zhang, R. & Mann, M. E. (2007). Decadal to centennial variability of the Atlantic from observations and models. *AGU monograph Past and Future Changes of the Ocean's Meridional Overturning Circulation: Mechanisms and Impacts*, accepted.

[601] Dengg, J., Beckmann, A. & Gerdes, R. (1996). The gulf stream separation problem. In: *The warmwatersphere of the North Atlantic Ocean.* [Krauss, W. (ed.)]. Gebruder-Borntrager, *25*, 3-290.

[602] Denton, G. H., Alley, R. B., Comer, G. C. & Broecker, W. S. (2005). The role of seasonality in abrupt climate change. *Quat. Sci. Rev., 24*, 1159-1182.

[603] Dickson, R. R. & Brown, J. (1994). The production of North Atlantic deep water: Sources, rates, and pathways. *J. Geophys. Res., 99*, C6, 12319-12341.

[604] Dickson, R. R., Lazier, J., Meincke, J., Rhines, P. & Swift, J. (1996). Long-term coordinated changes in the convective activity of the North Atlantic. *Prog. Oceanogr., 38*, 241-295

[605] Dixon, K., Delworth, T., Spelman, M. & Stouffer, R. (1999). The influence of transient surface fluxes on North Atlantic overturning in a coupled GCM climate change experiment. *Geophysical Research Letters, 26*, 2749–2752.

[606] Dong, B.W. & Sutton, R. T. (2007). Enhancement of ENSO variability by a weakened Atlantic thermohaline circulation in a coupled GCM. *Journal of Climate*, in press.

[607] Dong, B. W., Sutton, R. T. & Scaife, A. A. (2006). Multidecadal modulation of El Nino Southern Oscillation (ENSO) variance by Atlantic Ocean sea surface temperatures. *Geophys. Res. Letters, 3*, doi: 10.1 029/2006GL025766.

[608] Douville, H. & Royer, J. F. (1996). Sensitivity of the Asian summer monsoon to an anomalous Eurasian snow cover with the Meteo-France GCM. *Climate Dynamics, 12*, 449-466.

[609] Duplessy J. C., Shackleton, N. J., Fairbanks, R. G., Labeyrie, L., Oppo, D. & Kallel, N. (1988). Deepwater source variations during the last climatic cycle and their impact on the global deepwater circulation. *Paleoceanography, 3*, 343-3 60.

[610] Elderfield, H., Yu, J., Anand, P., Kiefer, T. & Nyland, B. (2006). Calibrations for benthic foraminiferal Mg/Ca paleothermometry and the carbonate ion hypothesis. *Earth Planet. Sci. Let., 250*, 633-649.

[611] Elliot, M., Labeyrie, L. D. & Duplessy, J. C. (2002). Changes in North Atlantic deep-water formation associated with the Dansgaard-Oeschger temperature oscillations (60-10 ka). *Quat. Sci. Rev., 21*, 1153-1165.

[612] Ellison, C. R. W., Chapman, M. R. & Hall, I. R. (2006). Surface and deep ocean interactions during the cold climate event 8200 years ago. *Science, 312*, 1929- 1932.

[613] Enfield, D. B., Mestas-Nuñez, A. M. & Trimble, P. J. (2001). The Atlantic multidecadal oscillation and its relation to rainfall and river flows in the continental U.S. *GRL, 28*, 2077-2080.

[614] EPICA Community Members, (2006). One-to-one coupling of glacial climate variability in Greenland and Antarctica. *Nature, 444*, 195-198.

[615] Flower, B. P., Hastings, D. W., Hill, H. W. & Quinn, T. M. (2004). Phasing of deglacial warming and Laurentide Ice Sheet meltwater in the Gulf of Mexico. *Geology, 32*, 597-600.

[616] Folland, C. K., Palmer, T. N. & Parker, D. E. (1986). Sahel rainfall and worldwide sea temperatures. *Nature, 320*, 602-607.

[617] Ganachaud, A. (2003a). Large-scale mass transports, water mass formation, and diffusivities estimated from World Ocean Circulation Experiment (WOCE) hydrographic data. *Journal of Geophysical Research-Oceans, 108*, 24.

[618] Ganachaud, A. (2003b). Error budget of inverse box models: The North Atlantic. *Journal of Atmospheric and Oceanic Technology, 20*, 1641-1655.

[619] Ganachaud, A. & Wunsch, C. (2000). Improved estimates of global ocean circulation, heat transport, and mixing from hydrographic data. *Nature, 408*, 453-457

[620] Giannini, A., Saravanan, R. & Chang, P. (2003). Oceanic Forcing of Sahel Rainfall on Interannual to Interdecadal Time Scales. *Science, 302(5647)*, 1027-1030, doi: 10.1 126/science.1089357.

[621] Gildor, H. & Tziperman, E. (2001). A sea ice climate switch mechanism for the 100-kyr glacial cycles. *J. Geophys. Res., 106*, C5, 9117-9133.

[622] Girton, J. G. & Sanford, T. B. (2003). Descent and modification of the overflow plume in the Denmark Strait. *J. Phys. Oceanogr., 33*, 7, 135 1-1364

[623] Gnanadesikan, A., Slater, R. D., Swathi, P. S. & Vallis, G. K. (2005). The energetics of ocean heat transport. *J. Clim., 18*, 2604-2616.

[624] Goldenberg, S. B., Landsea, C. W., Mestas-Nuñez, A. M. & Gray, W. M. (2001). The recent increase in Atlantic hurricane activity: Causes and implications. *Science, 293*, 474-479.

[625] Gordon, A. L. (1986). Inter-ocean exchange of thermocline water. *Journal of Geophysical Research-Oceans, 91*, 5037-5046.

[626] Gregory, J. M., et al. (2005). A model intercomparison of changes in the Atlantic thermohaline circulation in response to increasing atmospheric CO_2 concentration. *Geophys. Res. Lett., 32*, L12703, doi: 10.1029/2005GL023209.

[627] Grootes, P. M., Stuiver, M., White, J. W. C., Johnsen, S. J. & Jouzel, J. (1993). Comparison of oxygen isotope records from the GISP2 and GRIP Greenland ice cores. *Nature, 366*, 552-554.

[628] Gupta, A. K., Anderson, D. M. & Overpeck, J. T. (2003). Abrupt changes in the Asian southwest monsoon during the Holocene and their links to the North Atlantic Ocean. *Nature, 421*, 354-357.

[629] Hagen, S. & Keigwin, L. D. (2002). Sea-surface temperature variability and deep water reorganisation in the subtropical North Atlantic during Isotope Stage 2-4. *Mar. Geol., 189*, 145-162.

[630] Hall, I. R., Bianchi, G. G. & Evans, J. R. (2004). Centennial to millennial scale Holocene climate-deep water linkage in the North Atlantic. *Quat. Sci. Rev., 23*, 1529–1536.

[631] Hall, M. M. & Bryden, H. L. (1982). Direct estimates and mechanisms of ocean heat transport. *Deep-Sea Research Part A-Oceanographic Research Papers, 29*, 339- 359.

[632] Hallberg, R. & Gnanadesikan, A. (2006). The role of eddies in determining the structure and response of the wind-driven Southern Hemisphere overturning: Results from the Modeling Eddies in the Southern Ocean (MESO) Project. *J. Phys. Oceanogr., 36*, 2232-2252.

[633] Hammer, C. U., Clausen, H. B. & Langway, C. C. Jr., (1994). Electrical conductivity method (ECM) stratigraphic dating of the Byrd Station ice core, Antarctica. *Ann. Glaciol., 20*, 115–120.

[634] Haug, G. H., Hughen, K. A., Sigman, D. M., Peterson, L. C. & Röhl, U. (2001). Southward migration of the Intertropical Convergence Zone through the Holocene. *Science, 293*, 1304-1308.

[635] Held, I. M., Delworth, T. L., Lu, J., Findell, K. L. & Knutson, T. R. (2005). Simulation of Sahel drought in the 20th and 21st centuries. *Proceedings of the National Academy of Sciences, 102(50)*, 17891-17896.

[636] Hemming, S. R. (2004). Heinrich events: Massive late Pleistocene detritus layers of the North Atlantic and their global climate imprint. *Rev. Geophysics, 42*, RG1005, doi: 10.1 029/2003RG0001 28.

[637] Hendy, I. L. & Kennett, J. P. (2000). Dansgaard-Oeschger cycles and the California Current System: Planktonic foraminiferal response to rapid climate change in Santa Barbara Basin, Ocean Drilling Program hole 893A. *Paleoceanography, 15*, 30-42.

[638] Hewitt, C. D., Broccoli, A. J., Crucifix, M., Gregory, J. M., Mitchell, J. F. B. & Stouffer, R. J. (2006). The effect of a large freshwater perturbation on the glacial North Atlantic ocean using a coupled general circulation model. *Journal of Climate, 19*, 4436-4447.

[639] Hillaire-Marcel, C., de Vernal, A., Bilodeau, G. & Weaver, A. J. (2001). Absence of deep- water formation in the Labrador Sea during the last interglacial period. *Nature, 410*, 1073-1077.

[640] Holland, M. M., Bitz, C. M. & Tremblay, B. (2006). Future abrupt reductions in the summer Arctic sea ice. *Geophys. Res. Lett., 33*, L23 503, doi: 10.1 029/2006GL028024.

[641] Hu, A., Meehl, G. A., Washington, W. M. & Dai. A. (2004). Response of the Atlantic thermohaline circulation to increased atmospheric CO_2 in a coupled model. *Journal of Climate, 17*, 4267-4279.

[642] Huber, C., Leuenberger, M., Spahni, R., Fluckiger, J., Schwander, J., Stocker, T. F., Johnsen, S., Landais, A. & Jouzel, J. (2006). Isotope calibrated Greenland temperature record over Marine Isotope Stage 3 and its relation to CH_4. *Earth and Planetary Science Letters, 243*, 504-519.

[643] Hughen, K. A., Southon, J. R., Lehman, S. J. & Overpeck, J. T. (2000). Synchronous radiocarbon and climate shifts during the last deglaciation. *Science, 290,* 1951- 1954.

[644] Hughes, G. O. & Griffiths, R. W. (2006). A simple convective model of the global overturning circulation, including effects of entrainment into sinking regions. *Ocean Modelling, 12,* 46-79.

[645] Jia, Y, (2003). Ocean heat transport and its relationship to ocean circulation in the CMIP coupled models. *Climate Dynamics, 20,* 153-174.

[646] Johnsen, S. J., Dansgaard, W., Clausen, H. B. & Langway, C. C. Jr., (1972). Oxygen isotope profiles through the Antarctic and Greenland ice sheets. *Nature, 235,* 429- 434.

[647] Johnsen, S. J., et al. (1992). Irregular glacial interstadials recorded in a new Greenland ice core. *Nature, 359,* 311-313.

[648] Johnson, R. G. & McClure, B. T. (1976). A model for Northern Hemisphere continental ice sheet variation. *Quat. Res., 6,* 325-3 53.

[649] Jungclaus, J. H., Haak, H., Esch, M., Roeckner, E. & Marotzke, J. (2006). Will Greenland melting halt the thermohaline circulation? *Geophysical Research Letters, 33,* L17708, doi:10. 1029/2006GL026815.

[650] Kanzow, T., Cunningham, S., Rayner, D., Hirschi, J., Johns, W. E., Baringer, M., Bryden, H., Beal, L., Meinen, C. & Marotzke, J. (2007). Flow compensation associated with the meridional overturning circulation. *Science,* in press.

[651] Keigwin, L. D. (2004). Radiocarbon and stable isotope constraints on Last Glacial Maximum and Younger Dryas ventilation in the western North Atlantic. *Paleoceanography, 19,* 4.

[652] Keigwin, L. D. & Boyle, E. A. (2000). Detecting Holocene changes in thermohaline circulation. *Proc. Natl. Acad. Sci., 97,* 1343-1346.

[653] Keigwin, L. D., Sachs, J. P., Rosenthal, Y. & Boyle, E. A. (2005). The 8200 year BP event in the slope water system, western subpolar North Atlantic. *Paleocean., 20,* PA2003, doi: 10.1029/2004PA001074.

[654] Keigwin, L. D. & Schlegel, M. A. (2002). Ocean ventilation and sedimentation since the glacial maximum at 3 km in the western North Atlantic. *Geochem. Geophy. Geosy., 3.*

[655] Kissel, C., Laj, C., Labeyrie, L., Dokken, T., Voelker, A. & Blamart, D. (1999). Rapid climatic variations during marine isotopic stage 3: Magnetic analysis of sediments from Nordic Seas and North Atlantic. *Earth Planet. Sci. Let., 171,* 489-502.

[656] Kleiven, H. F., Kissel, C., Laj, C., Ninnemann, U. S., Richter, T. O. & Cortijo, E. (2008). Reduced North Atlantic deep water coeval with the glacial lake Agassiz freshwater outburst. *Science, 319,* 60-64.

[657] Knight, J. R., Allan, R. J., Folland, C. K., Vellinga, M. & Mann, M. E. (2005). A signature of persistent natural thermohaline circulation cycles in observed climate. *Geophysical Research Letters, 32,* 4, doi: 10.1 029/2005GL02423 3.

[658] Knight, J. R., Folland, C. K. & Scaife, A. A. (2006). Climate impacts of the Atlantic Multidecadal Oscillation. *GRL, 33,* doi: 10.1 029/2006GL026242.

[659] Koltermann, K. P., Sokov, A. V., Tereschenkov, V. P., Dobroliubov, S. A., Lorbacher, K. & Sy, A. (1999). Decadal changes in the thermohaline circulation of the North Atlantic, deep-sea research part II. *Topical Studies in Oceanography, 46,* 109- 138.

[660] Koutavas, A., Lynch-Stieglitz, J., Marchitto, T. M. Jr. & Sachs, J. P. (2002). El Niño–like pattern in Ice Age tropical Pacific sea surface temperature. *Science, 297,* 226-230.

[661] Kucera, M., et al., (2005). Reconstruction of sea-surface temperatures from assemblages of planktonic foraminifera: multi-technique approach based on geographically constrained calibration data sets and its application to glacial Atlantic and Pacific Oceans. *Quat. Sci. Rev.*, *24*, 7-9, 95 1-998.

[662] Kuhlbrodt, T., Griesel, A., Montoya, M., Levermann, A., Hofmann, M. & Rahmstorf, S. (2007). On the driving processes of the Atlantic meridional overturning circulation. *Rev. Geophys.*, *45*, RG200 1, doi: 10.1 029/2004RG000 *166*.

[663] Landsea, C. W. (2005), Hurricanes and global warming. *Nature*, *438*, 11-13.

[664] Latif, M., Böning, C., Willebrand, J., Biastoch, A., Dengg, J., Keenlyside, N. & Schweckendiek, U. (2006). Is the thermohaline circulation changing? *Journal of Climate*, *19*, 4631-4637.

[665] Lavin, A., Bryden, H. L. & Parilla, G. (1998). Meridional transport and heat flux variations in the subtropical North Atlantic. *Global Atmos. Ocean Sys.*, *6*, 269- 293.

[666] Lea, D. W., Pak, D. K. & Spero, H. J. (2000). Climate impact of late Quaternary equatorial Pacific sea surface temperature variations. *Science*, *289*, 17, 19-1724.

[667] LeGrande, A. N., Schmidt, G. A., Shindell, D. T., Field, C. V., Miller, R. L., Koch, D. M., Faluvegi, G. & Hoffmann, G. (2006). Consistent simulations of multiple proxy responses to an abrupt climate change event. *Proceedings of the National Academy of Sciences*, *103*, 837-842.

[668] Levermann, A., Griesel, A., Hofmann, M., Montoya, M. & Rahmstorf, S. (2005). Dynamic sea level changes following changes in the thermohaline circulation. *Clim. Dynamics*, *24*, 347-354, doi: 10.1007/s003 82-004-0505-y.

[669] Levitus, S., Antonov, J. I., Wang, J., Delworth, T. L., Dixon, K. W. & Broccoli, A. J. (2001). Anthropogenic warming of Earth's climate system. *Science*, *292(5515)*, 267-270.

[670] Li, C., Battisti, D. S., Schrag, D. P. & Tziperman, E. (2005). Abrupt climate shifts in Greenland due to displacements of the sea ice edge. *Geophys. Res. Let.*, *32*, doi: 10.1 029/2005GL023492.

[671] Lozier, S., Kelly, K., Baringer, M. & Delworth, T., et al. (2007). Implementation strategy for a JSOST near-term priority assessing meridional overturning circulation variability: Implications for rapid climate change, October 24, at http://www.usclivar.org/science_status/AMOC/AMOC_Strategy_Document.pdf.

[672] Lu, J. & Delworth, T. L. (2005). Oceanic forcing of the late 20th century Sahel drought. *Geophysical Research Letters*, *32*, L22706, doi: 10.1 029/2005GL023 316.

[673] Lumpkin, R. & Speer, K. (2003). Large-scale vertical and horizontal circulation in the North Atlantic Ocean. *Journal of Physical Oceanography*, *33*, 1902-1920.

[674] Lumpkin, R. & Speer, K. (2007). Global ocean meridional overturning. *J. Phys. Oceanogr.*, *37*, 2550-2562.

[675] Lund, D. C., Lynch-Stieglitz, J. & Curry, W. B. (2006). Gulf Stream density structure and transport during the past millennium. *Nature*, *444*, 601-604.

[676] Lynch-Stieglitz, J., Curry, W. B. & Slowey, N. (1999). Weaker Gulf Stream in the Florida Straits during the Last Glacial Maximum. *Nature*, *402*, 644-648.

[677] Lynch-Stieglitz, J., et al., (2006). Meridional overturning circulation in the South Atlantic at the last glacial maximum. *Geochem. Geophy. Geosy.*, *7*, Q10N03, doi: 10.1 029/2005GC001 226.

[678] Lynch-Stieglitz, J., et al. (2007). Atlantic meridional overturning circulation during the Last Glacial Maximum. *Science, 316(5821)*, 66-69.

[679] Macrander, A., Send, U., Vadimarsson, H., Jónsson, S. & Käse, R. H. (2005). Interannual changes in the overflow from the Nordic Seas into the Atlantic Ocean through Denmark Strait. *J. Geophys. Res., 32*, L06606, doi:10.1029/2004GL021463.

[680] Manabe, S. & Stouffer, R. J. (1988). Two stable equilibria of a coupled ocean-atmosphere model. *Journal of Climate, 1*, 841-866.

[681] Manabe, S. & Stouffer, R. J. (1994). Multiple-century response of a coupled ocean-atmosphere model to an increase of atmospheric carbon dioxide. *Journal of Climate, 7*, 5-23.

[682] Manighetti, B. & McCave, I. N. (1995). Late glacial and Holocene palaeocurrents through South Rockall Gap, NE Atlantic Ocean. *Paleocean., 10*, 611–626.

[683] Mann, M. E. & Emanuel, K. A. (2006). Atlantic hurricane trends linked to climate change. *Eos, 87*, 24, p 233, 238, 241.

[684] Marchal, O., Francois, R., Stocker, T. F. & Joos, F. (2000). Ocean thermohaline circulation and sedimentary Pa-231/Th-230 ratio. *Paleocean., 15*, 625–641.

[685] Marchitto, T. M., et al., (1998). Millennial-scale changes in North Atlantic circulation since the last glaciation. *Nature, 393*, 6685, 557-561.

[686] Marchitto, T. M. & Broecker, W. S. (2006). Deep water mass geometry in the glacial Atlantic Ocean: A review of constraints from the paleonutrient proxy Cd/Ca. *Geochemistry Geophysics Geosystems, 7*, Q12003.

[687] Masson-Delmotte, V., et al., (2005). GRIP deuterium excess reveals rapid and orbital-scale changes in Greenland moisture origin. *Science, 309*, 118-121.

[688] McCabe, G. J., Palecki, M. A. & Betancourt, J. L. (2004). Pacific and Atlantic Ocean influences on multidecadal drought frequency in the United States. *PNAS, 101*, 4136-4141.

[689] McCave, I. N. & Hall, I. R. (2006). Size sorting in marine muds: Processes, pitfalls, and prospects for paleoflow-speed proxies. *Geochemistry, Geophysics, Geosystems, 7*, Q10N05.

[690] McCave, I. N., Manighetti, B. & Beveridge, N. A. S. (1995). Circulation in the glacial North Atlantic inferred from grain-size measurements. *Nature, 374*, 149-15 1.

[691] McManus, J. F., Francois, R. Gherardi, J. M., Keigwin, L. D. & Brown-Leger, S. (2004). Collapse and rapid resumption of the Atlantic meridional circulation linked to deglacial climate changes. *Nature, 428*, 834-837.

[692] Meehl, G. A., Stocker, T. F., Collins, W. D., Friedlingstein, P., Gaye, A. T., Gregory, J. M., Kitoh, A., Knutti, R., Murphy, J. M., Noda, A., Raper, S. C. B., Watterson, I. G., Weaver, A. J. & Zhao, Z. C. (2007). Global climate projections. In: *Climate Change 2007: The Physical Science Basis. Contribution of Working Group 1 to the Fourth Assessment Report of the Intergovernmental Panel on Climate Change*. In: S., Solomon, D., Qin, M., Manning, Z., Chen, M., Marquis, K. B. Averyt, M. Tignor, & H. L. Miller (eds.). Cambridge University Press, Cambridge, United Kingdom, and New York, 996.

[693] Meissner, K. J. & Clark, P. U. (2006). Impact of floods versus routing events on the thermohaline circulation. *Geophysical Research Letters, 33*, L26705.

[694] Mikolajewicz, U., Crowley, T. J., Schiller, A. & Voss, R. (1997). Modelling teleconnections between the North Atlantic and North Pacific during the Younger Dryas. *Nature, 387*, 384-3, 87.

[695] Mikolajewicz, U. & Voss, R. (2000). The role of the individual air-sea flux components in CO_2-induced changes of the ocean's circulation and climate. *Climate Dynamics, 16*, 627-642.

[696] Munk, W. (1966). Abyssal recipes I. *Deep Sea Res. Oceanogr. Abstr., 13*, 707-730.

[697] Munk, W. & Wunsch, C. (1998). Abyssal recipes II: Energetics of tidal and wind mixing. *Deep Sea Res., Part I, 45*, 1977-2010.

[698] North, G. R. (1984). The small ice cap instability in diffusive climate models. *J. Atmos. Sci., 41*, 3390-3395.

[699] Oort, A., Anderson, L. & Peixoto, J. (1994). Estimates of the energy cycle of the Oceans. *J. Geophys. Res., 99*, 7665-7688.

[700] Oppo, D. W., McManus, J. F. & Cullen, J. L. (2003). Deepwater variability in the Holocene epoch. *Nature, 422*, 277-278.

[701] Otto-Bliesner, B. L., et al., (2007). *Last glacial maximum ocean thermohaline circulation*, PMIP2 model intercomparison and data constraints. *Geophysical Research Letters, 34*, L1 2706, doi: 10.1 029/2007GL029475.

[702] Paul, A. & Schafer-Neth, C. (2003). Modeling the water masses of the Atlantic Ocean at the last glacial maximum. *Paleoceanography, 18*, 3.

[703] Peltier, W. R., Vettoretti, G. & Stastna, M. (2006). Atlantic meridional overturning and climate response to Arctic Ocean freshening. *Geophysical Research Letters, 33*, 10.1029/2005GL025251.

[704] Peterson, Bruce, R. M., Holmes, J. W., McClelland, C. J., Vörösmarty, R. B., Lammers, A. I. & Shiklomanov, I. A. (2002). Shiklomanov, Stefan Rahmstorf, 2002: Increasing river discharge to the Arctic Ocean. *Science, 298(5601)*, 2171-2173, doi: 10.1126/science. 1077445.

[705] Peterson, B. J., McClelland, J., Curry, R., Holmes, R. M., Walsh, J. E. & Aagaard, K. (2006). Trajectory shifts in the Arctic and Subarctic freshwater cycle. *Science, 313*, 1061- 1066

[706] Peterson, L. C., Haug, G. H., Hughen, K. A. & Rohl, U. (2000). Rapid changes in the hydrologic cycle of the tropical Atlantic during the last glacial. *Science, 290*, 1947-195, 1.

[707] Pflaumann, U., et al., (2003). Glacial North Atlantic: Sea-surface conditions reconstructed by GLAMAP 2000. *Paleoceanography, 18*, 1065.

[708] Piotrowski, A. M., Goldstein, S. L., Hemming, S. R. & Fairbanks, R. G. (2004). Intensification and variability of ocean thermohaline circulation through the last deglaciation. *Earth Planet. Sci. Let., 225*, 205-220.

[709] Piotrowski, A. M., Goldstein, S. L., Hemming, S. R. & Fairbanks, R. G. (2005). Temporal relationships of carbon cycling and ocean circulation at glacial boundaries. *Science, 307*, 5717, 1933-1938.

[710] Rahmstorf, S., (2002). Ocean circulation and climate during the past 120,000 years. *Nature, 419*, 207-214.

[711] Rahmstorf, S. & Ganopolski, A. (1999). Long-term global warming scenarios computed with an efficient coupled climate model. *Clim. Change, 43*, 353-367.

[712] Randall, D. A., et al., (2007). Climate models and their evaluation. In: *Climate Change 2007: The Physical Science Basis. Contribution of Working Group 1 to the Fourth Assessment Report of the Intergovernmental Panel on Climate Change.* [Solomon, S., Qin, D., Manning, M., Chen, Z., Marquis, M., Averyt, K. B., Tignor, M. & Miller, H. L. (eds.). Cambridge University Press, Cambridge, United Kingdom, and New York.

[713] Rasmussen, T.L., Thomsen, E., Troelstra, S. R., Kuijpers, A. & Prins, M. A. (2002). Millennial-scale glacial variability versus Holocene stability: Changes in planktic and benthic foraminifera faunas and ocean circulation in the North Atlantic during the last 60000 years. *Marine Micropaleontology, 47*, 143-176.

[714] Rickaby, R. E. M. & Elderfield, H. (2005). Evidence from the high-latitude North Atlantic for variations in Antarctic Intermediate water flow during the last deglaciation. *Geochem. Geophys. Geosys.*, 6, Q05001, doi: 10.1 029/2004GC00085 8.

[715] Ridley, J. K., Huybrechts, P., Gregory, J. M. & Lowe, J. A. (2005). Elimination of the Greenland ice sheet in a high CO_2 climate. *Journal of Climate, 17*, 3409-3427.

[716] Roberts, M. J., Banks, H., Gedney, N., Gregory, J., Hill, R., Mullerworth, S., Pardaens, A., Rickard, G., Thorpe, R. & Wood, R. (2004). Impact of an eddy-permitting ocean resolution on control and climate change simulations with a global coupled GCM. *Journal of Climate, 17*, 3-20.

[717] Roberts, M. J. & Wood, R. A. (1997). Topography sensitivity studies with a Bryan-Cox type ocean model. *Journal of Physical Oceanography, 27*, 823-83 6.

[718] Robinson, L. F., et al. (2005). Radiocarbon variability in the western North Atlantic during the last deglaciation, Science, *310*, 5753, 1469-1473.

[719] Roemmich, D. & Wunsch, C. (1985). Two transatlantic sections: meridional circulation and heat flux in the subtropical North Atlantic Ocean. *Deep Sea Research, 32*, 619-664.

[720] Rooth, C. (1982). Hydrology and ocean circulation. *Prog. Ocean., 11*, 13 1–149.

[721] Rossby, T. (1996). The North Atlantic current and surrounding waters: At the crossroads. *Reviews of Geophysics, 34*, 463-481.

[722] Ruddiman, W. F. & McIntyre, A. (1981). The mode and mechanism of the last deglaciation: Oceanic evidence. *Quaternary Research, 16*, 125-134.

[723] Rumford, B., Count of., (1800). Essay VII, The propagation of heat in fluids. In: *Essays, political, economical, and philosophical, A new edition, 2.* T. Cadell, Jr. & W. Davies, London, p. 197-3 86.

[724] Rutberg, R. L., et al. (2000). Reduced North Atlantic deep water flux to the glacial Southern Ocean inferred from neodymium isotope ratios. *Nature, 405*, 6789, 935- 938.

[725] Samelson, R. (2004). Simple mechanistic models of mid-depth meridional overturning. *J. Phys. Oceanogr., 34*, 2096-2103

[726] Sandström, J. W. (1908). Dynamische versuche mit meerwasser. *Ann. Hydrogr. Mar. Meteorol.*, 36, 6-23.

[727] Sarnthein, M., Winn, K., Jung, S. J. A., Duplessy, J. C., Erlenkeuser, H. & Ganssen, G. (1994). Changes in east Atlantic deepwater circulation over the last 30,000 years: Eight time slice reconstructions. *Paleoceanography, 9*, 209-267.

[728] Schaeffer, M., Selten, F. M., Opsteegh, J. D. & Goosse, H. (2002). Intrinsic limits to predictability of abrupt regional climate change in IPCC SRES scenarios. *Geophys. Res. Lett.*, 29, doi: 10.1 029/2002GL0 15254.

[729] Schmittner, A. (2005). Decline of the marine ecosystem caused by a reduction in the Atlantic overturning circulation. *Nature, 434*, 628-63 3.

[730] Schmittner, A., Galbraith, E. D., Hostetler, S. W., Pedersen, T. F. & Zhang, R. (2007). Large fluctuations of dissolved oxygen in the Indian and Pacific oceans during Dansgaard-Oeschger oscillations caused by variations of North Atlantic deep water subduction. *Paleooceanography*, *22*, PA3207, doi: 10.1 029/2006PA00 1384.

[731] Schmittner, A., Latif, M. & Schneider, B. (2005). Model projections of the North Atlantic thermohaline circulation for the 21st century assessed by observations. *Geophysical Research Letters*, *32*, L23710, doi: 10. 1029/2005GL024368.

[732] Schmittner, A., Meissner, K. J., Eby, M. & Weaver, A. J. (2002). Forcing of the deep ocean circulation in simulations of the Last Glacial Maximum. *Paleoceanography*, *17*, 1015.

[733] Schneider, B., Latif, M. & Schmittner, A. (2007). Evaluation of different methods to assess model projections of future evolution of the Atlantic meridional overturning circulation. *Journal of Climate*, *20*, 2121-2132.

[734] Schweckendiek, U. & Willebrand, J. (2005). Mechanisms for the overturning response in global warming simulations. *Journal of Climate*, *18*, 4925-4936.

[735] Seager, R. & Battisti, D. S. (2007). Challenges to our understanding of the general circulation: abrupt climate change. In: *Global Circulation of the Atmosphere*. In: T. Schneider, & A. H. Sobel (eds.). Princeton University Press, Princeton, NJ, in press.

[736] Shackleton, N. J., Hall, M. A. & Vincent, E. (2000). Phase relationships between millennial scale events 64,000 to 24,000 years ago. *Paleoceanography*, *15*, 565- 569.

[737] Siddall, M., et al., (2007). Modeling the relationship between 23 1Pa/230Th distribution in North Atlantic sediment and Atlantic meridional overturning circulation. *Paleocean.*, *22*, PA2214, doi: 10. 1029/2006PA001358.

[738] Skinner, L. C. & Elderfield, H. (2007). Rapid fluctuations in the deep North Atlantic heat budget during the last glacial period. *Paleoceanography*, *22*, PA1205, doi:10.1029/ 2006PA001338.

[739] Sloyan, B. M. & Rintoul, S. R. (2001). The southern Ocean limb of the global deep overturning circulation. *J. Phys. Oceanogr.*, *31*, 143-173.

[740] Smethie, W. M., Jr. & Fine, R. A. (2001). Rates of North Atlantic deep water formation calculated from chlorofluorocarbon inventories. *Deep Sea Research, Part I*, *48*, 189-2 15.

[741] Smith, R. D., Maltrud, M. E., Bryan, F. O. & Hecht, M. W. (2000). Numerical simulation of the North Atlantic Ocean at 1/10°. *Journal of Physical Oceanography*, *30*, 1532- 1561.

[742] Sowers, T. & Bender, M. (1995). Climate records covering the last deglaciation. *Science*, *269*, 210-214.

[743] Stammer, D., Wunsch, C., Giering, R., Eckert, C., Heimbach, P., Marotzke, J., Adcroft, A., Hill, C. N. & Marshall, J. (2003). Volume, heat, and freshwater transports of the global ocean circulation 1993-2000, estimated from a general circulation model constrained by World Ocean Circulation Experiment (WOCE) data. *Journal of Geophysical Research*, *108*, 3007, doi: 10. 1029/2001JC001 115.

[744] Stocker, T. F. & Johnsen, S. J. (2003). A minimum thermodynamic model for the bipolar seesaw. *Paleoceanography*, *18*, doi: 10.1 029/2003PA000920.

[745] Stommel, H. (1958). The abyssal circulation. *Deep-Sea Research*, *5*, 80-82.

[746] Stoner, J. S., Channell, J. E. T., Hillaire-Marcel, C. & Kissel, C. (2000). Geomagnetic paleointensity and environmental record from Labrador Sea core MD95-2024: Global

marine sediment and ice core chronostratigraphy for the last 110 kyr. *Earth Planet. Sci. Let., 183*, 161-177.

[747] Stott, L., Poulsen, C., Lund, S. & Thunell, R. (2002). Super ENSO and global climate oscillations at millennial time scales. *Science, 297*, 222-226.

[748] Stouffer, R. J. & Manabe, S. (1999). Response of a coupled ocean-atmosphere model to increasing atmospheric carbon dioxide: Sensitivity to the rate of increase. *Journal of Climate, 12*, 2224-2237.

[749] Stouffer, R. J. & Manabe, S. (2003). Equilibrium response of thermohaline circulation to large changes in atmospheric CO_2 concentration. *Climate Dynamics, 20*, 759-773. Stouffer, R.J., et al., 2006: Investigating the causes of the response of the thermohaline circulation to past and future climate changes. *Journal of Climate, 19*, 1365-13 87.

[750] Stroeve, J., Holland, M. M., Meier, W., Scambos, T. & Serreze, M. (2007), Arctic sea ice decline: Faster than forecast, Geophys. Res. Lett., *34*, L09501, doi: 10.1 029/2007GL029703.

[751] Stroeve, J., Serreze, M., Drobot, S., Gearheard, S., Holland, M., Maslanik, J., Meier, W. & Scambos, T. (2008). Arctic sea ice extent plummets in 2007. *Eos, 89(2)*, 13-14.

[752] Stuiver, M. & Grootes, P. M. (2000). GISP2 oxygen isotope ratios. *Quat. Res., 53*, 277-284.

[753] Sutton, R. T. & Hodson, D. L. R. (2005). Atlantic Ocean forcing of North American and European summer climate. *Science, 309*, 115-118.

[754] Sutton, R. & Hodson, D. (2007). Climate response to basin-scale warming and cooling of the North Atlantic Ocean. *Journal of Climate, 20*, 89 1-907.

[755] Talley, L. D., Reid, J. L. & Robbins, P. E. (2003). Data-based meridional overturning streamfunctions for the global ocean. *Journal of Climate, 16*, 32 13-3226.

[756] Tang, Y. M. & Roberts, M. J. (2005). The impact of a bottom boundary layer scheme on the North Atlantic Ocean in a global coupled climate model. *Journal of Physical Oceanography, 35*, 202-2 17.

[757] Thorpe, R. B., Wood, R. A. & Mitchell, J. F. B. (2004). The sensitivity of the thermohaline circulation response to preindustrial and anthropogenic greenhouse gas forcing to the parameterization of mixing across the Greenland-Scotland ridge. *Ocean Modelling, 7*, 259-268.

[758] Timmermann, A., An, S. I., Krebs, U. & Goose, H. (2005a). ENSO suppression due to weakening of the Atlantic thermohaline circulation. *Journal of Climate, 18*, 3 122- 3 139.

[759] Timmermann, A., Krebs, U., Justino, F., Goosse, H. & Ivanochko, T. (2005b). Mechanisms for millennial-scale global synchronization during the last glacial period. *Paleoceanography, 20*, PA4008, doi: 10.1 029/2004PA00 1090.

[760] Timmermann, A., et al. (2007). The influence of a weakening of the Atlantic meridional overturning circulation on ENSO. *Journal of Climate*, in press.

[761] Toggweiler, J. R. & Samuels, B. (1993a). Is the magnitude of the deep outflow from the Atlantic Ocean actually governed by Southern Hemisphere winds? In: *The Global Carbon Cycle*. NATO ASI Ser., Ser. I, [Heimann, M. (ed.)]. Springer, New York, 333-366.

[762] Toggweiler, J. R. & Samuels, B. (1993b). New radiocarbon constraints on the upwelling of abyssal water to the ocean's surface. In: *The Global Carbon Cycle*. NATO ASI Ser., Ser. I, [Heimann, M. (ed.)]. Springer, New York, 303-331.

[763] Toggweiler, J. R. & Samuels, B. (1995). Effect of Drake passage on the global thermohaline circulation. *Deep Sea Res., Part 1, 42*, 477-500.

[764] Toggweiler, J. R. & Samuels, B. (1998). On the ocean's large scale circulation in the limit of no vertical mixing. *J. Phys. Oceanogr., 28*, 1832-1852.

[765] Toresen, R. & Østvedt, O. J. (2000). Variation in abundance of Norwegian spring-spawning herring (Clupea harengus, Clupeidae) throughout the 20th century and the influence of climatic fluctuations. *Fish and Fisheries, 1*, 231-256.

[766] Velicogna, I. & Wahr, J. (2006). Acceleration of Greenland ice mass loss in spring 2004. *Nature, 443*, 329-33 1.

[767] Vellinga, M. & Wood, R. A. (2002). Global climatic impacts of a collapse of the Atlantic thermohaline circulation. *Clim. Change, 54*, 25 1-267.

[768] Vellinga, M. A. & Wood, R. A, (2007). Impacts of thermohaline circulation shutdown in the twenty-first century. *Clim. Change*, doi: 10.1007/s10584-006-9146-y.

[769] Vikebo, F. B., Sundby, S., Adlandsvik, B. & Ottera, O. H. (2007). Impacts of a reduced thermohaline circulation on transport and growth of larvae and pelagic juveniles of Arcto-Norwegian cod (*Gadus morhua*). *Fish. Oceanogr., 16(3)*, 2 16-228.

[770] Voss, R. & Mikolajewicz, U. (2001). Long-term climate changes due to increased CO_2 concentration in the coupled atmosphere-ocean general circulation model ECHAM3/LSG. *Climate Dynamics, 17*, 45-60.

[771] Wadhams, P., Holfort, J., Hansen, E. & Wilkinson, J. P. (2002), A deep convective chimney in the winter greenland sea, Geophys. Res. Lett. *29(10)*, 1434, doi: 10.1029/2001GL014306.

[772] Wang, Y. J., Cheng, H., Edwards, R. L., An, Z. S., Wu, J. Y., Shen, C. C. & Dorale, J. A. (2001). A high-resolution absolute-dated late Pleistocene monsoon record from Hulu Cave, China. *Science, 294*, 2345-2348.

[773] Wang, X., Auler, A. S., Edwards, R. L., Cheng, H., Cristalli, P. S., Smart, P. L., Richards, D. A., Shen, C. C. (2004). Wet periods in northeastern Brazil over the past 210 kyr linked to distant climate anomalies. *Nature, 432*, 740-743.

[774] Weaver, A. J., Eby, M., Kienast, M. & Saenko, O. A. (2007). Response of the Atlantic meridional overturning circulation to increasing atmospheric CO_2: Sensitivity to mean climate state. *Geophysical Research Letters, 34*, L05708, doi: 10.1 029/2006GL028756.

[775] Weaver, A. J. & Hillaire-Marcel, C. (2004a). Ice growth in the greenhouse: A seductive paradox but unrealistic scenario. *Geoscience Canada, 31*, 77-85.

[776] Weaver, A. J. & Hillaire-Marcel, C. (2004b). Global warming and the next ice age. *Science, 304*, 400-402.

[777] Webb, D. J. & Suginohara, N. (2001). Vertical mixing in the ocean. *Nature, 409*, 37. Weber, S.L., et al., 2007: The modern and glacial overturning circulation in the Atlantic. ocean in PMIP coupled model simulations. *Climate of the Past, 3*, 51-64.

[778] Wiersma, A. P., Renssen, H., Goosse, H. & Fichefet, T. (2006). Evaluation of different freshwater forcing scenarios for the 8.2 ka BP event in a coupled climate model. *Climate Dynamics, 27*, 831-849.

[779] Willamowski, C. & Zahn, R. (2000). Upper ocean circulation in the glacial North Atlantic from benthic foraminiferal isotope and trace element fingerprinting. *Paleocean., 15*, 5 15-527.

[780] Willebrand, J., Barnier, B., Boning, C., Dieterich, C., Killworth, P. D., Le Provost, C., Jia, Y., Molines, J. M. & New, A. L. (2001). Circulation characteristics in three eddy-permitting models of the North Atlantic. *Progress in Oceanography, 48*, 123-161.

[781] Winguth, A. M. E., Archer, D., Duplessy, J. C., Maier-Reimer, E. & Mikolajewicz, U. (1999). Sensitivity of paleonutrient tracer distributions and deep-sea circulation to glacial boundary conditions. *Paleocean., 14*, 304-323.

[782] Winton, M., (2006). Does the Arctic sea ice have a tipping point? *Geophys. Res. Lett., 33*, L23 504, doi: 10. 1029/2006GL028017.

[783] Winton, M., Hallberg, R. & Gnanadesikan, A. (1998). Simulation of density-driven frictional downslope flow in z-coordinate ocean models. *Journal of Physical Oceanography, 28*, 2163-2174.

[784] Wood, R. A., Keen, A. B., Mitchell, J. F. B. & Gregory, J. M. (1999). Changing spatial structure of the thermohaline circulation in response to atmospheric CO_2 forcing in a climate model. *Nature, 399*, 572-575.

[785] Wood, R. A., Vellinga, M. & Thorpe, R. (2003). Global warming and thermohaline circulation stability. *Philosophical Transactions of the Royal Society of London Series A, 361*, 1961-1975.

[786] Wu, P., Wood, R. & Stott, P. (2004). Does the recent freshening trend in the North Atlantic indicate a weakening thermohaline circulation? *Geophys. Res. Lett., 31(2)*, L02301, doi:10.129/2003GL018584.

[787] Wunsch, C., (1996). *The ocean circulation inverse problem.* Cambridge University Press, Cambridge, United Kingdom, 458.

[788] Wunsch, C., (1998). The work done by the wind on the general circulation. *J. Phys. Oceanogr., 28*, 2332-2340.

[789] Wunsch, C., (2003). Determining paleoceanographic circulations, with emphasis on the Last Glacial Maximum. *Quat. Sci. Rev., 22*, 371-385.

[790] Wunsch, C. & Ferrari, R. (2004). Vertical mixing, energy and the general circulation of the oceans. *Annu. Rev. Fluid Mech., 36*, 281-3 14.

[791] Yoshida, Y., et al. (2005). Multi-century ensemble global warming projections using the Community Climate System Model (CCSM3). *Journal of the Earth Simulator, 3*, 2-10.

[792] Yu, E. F., Francois, R. & Bacon, P. (1996). Similar rates of modern and last-glacial ocean thermohaline circulation inferred from radiochemical data. *Nature, 379*, 689-694.

[793] Zahn, R., Schonfeld, J., Kudrass, H. R., Park, M. H., Erlenkeuser, H. & Grootes, P. (1997). Thermohaline instability in the North Atlantic during meltwater events: Stable isotope and ice-rafted detritus records from core S075-26KL, Portugese margin. *Paleoceanography, 12*, 696-7 10.

[794] Zhang, R., (2007). Anticorrelated multidecadal variations between surface and subsurface tropical North Atlantic. *Geophysical Research Letters*, 34, L12713, doi: 10.1 029/2007GL030225.

[795] Zhang, R. & Delworth, T. L. (2005). Simulated tropical response to a substantial weakening of the Atlantic thermohaline circulation. *Jour. Clim., 18*, 1853-1860.

[796] Zhang, R. & Delworth, T. L. (2006). Impact of Atlantic multidecadal oscillations on India/Sahel rainfall and Atlantic hurricanes. *Geophysical Research Letters, 33(5)*, doi: 10.1 029/2006GL026267.

[797] Zhang, R., Delworth, T. L. & Held. I. M. (2007a). Can the Atlantic Ocean drive the observed multidecadal variability in Northern Hemisphere mean temperature? *Geophysical Research Letters, 34*, L02709, doi: 10. 1029/2006GL028683.

[798] Zhang, X. & Walsh, J. E. (2006). Toward a seasonally ice-covered Arctic Ocean: scenarios from the IPCC AR4 model simulations. *J. Clim., 19*, 1730-1747.

[799] Zhang, X., et al., (2007b). Detection of human influence on twentieth-century precipitation trends. *Nature, 468*, 461-465.

[800] Zickfeld, K., Levermann, A., Granger Morgan, M., Kuhlbrodt, T., Rahmstorf, S. & Keith, D. (2007). Expert judgements on the response of the Atlantic meridional overturning circulation to climate change. *Clim. Change, 82*, doi:10.1007/s10584- 007- 9246-3.

[801] ACIA, (2004). Impacts of a warming Arctic: Arctic climate impact assessment. Cambridge University Press, 144.

[802] Alley, R. B. & Agustsdottir, A. M. (2005). The 8k event: cause and consequences of a major Holocene abrupt climate change. *Quaternary Science Reviews, 24(10-1 1)*, 1123-1149.

[803] Anisimov, O. A. & Nelson, F. E. (1997). Permafrost zonation and climate change in the northern hemisphere: Results from transient general circulation models. *Clim. Change, 35*, 241-258.

[804] Aoki, Y., Shimizu, S., Yamane, T., Tanaka, T., Nakayama, K., Hayashi, T. & Okuda, Y. (2000). Methane hydrate accumulation along the western Nankai Trough. *Gas Hydrates: Challenges for the Future*, 136-145.

[805] Archer, D. (2005). Fate of fossil-fuel CO_2 in geologic time. *J. Geophysical Res. Oceans*, doi: 10.1 029/2004JC002625.

[806] Archer, D. (2007). Methane hydrate stability and anthropogenic climate change. *Biogeosciences, 4*, 521-544,.

[807] Archer, D. E. & Buffett, B. (2005). Time-dependent response of the global ocean clathrate reservoir to climatic and anthropogenic forcing. *Geochem., Geophys., Geosys., 6(3)*, doi: 10.1029/2004GC000854.

[808] Archer, D., Kheshgi, H. & Maier-Riemer, E. (1997). Multiple timescales for neutralization of fossil fuel CO2. *Geophys. Res. Letters, 24*, 405-408.

[809] Archer, D., Martin, P., Buffett, B., Brovkin, V., Rahmstorf, S. & Ganopolski, A. (2004). The importance of ocean temperature to global biogeochemistry. *Earth and Planetary Science Letters, 222*, 333-348.

[810] Bartlett, K. B. & Harriss, R. C. (1993). Review and assessment of methane emissions from wetlands. *Chemosphere, 26*, 261-320.

[811] Belyea L. R. & Baird, A. J. (2007). Beyond "the limits to peat bog growth": Cross-scale feedback in peatland development. *Ecol. Mongr., 76*, 299–322.

[812] Bergamaschi, P., et al. (2007). Satellite chartography of atmospheric methane from SCIAMACHY on board ENVISAT: 2. Evaluation based on inverse model simulations. *J. Geophys. Res., 112*, D02304, doi:10.1029/2006JD007268.

[813] Berner, R. A., Lasaga, A. C. & Garrels, R.M. (1983). The carbonate-silicate geochemical cycle and its effect on atmospheric carbon dioxide over the past 100 million years. *Am. J. Sci., 283*, 641-683.

[814] Blake, D. R. & Rowland, F. S. (1988). Continuing worldwide increase in tropospheric methane, 1978-1987. *Science, 239*, 1129-113 1.

[815] Blunier, T. & Brook, E. J. (2001). Timing of millennial-scale climate change in Antarctica and Greenland during the last glacial period. *Science*, *291*, 109-112.

[816] Boswell, R. (2005). Changing perspectives on the resource potential of methane hydrates. In: Fire in the Ice, newsletter of the U.S. Department of Energy, Office of Fossil Energy, National Energy Technology Laboratory, summer http://www.netl.doe.gov/technologies/ oil

[817] Boswell, R. (2007). Resource potential of methane hydrate coming into focus. *J. Petrol. Sci. Engin.*, *56*, 9-13.

[818] Boudreau, B. P., Algar, C., Johnson, B. D., Croudace, I., Reed, A., Furukawa, Y., Dorgan, K. M., Jumars, P. A., Grader, A. S. & Gardiner, B. S. (2005). Bubble growth and rise in soft sediments. *Geology*, *33(6)*, 517-520.

[819] Bousquet, P., Ciais, P., Miller, J. B., Dlugokencky, E. J., Hauglustaine, D. A., Prigent, C., Van der Werf, G. R., Peylin, P., Brunke, E. G., Carouge, C., Langenfelds, R. L., Lathiere, J., Papa, F., Ramonet, M.; Schmidt, M., Steele, L. P., Tyler, S. C. & White, J. (2006). Contribution of anthropogenic and natural sources to atmospheric methane variability. *Nature*, *443*, 439-443.

[820] Bowen, G. J., Beerling, D. J., Koch, P. L., Zachos, J. C. & Quattlebaum, T. (2004). A humid climate state during the Palaeocene/Eocene thermal maximum. *Nature*, *432(7016)*, 495-499.

[821] Bralower, T. J., Thomas, D. J., Zachos, J. C., Hirschmann, M. M., Rohl, U., Sigurdsson, H., Thomas, E. & Whitney, D. L. (1997). High-resolution records of the late Paleocene thermal maximum and circum-Caribbean volcanism: Is there a causal link? *Geology*, *25(11)*, 963-966.

[822] Brewer, P. G., Paull, C., Peltzer, E. T., Ussler, W., Rehder, G. & Friederich, G. (2002). Measurements of the fate of gas hydrates during transit through the ocean water column. *Geophysical Research Letters*, *29(22)*.

[823] Brook, E. & Mitchell, L. (2007). Timing and trends in Northern and Southern. Hemisphere Atmospheric Methane during the Holocene: new results from Antarctic and Greenlandic ice cores, *Eos Trans. AGU, 88*(52), *Fall Meet. Suppl.*, Abstract U21F-06.

[824] Brook, E. & Wolff, E. (2005). The future of ice core science. *Eos*, *87*, 39.

[825] Brook, E. J., Harder, S., Severinghaus, J. P., Steig, E. & Sucher, C. (2000). On the origin and timing of rapid changes in atmospheric methane during the last glacial period. *Global Biogeochemical Cycles*, *14*, 559-572.

[826] Brook, E. J., Sowers, T. & Orchardo, J. (1996). Rapid variations in atmospheric methane concentration during the past 110 ka. *Science*, *273*, 1087-1091.

[827] Brook, E., White, J. W. C., Schilla, A., Bender, M., Barnett, B. A., Serveringhaus, J., Taylor, K. C., Alley, R. B. & Steig, E. J. (2005). Timing of millennial-scale climate change at Siple Dome, West Antarctica, during the last glacial period. *Quaternary Science Reviews*, *24*, 1333-1343.

[828] Brooks, J. M., Bryant, W. R., Bernard, B. B. & Cameron, N. R. (2000). The nature of gas hydrates on the Nigerian continental slope. *Gas Hydrates: Challenges for the Future*, 76-93.

[829] Bryn, P., Berg, K., Forsberg, C. F., Solheim, A. & Kvalstad, T. J. (2005). Explaining the Storegga slide. *Marine and Petroleum Geology*, *22(1-2)*, 11-19.

[830] Bubier, J. L., Moore, T. R., Bellisario, L., Comer, N. T. & Crill, P. M. (1995). Ecological Controls on Methane Emissions from a Northern Peatland Complex in the Zone of Discontinuous Permafrost, Manitoba, Canada. *Glob. Biogeochem. Cyc.*, *9*, 455-470.

[831] Bubier, J. L., Moore, T.. R., Savage, K. & Crill, P. (2005). A comparison of methane flux in a boreal landscape between a dry and a wet year. *Global Biogeochemical Cycles*, 19, GB1*023*, doi:10.1029/2004GB002351.

[832] Buffett, B. & D.E. Archer, 2004: Global inventory of methane clathrate: Sensitivity to changes in environmental conditions. *Earth and Planetary Science Letters*, *227*, 185-199.

[833] Bunz, S. & J. Mienert, 2004: Acoustic imaging of gas hydrate and free gas at the Storegga slide. *Journal of Geophysical Research-Solid Earth*, 109(B4).

[834] Buteau, S., R. Fortier, G. Delisle, and M. Allard. 2004: Numerical simulation of the impacts of climate warming on a permafrost mound. *Permafr. Periglac. Proc.*, 15, 41-57.

[835] Camill, P, (2005). Permafrost thaw accelerates in boreal peatlands during late-20th century climate warming. *Clim. Change*, *68*, 135-152.

[836] Cannariato, K. G. & Stott, L. D. (2004). Evidence against clathrate-derived methane release to Santa Barbara Basin surface waters? *Geochemistry Geophysics Geosystems*, *5*.

[837] Chapin, F. S., Sturm, M., Serreze, M. C., McFadden, J. P., Key, J. R., Lloyd, A. H., McGuire, A. D., Rupp, T. S., Lynch, A. H., Schimel, J. P., Beringer, J., Chapman, W. L., Epstein, H. E., Euskirchen, L. D., Hinzman, L. D., Jia, G., Ping, C. L., Tape, K. D., Thompson, C. D. C., Walker, D. A. & Welker, J. M. (2005). Role of land-surface changes in arctic summer warming. *Science*, doi: 10.1126/science. 1117368.

[838] Chapman, W. L. & Walsh, J. E. (2007). Simulations of Arctic temperature and pressure by global coupled models. *J. Clim.*, 20, 609-632.

[839] Chappellaz, J., Barnola, J. M., Raynaud, D., Korotkevich, Y. S. & Lorius, C. (1990). Ice- core record of atmospheric methane over the past *160,*000 years. *Nature*, *345*, 127-13 1.

[840] Chappellaz, J., Blunier, T., Kints, S., Dällenbach, A., Barnola, J. M., Schwander, J., Raynaud, D. & Stauffer, B. (1997). Changes in the atmospheric CH4 gradient between Greenland and Antarctica during the Holocene. *Journal of Geophysical Research, 102,* 15987-15997.

[841] Chappellaz, J., Blunier, T., Raynaud, D., Barnola, J. M., Schwander, J. & Stauffer, B. (1993a). Synchronous changes in atmospheric CH_4 and Greenland climate between 40 and 8 kyr BP. *Nature*, *366*, 443- 445.

[842] Chappellaz, J. A., Fung, I. Y. & Thompson, A. M. (1993b). The atmospheric CH4 increase since the Last Glacial Maximum (1). Source estimates. *Tellus*, *45B(3)*, 228-241.

[843] Christensen, T. R., Ekberg, A., Strom, L., Mastepanov, M., Panikov, N., Mats, O., Svensson, B. H., Nykanen, H., Martikainen, P. J. & Oskarsson, H. (2003). Factors controlling large scale variations in methane emissions from wetlands. *Geophys. Res. Lett.*, *30(7)*, 1*414,* doi:10.1029/2002GL016848.

[844] Christensen, T. R., Johansson, T. R., Akerman, H. J., Mastepanov, M., Malmer, N., Friborg, T., Crill, P. & Svensson, B. H. (2004). Thawing sub-arctic permafrost: Effects

on vegetation and methane emissions. *Geophys. Res. Lett.*, *31*, L04501, doi: 10.1 029/2003GL01 8680.

[845] Christensen, T. R., Prentice, I. C., Kaplan, J., Haxeltine, A. & Sitch, S. (1996). Methane flux from northern wetlands and tundra: an ecosystem modeling approach. *Tellus*, *48B*, 652-661.

[846] Collett T. S. & Ginsburg, G. D. (1998). Gas hydrates in the Messoyakha Gas Field of the West Siberian Basin—A re-examination of the geologic evidence. *mnt. J. Offshore and Polar Engineering*, *8*, 22-2 9.

[847] Collett, T. S. & Kuuskraa, V. A. (1998). Hydrates contain vast store of world gas resources. *Oil and Gas Journal*, *96*, 90-95.

[848] Dallenbach, A., Blunier, T., Fluckiger, J., Stauffer, B., Chappellaz, J. & Raynoud, D. (2000). Changes in the atmospheric CH_4 gradient between Greenland and Antarctica during the Last Glacial and the transition to the Holocene. *Geophys. Res. Lett.*, *27*, 1005-1008.

[849] Dansgaard, W., White, J. W. C. & Johnson, S. J. (1989). The abrupt termination of the Younger Dryas climate event. *Nature*, *339*, 532-534.

[850] Davidson, E. A., & Janssens, I. A. (2006). Temperature sensitivity of soil carbon decomposition and feedbacks to climate change. *Nature*, *440*, 165-173.

[851] De Batist M, Klerkx, J., Van Rensbergen, P., Vanneste, M., Poort, J., Golmshtok, A. Y., Kremlev, A. A., Khlystov, O. M. & Krinitsky, P. (2002). Active hydrate destabilization in Lake Baikal, Siberia? *Terra Nova*, *14*, 436-442.

[852] Delmotte, M., Chappellaz, J., Brook, E., Yiou, P., Barnola, J. M., Goujon, C., Raynaud, D. & Lipenkov, V. I. (2004). Atmospheric methane during the last four glacial-interglacial cycles: Rapid changes and their link with Antarctic temperature. *Journal of Geophysical Research*, *109*, D1 2104, doi: 10.1 029/2003JD0044 17.

[853] Denman, K. L., Brasseur, G., Chidthaisong, A., Ciais, P., Cox, P. M., Dickinson, R. E., Hauglustaine, D., Heinze, C., Holland, E., Jacob, D., Lohmann, U., Ramachandran, S., da Silva Dias, P. L., Wofsy, S. C. & Zhang, X. (2007). Couplings between changes in the climate system and biogeochemistry. *In: Climate Change 2007: The Physical Science Basis. Contribution of Working Group 1 to the Fourth Assessment Report of the Intergovernmental Panel on Climate Change*. In: S., Solomon, D., Qin, M., Manning, Z., Chen, M., Marquis, K. B., Averyt, M. Tignor, & H. L. Miller, (eds.). Cambridge University Press, Cambridge, United Kingdom, and New York.

[854] Devol, A. H., Richey, J. E., Forsberg, B. R. & Martinelli, L. A. (1990). Seasonal dynamics in methane emissions from the Amazon River floodplain to the troposphere. *J. Geophys. Res.*, *95*, 16417-16426.

[855] D'Hondt, S., Jorgensen, B. B., Miller, D. J., Batzke, A., Blake, R., Cragg, B. A., Cypionka, H., Dickens, G. R., Ferdelman, T., Hinrichs, K. U., Holm, N. G., Mitterer, R., Spivack, A., Wang, G. Z., Bekins, B., Engelen, B., Ford, K., Gettemy, G., Rutherford, S. D., Sass, H., Skilbeck, C. G., Aiello, I. W., Guerin, G., House, C. H., Inagaki, F., Meister, P., Naehr, T., Niitsuma, S., Parkes, R. J., Schippers, A., Smith, D. C., Teske, A., Wiegel, J., Padilla, C. N. & Acosta, J. L. S. (2004). Distributions of microbial activities in deep subseafloor sediments. *Science*, *306(5705)*, 2216-2221.

[856] Dickens, G. R., (2001a). Modeling the global carbon cycle with a gas hydrate capacitor: significance for the latest Paleocene thermal maximum. In: *Natural Gas Hydrates.*

[Paull, C.K. & W.P. Dillon (eds.)]. American Geophysical Union *Occurrence, Distribution, and Detection, Geophysical Monographs, 124,* 19-38.

[857] Dickens, G. R., (2001b). On the fate of past gas: what happens to methane released from a bacterially mediated gas hydrate capacitor? *Geochem. Geophys. Geosyst.,* 2, 2000GC0001 31.

[858] Dickens, G. R., (2001c). The potential volume of oceanic methane hydrates with variable external conditions. *Org. Geochem., 32,* 1179-1193.

[859] Dickens, G. R., O'Heill, J. R., Rea, D. K. & Owens, R. M. (1995). Dissociation of oceanic methane hydrate as a cause of the carbon isotope excursion at the end of the Paleocene. *Paleoceanography, 19,* 965-971.

[860] Dimitrov, L. & Woodside, J. (2003). Deep sea pockmark environments in the eastern Mediterranean. *Marine Geology, 195(1-4),* 263-276.

[861] Dlugokencky, E. J., Dutton, E. G., Novelli, P. C., Tans, P. P., Masarie, K. A., Lantz, K. O. & Madronich, S. (1996). Changes in CH4 and CO growth rates after the eruption of Mt Pinatubo and their link with changes in tropical tropospheric UV flux. *Geophys. Res. Lett., 23(20),* 2761-2764.

[862] Dlugokencky, E. J., Houweling, S., Bruhwiler, L., Masarie, K. A., Lang, P. M., Miller, J. B. & Tans, P. P. (2003). Atmospheric methane levels off: Temporary pause or a new steady state? *Geophys. Res. Lett., 30(19),* doi:10.1029/2003GL018126.

[863] Dlugokencky, E. J., Masarie, K. A., Lang, P. M. & Tans, P. P. (1998). Continuing decline in the growth rate of the atmospheric methane burden. *Nature, 393,* 447-450.

[864] Dobrynin, V. M., Korotajev, Y. P. & Plyuschev, D. V. (1981). Gas hydrates: A possible energy resource. [Meyer, R.F. & J.C. Olson (eds.)]. *Long Term Energy Resources,* v. *1,* Pitman, Boston, 727-729.

[865] Driscoll, N. W., Weissel, J. K. & Goff, J. A. (2000). Potential for large-scale submarine slope failure and tsunami generation along the US mid-Atlantic coast. *Geology, 28(5),* 407-410.

[866] Dueck, T., et al., (2007). No evidence for substantial aerobic methane emission by terrestrial plants: A ^{13}C-labelling approach. *New Phytologist,* doi: 10.1111/j. 1469-8137.2007.02103.x.

[867] Dugan, B. & Flemings, P. B. (2000). Overpressure and fluid flow in the New Jersey continental slope: Implications for slope failure and cold seeps. *Science, 289,* 288- 291.

[868] Dykoski, C. A., Edwards, R. L., Cheng, H., Yuan, D. X., Cai, Y. J., Zhang, M. L., Lin, Y. S., An, Z. S. & Revenaugh, J. (2005). A high resolution, absolute-dated Holocene and deglacial Asian monsoon record from Dongge Cave, China. *Earth Planet. Sci. Lett., 233,* 71-86.

[869] EPICA Community Members, (2006). One-to-one coupling of glacial climate variability in Greenland and Antarctica. *Nature, 444,* doi: 10.103 8/nature05301.

[870] Etheridge, D. M., Steele, L. P., Francey, R. J. & Langenfelds, R. L. (1998). Atmospheric methane between 1000 A.D. and present: Evidence of anthropogenic emissions and climatic variability. *J. Geophys. Res., 103,* 15979-15993.

[871] Euskirchen, E. S., McGuire, A. D., Kicklighter, D. W., Zhuang, Q., Clein, J. S., Dargaville, R. J., Dye, D. G., Kimball, J. S., McDonald, K. C., Melillo, J. M., Romanovsky, V. E. & Smith, N. V. (2006). Importance of recent shifts in soil thermal dynamics on growing season length, productivity, and carbon sequestration in terrestrial high-latitude ecosystems. *Glob. Change Biol., 12,* 731-750.

[872] Fiore A. M., Horowitz, L. W., Dlugokencky, E. J. & West, J. J. (2006). Impact of meteorology and emissions on methane trends, 1990\u20132004. *Geophys. Res. Lett.*, *33*, L12809, doi:10.1029/2006GL026199.

[873] Flemings, B. P., Liu, X. & Winters, W. J. (2003). Critical pressure and multiphase flow in Blake Ridge gas hydrates. *Geology*, *31*, 1057-1060.

[874] Forster, P., Ramaswamy, V., Artaxo, P., Berntsen, T., Betts, R., Fahey, D. W., Haywood, J., Lean, J., Lowe, D. C., Myhre, G., Nganga, J., Prinn, R., Raga, G., Schulz, M. & Van Dorland, R. (2007). Changes in atmospheric constituents and in radiative forcing. *In: Climate Change 2007: The Physical Science Basis. Contribution of Working Group 1 to the Fourth Assessment Report of the Intergovernmental Panel on Climate Change.* In: S., Solomon, D., Qin, M., Manning, Z., Chen, M., Marquis, K. B., Averyt, M. Tignor, & H. L. Miller (eds.). Cambridge University Press, Cambridge, United Kingdom, and New York.

[875] Flückiger, J., Blunier, T., Stauffer, B., Chappellaz, J., Spahni, R., Kawamura, K., Schwander, J., Stocker, T. F. & Dahl-Jensen, D. (2004). N2O and CH4 variations during the last glacial epoch: Insight into global processes. *Glob. Biogeochem. Cycles*, *18(1)*, GB 1020, doi: 10.1 029/2003GB002 122.

[876] Frankenberg, C., Meirink, J. F., Bergamaschi, P., Goede, A. P. H., Heimann, M., Körner, S., Platt, U., van Weele, M. & Wagner, T. (2006). Satellite chartography of atmospheric methane from SCIAMACHY on board ENVISAT: Analysis of the years 2003 and 2004. *J. Geophys. Res.*, *111*, D07303, doi: 10.1029/2005JD006235.

[877] Frey, K. E. & Smith, L. C. (2007). How well do we know northern land cover? Comparison of four global vegetation and wetland products with a new ground- truth database for West Siberia. *Glob. Biogeochem. Cyc.*, *21*, GB 1016, doi: 10.1 029/2006GB002706.

[878] Friborg, T., Soegaard, H., Christensen, T. R., Lloyd, C. R. & Panikov, N. S. (2003). Siberian wetlands: Where a sink is a source. *Geophys. Res. Lett.*, *30*.

[879] Frolking S., Roulet, N. T., Moore, T. R., Richard, P. J. H., Lavoie, M. & Muller, S. D. (2001). Modeling northern peatland decomposition and peat accumulation. *Ecosystems*, *4*, 479-498.

[880] Gardiner, B. S., Boudreau, B. P. & Johnson, B. D. (2003). Growth of disk-shaped bubbles in sediments, *Geochimica et Cosmochimica Acta, 67(8)*, 1485-1494.

[881] Gauci,V., Matthews, E., Dise, N., Walter, B., Koch, D., Granberg, G. & Vile, M. (2004). Sulfur pollution suppression of the wetland methane source in the 20th and 21st centuries. *Proc. Nat. Acad. Sci.*, *101*, 12583-12587.

[882] Gedney, N., Cox, P. M. & Huntingford, C. (2004). Climate feedback from wetland methane emissions. *Geophys. Res. Lett.*, 31, L20503, doi: 10.1 029/2004GL0209 19.

[883] Gorman, A. R., Holbrook, W. S., Hornbach, M. J., Hackwith, K. L., Lizarralde, D. & Pecher, I. (2002). Migration of methane gas through the hydrate stability zone in a low-flux hydrate province. *Geology*, *30(4)*, 327-330.

[884] Grachev A. M., Brook, E. J. & Severinghaus, J. P. (2007). Abrupt changes in atmospheric methane at the MIS 5b-5a transition. *Geophys. Res. Lett.*, *34*, L20703, doi: 10.1 029/2007GL029799.

[885] Grant, N. J. & Whiticar, M. J. (2002). Stable carbon isotopic evidence for methane oxidation in plumes above Hydrate Ridge, Cascadia Oregon margin. *Global Biogeochemical Cycles, 16(4)*, 1124.

[886] Hampton, M. A., Lee, H. J. & Locat, J. (1996). Submarine landslides. *Reviews of Geophysics*, *34(1)*, 33-59.

[887] Hansen, J. & Sato, M. (2001). Trends of measured climate forcing agents. *Proc. Natl. Acad. Sci.*, 98, 14778-14783, doi:10.1073/pnas.261553698.

[888] Hansen, J., Sato, M., Ruedy, R., Lacis, A. & Oinas, V. (2000). Global warming in the twenty-first century: An alternative scenario. *Proc. Natl. Acad. Sci.*, *97*, 9875-9880, doi:10.1073/pnas.170278997.

[889] Harvey, L. D. D. & Huang, Z. (1995). Evaluation of the potential impact of methane clathrate destabilization on future global warming. *J. Geophysical Res.*, *100*, 2905-2926.

[890] Heeschen, K. U., Collier, R. W., de Angelis, M. A., Suess, E., Rehder, G., Linke, P. & Klinkhammer, G. P. (2005). Methane sources, distributions, and fluxes from cold vent sites at Hydrate Ridge, Cascadia margin. *Global Biogeochemical Cycles*, *19(2)*.

[891] Heeschen, K. U., Trehu, A. M., Collier, R. W., Suess, E. & Rehder, G. (2003). Distribution and height of methane bubble plumes on the Cascadia margin characterized by acoustic imaging. *Geophysical Research Letters*, *30(12)*.

[892] Hesselbo, S. P., Grocke, D. R., Jenkyns, H. C., Bjerrum, C. J., Farrimond, P., Bell, H. S. M. & Green, O. R. (2000). Massive dissociation of gas hydrate during a Jurassic oceanic anoxic event. *Nature*, *406(6794)*, 392-395.

[893] Hill, J. C., Driscoll, N. W., Weissel, J. K. & Goff, J. A. (2004). Large-scale elongated gas blowouts along the US Atlantic margin. *Journal of Geophysical Research-Solid Earth*, 109(B9).

[894] Hill, T. M., Kennett, J. P., Valentine, D. L., Yang, Z., Reddy, C. M., Nelson, R. K., Behl, R. J., Robert, C. & Beaufort, L. (2006). Climatically driven emissions of hydrocarbons from marine sediments during deglaciation. *Proc. Natl. Acad. Sci.*, *103*, 13570- 13574.

[895] Hinrichs K. U., Hayes, J. M., Sylva, S. P., Brewer, P. G. & Delong, E. F. (1999). Methane- consuming archaebacteria in marine sediments. *Nature*, *398*, 802-805.

[896] Hinrichs, K. U., Hmelo, L. R. & Sylva, S. P. (2003). Molecular fossil record of elevated methane levels in late pleistocene coastal waters. *Science*, *299(5610)*, 1214-1217.

[897] Hinzman, L. D., Bettez, N. D., Bolton, W. R., Chapin, F. S., Dyurgerov, M. B., Fastie, C. L., Griffith, B., Hollister, R. D., Hope, A., untington, H. P., Jensen, H. A. M., Jia, G. J., Jorgenson, T., Kane, D. L., Klein, D. R., Kofinas, G., Lynch, A. H., Lloyd, A. H., McGuire, A. D., Nelson, F. E., Oechel, W. C., Osterkamp, T. E., Racine, C. H., Romanovsky, V. E., Stone, R. S., Stow, D. A., Sturm, M., Tweedie, C. E., Vourlitis, G. L., Walker, M. D., Walker, D. A., Webber, P. J., Welker, J. M., Winker, K. & Yoshikawa, K. (2005). Evidence and implications of recent climate change in northern Alaska and other arctic regions. *Clim. Change*, *72*, 25 1-298.

[898] Hofmann, D. J., Butler, J. H., Dlugokencky, E. J., Elkins, J. W., Masarie, K., Montzka, S. A. & Tans, P. (2006). The role of carbon dioxide in climate forcing from 1979 to 2004: Introduction of the annual greenhouse gas index. *Tellus B*, *58*, 614-619.

[899] Hornbach, M. J., Saffer, D. M. & Holbrook, W. S. (2004). Critically pressured free-gas reservoirs below gas-hydrate provinces. *Nature*, *427(6970)*, 142-144.

[900] Hovland, M. & Judd, A. G. (1998). Seabed pockmarks and seepages. Graham & Trotman, London.

[901] Hovland, M., Svensen, H., Forsberg, C. F., Johansen, H., Fichler, C., Fossa, J. H., Jonsson, R. & Rueslatten, H. (2005). Complex pockmarks with carbonate-ridges off mid- Norway: Products of sediment degassing. *Marine Geology, 218(1-4)*, 191-206.

[902] Huber, C., Leuenberger, M., Spahni, R., Flükiger, J., Schwander, J., Stocker, T. F., Johnsen, S., Landais, A. & Jouzel, J. (2006). Isotope calibrated Greenland temperature record over Marine Isotope Stage 3 and its relation to CH$_4$. *Earth and Planet. Sci. Lett., 243*, 504-519.

[903] Huber, M., Sloan, L. & Shellito, C. (2002). Early Paleogene oceans and climate: A fully coupled modeling approach using the National Center for Atmospheric Research Community Climate System Model. In: *Causes and consequences of globally warm climates in the Early Paleogene.* In: S. L., Wing, P. D., Gingerich, B. Schmitz, & E. Thomas (eds.). Geological Society of America, Boulder, CO, 25-47.

[904] Jenkyns, H. C. (2003). Evidence for rapid climate change in the Mesozoic-Palaeogene greenhouse world. *Philosophical Transactions of the Royal Society of London Series a-Mathematical Physical and Engineering Sciences, 361(1810)*, 1885- 1916.

[905] Joabsson, A. & Christensen, T. R. (2001). Methane emissions from wetlands and their relationship with vascular plants: An Arctic example. *Global Change Biol., 7*, 919-932.

[906] Jorgenson, M. T., Racine, C. H., Walters, J. C. & Osterkamp, T. E. (2001). Permafrost degradation and ecological changes associated with a warming climate in central Alaska. *Clim. Change, 48*, 55 1-579.

[907] Jorgenson, M. T., Shur, Y. L. & Pullman, E. R. (2006). Abrupt increase in permafrost degradation in Arctic Alaska. *Geophys. Res. Lett., 33*, L02503, doi: 10.1 029/2005GL024960

[908] Kaplan, J. O., (2002). Wetlands at the last glacial maximum: Distribution and methane emissions. *Geopysical Research Letters, 29*, doi: 10.1029/2001 GL0 13366.

[909] Kaplan. J. O., Folberth, G. & Hauglustaine, D. A. (2006). Role of methane and volatile biogenic compound sources in late glacial and Holocene fluctuations of atmospheric methane concentrations. *Global Biogeochemical Cycles, 20*, GB2016, doi: 10. 1029/2005GB002590.

[910] Kelly, M., Edwards, R. L., Cheng, H., Yuan, D., Cai, Y., Zhang, M., Lin, Y. & An, Z. (2006). High resolution characterization of the Asian Monsoon between *146,*000 and 99,000 years B.P. from Dongge Cave, China and correlation of events surrounding Termination II. *Palaeogeography, Palaeoclimatology, Palaeoecology, 236*, 20-3 8.

[911] Kennett, J. P., Cannariato, K. G., Hendy, I. L. & Behl, R. J. (2000). Carbon isotopic evidence for methane hydrate instability during quaternary interstadials. *Science, 288(5463)*, 128-133.

[912] Kennett, J. P., Cannariato, K. G., Hendy, I. L. & Behl, R. J. (2003). *Methane hydrates in Quaternary climate change: The clathrate gun hypothesis.* AGU Press.

[913] Kennett, J. P. & Stott, L. D. (1991). Abrupt deep sea warming, paleoceanographic changes and benthic extinctions at the end of the Paleocene. *Nature, 353*, 319- 322.

[914] Keppler, F., Hamilton, J., Braß, M. & Rockman, T. (2006). Methane emissions from terrestrial plants under aerobic conditions. *Nature, 439*, 187-191.

[915] Kent, D. V., Cramer, B. S., Lanci, L., Wang, D., Wright, J. D. & Van der Voo, R. (2003). A case for a comet impact trigger for the Paleocene/Eocene thermal maximum and carbon isotope excursion. *Earth and Planetary Science Letters, 211(1-2)*, 13-26.

[916] King, J. Y. & Reeburgh, W. S. (2002). A pulse-labeling experiment to determine the contribution of recent plant photosynthates to net methane emission in arctic wet sedge tundra. *Soil Biology & Biochemistry*, *34*, 173-180.

[917] King, J. Y., Reeburgh, W. S. & Regli, S. K. (1998). Methane emission and transport by arctic sedges in Alaska: Results of a vegetation removal experiment. *J. Geophys. Res.*, *103*, 29083-29092.

[918] Kobashi, T., Severinghaus, J., Brook, E., Barnola, J. M. & Grachev, A. (2007). Precise timing and characterization of abrupt climate change at 8.2k B .P. from air trapped in polar ice. *Quaternary Science Reviews*, *26*, 12, 12-1222.

[919] Koch, P. L., Zachos, J. C. & Gingerich. P. D. (1992). Coupled isotopic change in marine and continental carbon reservoirs near the Paleocene/Eocene boundary. *Nature*, *358*, 3 19-322.

[920] Krason J, (2000). Messoyakh gas field (W. Siberia). A model for development of the methane hydrate deposits of Mackenzie Delta. In: *Gas Hydrates: Challenges for the Future*. [Holder, G.D. & P.R. Bishnoi (eds.)]. New York Academy of Sciences, 173-188.

[921] Kump, L. R. & Arthur, M. A. (1999). Interpreting carbon-isotope excursions: Carbonates and organic matter. *Chem. Geol.*, *161*, 181-198.

[922] Kvenvolden, K. A. (1993). Gas hydrates—Geological perspective and global change. *Reviews of Geophysics*, *31*, 173-187.

[923] Kvenvolden, K. A. (1999). Potential effects of gas hydrate on human welfare. *Proc. Natl. Acad. Sci.*, *96*, 3420-3426.

[924] Kvenvolden, K. A. (2000). Gas hydrate and humans. In: *Gas Hydrates: Challenges for the Future*. [Holder, G.D. & P.R. Bishnoi (eds.)]. New York Academy of Sciences, 17-22.

[925] Kvenvolden, K. A. & T. D. Lorenson, (2001). The global occurrence of natural gas hydrates. In: Natural Gas Hydrates: Occurrence, Distribution, and Detection. [Paull, C.K. & W.P. Dillon (eds.)]. *American Geophysical Union Geophysical Monographs*, *124*, 3-18.

[926] Larson, R. L. (1991). Geological consequences of superplumes. *Geology*, *19*, 963-966.

[927] Lawrence, D. M. & Slater, A. G. (2005). A projection of severe near-surface permafrost degradation during the 21st century. *Geophys. Res. Lett.*, *24*, L24401, doi: 10.1 029/2005GL025080.

[928] Lawrence, D. M., Slater, A. G., Romanovsky, V. E. & Nicolsky, D. J. (2008). The sensitivity of a model projection of near-surface permafrost degradation to soil column depth and inclusion of soil organic matter. *J. Geophys. Res.*, *113*, F0201 1, doi: 10.1 029/2007JF000883.

[929] Leifer, I., Clark, J. F. & Chen, R. F. (2000). Modifications of the local environment by natural marine hydrocarbon seeps. *Geophysical Research Letters*, *27(22)*, 3711- 3714.

[930] Leifer, I. & MacDonald, I. (2003). Dynamics of the gas flux from shallow gas hydrate deposits: interaction between oily hydrate bubbles and the oceanic environment. *Earth and Planetary Science Letters*, *210(3-4)*, 411-424.

[931] Li, C., Qiu, J. J., Frolking, S., Xiao, X., Salas, W., Moore, B., Boles, S., Huang, Y. & Sass, R. (2002). Reduced methane emissions from large-scale changes in water management of China's rice paddies during 1980-2000. *Geophys. Res. Lett.*, *29(20)*, doi: 10. 1029/2002GL015370.

[932] Loulergue, L., Schilt, A., Spahni, R., Masson-Delmotte, V., Blunier, T., Lemieux, B., Barnola, J. M., Raynaud, D., Stocker, T. F. & Chappellaz, J. (2008). Orbital and millennial- scale features of atmospheric CH4 over the past 800,000 years. *Nature, 453,* 383- 386.

[933] Luyendyk, B., Kennett, J. & Clark, J. F. (2005). Hypothesis for increased atmospheric methane input from hydrocarbon seeps on exposed continental shelves during glacial low sea level. *Marine and Petroleum Geology, 22(4)*, 59 1-596.

[934] Luyten, J. R., Pedlosky, J. & Stommel, H. (1983). The ventilated thermocline, *J. Phys. Ocean., 13*, 292-309.

[935] MacDonald, G. J. (1990). Role of methane clathrates in past and future climates. *Clim. Change, 16*, 247-281.

[936] MacDonald, G. M., Berilman, D. W., Kremenetski, K. V., Sheng, Y., Smith, L. & Velichko, A. (2006). Rapid development of circumarctic peatlands and atmospheric CH_4 and CO_2 variations. *Science, 314,* 285-288.

[937] MacDonald, I. R., Bohrmann, G., Escobar, E., Abegg, F., Blanchon, P., Blinova, V., Bruckmann, W., Drews, M., Eisenhauer, A., Han, X., Heeschen, K., Meier, F., Mortera, C., Naehr, T., Orcutt, B., Bernard, B., Brooks, J. & de Farago, M. (2004). Asphalt volcanism and chemosynthetic life in the Campeche Knolls, Gulf of Mexico. *Science, 304(5673)*, 999-1002.

[938] MacDonald, I. R., Guinasso, N. L., Sassen, R., Brooks, J. M., Lee, L. & Scott, K. T. (1994). Gas hydrate that breaches the sea-floor on the continental-slope of the Gulf of Mexico. *Geology, 22(8)*, 699-702.

[939] MacDonald, I. R., Leifer, I., Sassen, R., Stine, P., Mitchell, R. & Guinasso, N. (2002). Transfer of hydrocarbons from natural seeps to the water column and atmosphere. *Geofluids, 2(2)*, 95-107.

[940] MacDonald, J. A., Fowler, D., Hargreaves, K. J., Skiba, U., Leith, I. D. & Murray, M. B. (1998). Methane emission rates from a northern wetland; response to temperature, water table and transport. *Atmospheric Environment, 32*, 3219.

[941] MacFarling-Meure, C., Etheridge, D., Trudinger, C., Steele, P. S., Langenfelds, R., van Ommen, T., Smith, A. & Elkins, J. (2006). Law dome CO_2, CH_4, and N_2O records extended to 2000 years BP. *Geophysical Research Letters, 33*, L14810, doi: 10.1 029/2006GL0261 52.

[942] Martin, P., Archer, D. & Lea, D. (2005). Role of deep sea temperatures in the carbon cycle during the last glacial. *Paleoceanography*, doi:10.1029/2003PA000914.

[943] Maslin, M., Owen, M., Day, S. & Long, D. (2004). Linking continental-slope failures and climate change: Testing the clathrate gun hypothesis. *Geology, 32(1)*, 53-56.

[944] Masson, D. G., Harbitz, C. B., Wynn, R. B., Pedersen, G. & Løvholt, F. (2006). Submarine landslides: processes, triggers, and hazard prediction. *Philosophical Transactions of the Royal Society, 364,* 2009-203 9.

[945] McGuire, A. D., Chapin, F. S., Walsh, J. E. & Wirth, C. (2006). Integrated regional changes in arctic climate feedbacks: Implications for the global climate system. *Annual Review of Environment and Resources, 31*, 61-91.

[946] Meehl, G. A., Stocker, T. F., Collins, W. D., Friedlingstein, P., Gaye, A. T., Gregory, J. M., Kitoh, A., Knutti, R., Murphy, J. M., Noda, A., Raper, S. C. B., Watterson, I. G., Weaver, A. J. & Zhao, Z. C. (2007). Global climate projections. In: *Climate Change The Physical Science Basis. Contribution of Working Group I to the Fourth Assessment*

Report of the Intergovernmental Panel on Climate Change. [Solomon, S., Qin, D., Manning, M., Chen, Z., Marquis, M., Averyt, K. B., Tignor, M. & Miller, H. L. (eds.). Cambridge University Press, Cambridge, United Kingdom, and New York.

[947] Melack, J. M., Hess, L. L., Gastil, M., Forsberg, B. R., Hamilton, S. K., Lima, l. B. T. & Novo, E. M. L. M. (2004). Regionalization of methane emissions in the Amazon Basin with microwave remote sensing. *Global Change Biology, 10*, 530-544.

[948] Meyer, R. F. (1981). Speculations on oil and gas resources in small fields and unconventional deposits. In: *Long-term Energy Resources*, v. *1*, 49-72 In: R. F. Meyer, & J. C. Olson (eds.). Pitman, Boston.

[949] Mienert, J., Andreassen, K., Posewang, J. & Lukas, D. (2000). Changes of the hydrate stability zone of the Norwegian margin from glacial to interglacial times. *Gas Hydrates: Challenges for the Future*, 200-2 10.

[950] Mienert, J., Vanneste, M., Bunz, S., Andreassen, K., Haflidason, H. & Sejrup, H. P. (2005). Ocean warming and gas hydrate stability on the mid-Norwegian margin at the Storegga slide. *Marine and Petroleum Geology, 22(1-2)*, 233-244.

[951] Milkov, A. V. (2000). Worldwide distribution of submarine mud volcanoes and associated gas hydrates. *Marine Geology, 167(1-2)*, 29-42.

[952] Milkov, A. V. (2004). Global estimates of hydrate-bound gas in marine sediments: How much is really out there? *Earth-Science Reviews, 66(3-4)*, 183-197.

[953] Milkov, A. V. & Sassen, R. (2000). Thickness of the gas hydrate stability zone, Gulf of Mexico continental slope. *Marine and Petroleum Geology, 17(9)*, 981-991.

[954] Milkov, A. V. & Sassen, R. (2001). Estimate of gas hydrate resource, northwestern Gulf of Mexico continental slope. *Marine Geology, 179(1-2)*, 7 1-83.

[955] Milkov, A. V. & Sassen, R. (2003). Two-dimensional modeling of gas hydrate decomposition in the northwestern Gulf of Mexico: Significance to global change assessment. *Global and Planetary Change, 36(1-2)*, 31-46.

[956] Milkov, A. V., Vogt, P. R., Crane, K., Lein, A. Y., Sassen, R. & Cherkashev, G. A. (2004). Geological, geochemical, and microbial processes at the hydrate-bearing Hakon Mosby mud volcano: A review. *Chemical Geology, 205(3-4)*, 347-366.

[957] Minkkinen, K., Laine, J., Nykanen, H., Marikainen, P. J. (1997). Importance of drainage ditches in emissions of methane from mires drained for forestry. *Can. J. For. Res., 27*, 949-952.

[958] Nisbet, E. G. (2002). Have sudden large releases of methane from geological reservoirs occurred since the last glacial maximum, and could such releases occur again? *Philosophical Transactions of the Royal Society of London Series A— Mathematical Physical and Engineering Sciences, 360(1793)*, 581-607.

[959] Nisbet, E. G. & Piper, D. J. W. (1998). Giant submarine landslides. *Nature, 392(6674)*, 329-330.

[960] Nouze, H., Henry, P., Noble, M., Martin, V. & Pascal, G. (2004). Large gas hydrate accumulations on the eastern Nankai Trough inferred from new high-resolution 2- D seismic data. *Geophysical Research Letters, 31(13)*, L1 3308, doi: 10.1 029/2004GL01 9848.

[961] Nykänen, H., Heikkinen, J. E. P., Pirinen, L., Tiilikainen, K. & Martikainen, P. J. (2003). Annual CO_2 exchange and CH_4 fluxes on a subarctic palsa mire during climatically different years. *Global Biogeochem. Cycles, 17(1)*, 1018, doi: 10.1029/2002GB001 861.

[962] Olivier, J. G. J. & Berdowski, J. J. M. (2001a). Global emissions sources and sinks. In: J., Berdowski, R. Guicherit, & B. J. Heij, (eds.) "The Climate System", 33-78. A.A. Balkema Publishers/Swets & Zeitlinger Publishers, Lisse, The Netherlands. ISBN 90 5809 255 0.

[963] Osterkamp, T. E. & Jorgenson, J. C. (2006). Warming of permafrost in the Arctic National Wildlife Refuge. *Alaska. Permafr. Periglac. Proc., 17*, 65-69. Osterkamp, T.E. & V.E. Romanovsky, 1999: Evidence for warming and thawing of discontinuous permafrost in Alaska. *Permafr. Periglac. Proc.*, 10, 17-37.

[964] Pagani, M., Caldeira, K., Archer, D. & Zachos, J. C. (2006). An ancient carbon mystery. *Science, 314(5805)*, 1556-1557.

[965] Paull, C. K., Brewer, P. G., Ussler, W., Peltzer, E. T., Rehder, G. & Clague, D. (2003). An experiment demonstrating that marine slumping is a mechanism to transfer methane from seafloor gas-hydrate deposits into the upper ocean and atmosphere. *Geo-Marine Letters, 22(4)*, 198-203.

[966] Paull, C. K., Ussler, W. & Dillon, W. P. (1991). Is the extent of glaciation limited by marine gas hydrates? *Geophys. Res. Lett., 18*, 432-434.

[967] Payette, S., Delwaide, A., Caccianiga, M. & Beauchemin, M. (2004). Accelerated thawing of subarctic peatland permafrost over the last 50 years. *Geophys. Res. Lett., 31*, L1 8208, doi:10. 1029/2004GL020358.

[968] Pecher, I. A., Kukowski, N., Huebscher, C., Greinert, J. & Bialas, J. (2001). The link between bottom-simulating reflections and methane flux into the gas hydrate stability zone—New evidence from Lima basin, Peru margin. *Earth and Planetary Science Letters, 185(3-4)*, 343-354.

[969] Peterson, L. C., Haug, G. H., Hughen, K. A. & Rohl, U. (2000). Rapid changes in the hydrologic cycle of the tropical Atlantic during the last glacial. *Science, 290*, 1947-1951.

[970] Ramaswamy, V., Boucher, O. & Haigh, J., et al. (2001). Radiative forcing of climate change. In: *Climate Change 2001: The Scientific Basis. Contribution of Working Group I to the Third Assessment Report of the Intergovernmental Panel on Climate Change.* In: J. T., Houghton, Y., Ding, & D. J. Griggs, et al. (eds.)]. Cambridge University Press, Cambridge, United Kingdom, 350-416.

[971] Reeburgh, W. S. (2004). Global methane biogeochemistry. In: *Treatise on Geochemistry. The Atmosphere.* [Keeling, R.F. (ed.)]. Elsevier, Amsterdam, v. *4*, 65-89.

[972] Rehder, G., Brewer, P. W., Peltzer, E. T. & Friederich, G. (2002). Enhanced lifetime of methane bubble streams within the deep ocean. *Geophysical Research Letters, 29(15)*, 1731, doi:10.1029/2001GL013966.

[973] Rehder, G., Keir, R. S., Suess, E. & Rhein, M. (1999). Methane in the northern Atlantic controlled by microbial oxidation and atmospheric history. *Geophysical Research Letters, 26(5)*, 587-590.

[974] Riedel, M., Spence, G. D., Chapman, N. R. & Hyndman, R. D. (2002). Seismic investigations of a vent field associated with gas hydrates, offshore Vancouver Island. *Journal of Geophysical Research-Solid Earth*, 107(B9).

[975] Riordan, B., Verbyla, D. & McGuire, A. D. (2006). Shrinking ponds in subarctic Alaska based on 1950-2002 remotely sensed images. *Journal of Geophysical Research-Biogeosciences*, 111, G04002, doi: 10. 1029/2005JG000150.

[976] Rogner, H. H, (1997). An assessment of world hydrocarbon resources. *Annu. Rev. Energy Environ., 22*, 217-262.

[977] Romanovsky, V. E., Burgess, M., Smith, S., Yoshikawa, K. & Brown, J. (2002). Permafrost temperature records: Indicator of climate change. *Eos, 589*, 593-594.

[978] Rothwell, R. G., Reeder, M. S., Anastasakis, G., Stow, D. A. V., Thomson, J. & Kahler, G. (2000). Low sea-level stand emplacement of megaturbidites in the western and eastern Mediterranean Sea. *Sedimentary Geology, 135(1-4)*, 75-88.

[979] Saito, K., Kimoto, M., Zhang, T., Takata, K. & Emori, S. (2007). Evaluating a high-resolution climate model: Simulated hydrothermal regimes in frozen ground regions and their change under the global warming scenario. *J. Geophys. Res., 112*, F02S1 1, doi:10.1029/2006JF000577.

[980] Sassen, R., Losh, S. L., Cathles, L., Roberts, H. H., Whelan, J. K., Milkov, A. V., Sweet, S. T. & DeFreitas, D. A. (2001a). Massive vein-filling gas hydrate: relation to ongoing gas migration from the deep subsurface in the Gulf of Mexico. *Marine and Petroleum Geology, 18(5)*, 551-560.

[981] Sassen, R. & MacDonald, I. R. (1994). Evidence of structure H hydrate, Gulf of Mexico continental slope. *Org. Geochem., 22*, 1029-1032.

[982] Sassen, R., Milkov, A. V., Roberts, H. H., Sweet, S. T. & DeFreitas, D. A. (2003). Geochemical evidence of rapid hydrocarbon venting from a seafloor-piercing mud diapir, Gulf of Mexico continental shelf. *Marine Geology, 198(3-4)*, 319- 329.

[983] Sassen, R., Sweet, S. T., Milkov, A. V., DeFreitas, D. A. & Kennicutt, M. C. (2001b). Thermogenic vent gas and gas hydrate in the Gulf of Mexico slope: Is gas hydrate decomposition significant? *Geology, 29(2)*, 107-110.

[984] Sazonova, T. S., Romanovsky, V. E., Walsh, J. E. & Sergueev, D. O. (2004). Permafrost dynamics in the 20th and 21st centuries along the East Siberian transect. *J. Geophys. Res., 109*, D01108, doi:10.1029/2003JD003680.

[985] Schaefer, H., Whiticar, M., Brook, E., Petrenko, V., Ferretti, D. & Severinghaus, J. (2006). Ice record of $\delta^{13}C$ for atmospheric methane across the Younger Dryas-Pre Boreal Transition. *Science, 313*, 1109-1112.

[986] Schlesinger, W. (1997). *Biogeochemistry: An analysis of global change*. Academic Press.

[987] Schmidt, G. A. & Shindell, D. T. (2003). Atmospheric composition, radiative forcing, and climate change as a consequence of a massive methane release from gas hydrates. *Paleoceanography, 18(1)*.

[988] Schmitz, B., Peucker-Ehrenbrink, B., Heilmann-Clausen, C., Aberg, G., Asaro, F. & Lee, C. T. A. (2004). Basaltic explosive volcanism, but no comet impact, at the Paleocene-Eocene boundary: high-resolution chemical and isotopic records from Egypt, Spain and Denmark. *Earth and Planetary Science Letters, 225(1-2)*, 1-17.

[989] Schwander, J. (2006). Ice core chronologies. In: *Encyclopedia of Quaternary Science*. [Elias, S. (ed.)]. Elsevier, Amsterdam.

[990] Schwander, J., Barnola, J. M., Andrie, C., Leuenberger, M., Ludin, A., Raynaud, D. & Stauffer, B. (1993). The age of the air in the firn and the ice at Summit, Greenland. *J. Geophys. Res., 98*, 2831-2838.

[991] Serreze, M. C. & Francis, J. A. (2006). The arctic amplification debate. *Clim. Change, 76*, 241-264.

[992] Serreze, M. C., Walsh, J. E., Chapin, F. S., Osterkamp, T., Dyurgerov, M., Romanovsky, V. Oechel, W. C., Morison, J., Zhang, T. & Barry, R. G. (2000). Observational evidence of recent change in the northern high-latitude environment. *Clim. Change, 46*, 159-207.

[993] Severinghaus, J. & Brook, E. (1999). Simultaneous tropical-abrupt climate change at the end of the last glacial period inferred from trapped air in polar ice. *Science, 286*, 930-934.

[994] Severinghaus, J. P., Sowers, T., Brook, E., Alley, R. B. & Bender, M. L. (1998). Timing of abrupt climate change at the end of the Younger Dryas interval from thermally fractionated gases in polar ice. *Nature, 391*, 141-148.

[995] Shakhova, N., Semiletov, I. & Panteleev, G. (2005). The distribution of methane on the Siberian Arctic shelves: Implications for the marine methane cycle. *Geophys. Res. Lett., 32(9)*, L09601, doi: 10.1 029/2005GL02275 1.

[996] Shindell, D. T., Walter, B. P. & Faluvegi, G. (2004). Impacts of climate change on methane emissions from wetlands. *Geophys. Res. Lett., 31*, L21202, doi: 10.1 029/2004GL02 1009.

[997] Sluijs, A., Schouten, S., Pagani, M., Woltering, M., Brinkhuis, H., Sinninghe Damsté, J. S,. Dickens, G. R., Huber, M., Reichart, G. J., Stein, R., Matthiessen, J., Lourens, L. J., Pedentchouk, N., Backman, J. & Moran, K. (2006). the Expedition 302 Scientists, Subtropical Arctic Ocean temperatures during the Palaeocene/Eocene thermal maximum. *Nature, 441*, 610-613.

[998] Sluijs, A., Brinkhuis, H., Schouten, S., Bohaty, S., John, C., Zachos, J., Reichart, G. R., Damste, J. S. S., Crouch, E. & Dickens, G. R. (2007). Environmental precursors to rapid light carbon injection at the Paleocene/Eocene boundary, *Nature, 450*, 1218-1222.

[999] Smith, D. E., Shi, S., Cullingford, R. A., Dawson, A. G., Dawson, S., Firth, C. R., Foster, I. D. L., Fretwell, P. T., Haggart, B. A., Holloway, L. K. & Long, D. (2004). The holocene storegga slide Tsunami in the United Kingdom. *Quaternary Science Reviews, 23(23-24)*, 2291-2321.

[1000] Smith, L. C., Sheng, Y., MacDonald, G. M. & Hinzman, L. D. (2005). Disappearing Arctic lakes. *Science, 308*, 1429-1429.

[1001] Smith, L. K., Lewis, W. M., Chanton, J. P., Cronin, G. & Hamilton, S. K. (2000). Methane emissions from the Orinoco River floodplain, Venezuela. *Biogeochemistry, 51*, 113-140.

[1002] Solheim, A., Berg, K., Forsberg, C. F. & Bryn, P. (2005). The Storegga slide complex: Repetitive large scale sliding with similar cause and development. *Marine and Petroleum Geology, 22(1-2)*, 97-107.

[1003] Sowers, T. (2006). Late quaternary atmospheric CH4 isotope record suggests marine clathrates are stable. *Science, 311(5762)*, 838-840.

[1004] Spahni, R., Schwander, J., Flückiger. J., Stauffer, B., Chappellaz, J. & Raynaud, D. (2003). The attenuation of fast atmospheric CH4 variations recorded in polar ice cores. *Geophys. Res. Lett., 30*, doi:10.1029/2003GL017093.

[1005] Spahni, R., et al. (2005). Atmospheric methane and nitrous oxide of the late Pleistocene from Antarctic ice cores. *Science, 310*, 13 17-1321.

[1006] St. Louis, V. I., Kelly, C. A., Duchemin, E., Rudd, J. W. M. & Rosenberg, D. M. (2000). Reservoir surfaces as sources of greenhouse gases to the atmosphere: A global estimate. *BioScience*, *50*, 766-775.

[1007] Steele, L. P., Dlugokencky, E. J., Lang, P. M., Tans, P. P., Martin, R. C. & Masarie, K. A. (1992). Slowing down of the global accumulation of atmospheric methane during the 1980s. *Nature*, *358(6384)*, 313-316.

[1008] Stendel, M. & Christensen, J. H. (2002). Impact of global warming on permafrost conditions in a coupled GCM. *Geophys. Res. Lett.*, *13*, doi: 10.1029/2001GL014345.

[1009] Storey, M., Duncan, R. & Swisher, C. C. (2007). Paleocene-Eocene thermal maximum and the opening of the northeast Atlantic. *Science*, *316*, 587-589.

[1010] Sturm, M., McFadden, J. P., Liston, G. E., Chapin, F. S. III, Racine, C. H. & Holmgren, J. (2001). Snow-shrub interactions in Arctic tundra: A hypothesis with climatic implications. *J. Clim.*, *14*, 336-344.

[1011] Sultan, N., Cochonat, P., Foucher, J. P. & Mienert, J. (2004). Effect of gas hydrates melting on seafloor slope instability. *Marine Geology*, *213(1-4)*, 379-401.

[1012] Svensen, H., Planke, S., Malthe-Sorenssen, A., Jamtveit, B., Myklebust, R., Eidem, T. R. & Rey, S. S. (2004). Release of methane from a volcanic basin as a mechanism for initial Eocene global warming. *Nature*, *429(6991)*, 542-545.

[1013] Taylor, M. H., Dillon, W. P. & Pecher, I. A. (2000). Trapping and migration of methane associated with the gas hydrate stability zone at the Blake Ridge Diapir: New insights from seismic data. *Marine Geology*, *164(1-2)*, 79-89.

[1014] Thomas, D. J., Zachos, J. C., Bralower, T. J., Thomas, E. & Bohaty, S. (2002). Warming the fuel for the fire: Evidence for the thermal dissociation of methane hydrate during the Paleocene-Eocene thermal maximum. *Geology*, *30(12)*, 1067-1070.

[1015] Thorpe, R. B., Law, K. S., Bekki, S., Pyle, J. A., Nisbet, E. G. (1996). Is methane-driven deglaciation consistent with the ice core record? *Journal of Geophysical Research*, *101*, 28627-28635.

[1016] Torres, M. E., McManus, J., Hammond, D. E., de Angelis, M. A., Heeschen, K. U., Colbert, S. L., Tryon, M. D., Brown, K. M. & Suess, E. (2002). Fluid and chemical fluxes in and out of sediments hosting methane hydrate deposits on Hydrate Ridge, OR, I: Hydrological provinces. *Earth and Planetary Science Letters*, *201(3-4)*, 525- 540.

[1017] Torres, M. E., Wallmann, K., Trehu, A. M., Bohrmann, G., Borowski, W. S. & Tomaru, H. (2004). Gas hydrate growth, methane transport, and chloride enrichment at the southern summit of Hydrate Ridge, Cascadia margin off Oregon. *Earth and Planetary Science Letters*, *226(1-2)*, 225-241.

[1018] Trehu, A. M., Flemings, P. B., Bangs, N. L., Chevallier, J., Gracia, E., Johnson, J. E., Liu, C. S., Liu, X. L., Riedel, M. & Torres, M. E. (2004a). Feeding methane vents and gas hydrate deposits at south Hydrate Ridge. *Geophysical Research Letters*, *31(23)*.

[1019] Trehu, A. M., Long, P. E., Torres, M. E., Bohrmann, G., Rack, F. R., Collett, T. S., Goldberg, D. S., Milkov, A. V., Riedel, M., Schultheiss, P., Bangs, N. L., Barr, S. R., Borowski, W. S., Claypool, G. E., Delwiche, M. E., Dickens, G. R., Gracia, E., Guerin, G., Holland, M., Johnson, J. E., Lee, Y. J., Liu, C. S., Su, X., Teichert, B., Tomaru, H., Vanneste, M., Watanabe, M. & Weinberger, J. L. (2004b). Three-dimensional distribution of gas hydrate beneath southern Hydrate Ridge: Constraints from ODP Leg 204. *Earth and Planetary Science Letters*, *222(3-4)*, 845-862.

[1020]   Turetsky, M. R., Wieder, R. K. & Vitt, D. H. (2002). Boreal peatland C fluxes under varying permafrost regimes. *Soil Biology & Biochemistry*, *34*, 907-9 12.

[1021]   Turunen, J., Tomppo, E., Tolonen, K. & Reinikainen, A. (2002). Estimating carbon accumulation rates of undrained mires in Finland—Application to boreal and subarctic regions. *The Holocene*, *1*, doi: 10.1191/0959683 602hl522rp.

[1022]   Uchida, T., Dallimore, S. & Mikami, J. (2002). Occurrences of natural gas hydrates beneath the permafrost zone in Mackenzie Delta: Visual and x-ray CT imagery. In: *Gas Hydrates: Challenges for the Future*. In: G. D. Holder, & P. R. Bishnoi (eds.). New York Academy of Sciences, 1021-1033.

[1023]   Uchida, M., Shibata, Y., Ohkushi, K., Ahagon, N. & Hoshiba, M. (2004). Episodic methane release events from Last Glacial marginal sediments in the western North Pacific. *Geochemistry Geophysics Geosystems*, 5.

[1024]   Valdes, P. J., Beerling, D. J. & Johnson, C. E. (2005). The ice age methane budget. *Geophysical Research Letters*, 32, L02704, doi. 10.1 029/2004GL02 1004.

[1025]   Valentine, D.L., Blanton, D. C., Reeburgh, W. S. & Kastner, M. (2001). Water column methane oxidation adjacent to an area of active hydrate dissociation, Eel River Basin. *Geochimica et Cosmochimica Acta*, *65(16)*, 263 3-2640.

[1026]   van der Werf, G. R., Randerson, J. T., Collatz, G. J., Giglio, L., Kasibhatla, P. S., Arellano, A. F. Jr., Olsen, S. C. & Kasischke, E. S. (2004). Continental-scale partitioning of fire emissions during the 1997 to 2001 El Nino/La Nina period. *Science*, *303*, 73-74.

[1027]   Vogt, P. R. & Jung, W. Y. (2002). Holocene mass wasting on upper non-Polar continental slopes - due to post-Glacial ocean warming and hydrate dissociation? *Geophysical Research Letters*, *29(9)*, 1341, doi:10.1029/2001GL013488.

[1028]   von Huissteten, J. (2004). Methane emission from northern wetlands in Europe during oxygen isotope stage 3. *Quaternary Science Reviews*, *23*, 1989-2005.

[1029]   Waddington, J. M., Roulet, N. T. & Swanson, R. V. (1996). Water table control of CH4 emission enhancement by vascular plants in boreal peatlands. *J. Geophys. Res.*, *101*, 22775-22785.

[1030]   Wagner, D., Gattinger, A., Embacher, A., Pfeiffer, E. M., Schloter, M. & Lipski, A. (2007). Methanogenic activity and biomass in Holocene permafrost deposits of the Lena Delta, Siberian Arctic and its implication for the global methane budge. *Glob. Change Biol.*, *13*, 1089-1099.

[1031]   Wakeham, S. G., Lewis, C. M., Hopmans, E. C., Schouten, S. & Sinninghe Damsté, J. S. (2003). Archaea mediate anaerobic oxidation of methane in deep euxinic waters of the Black Sea. *Geochimica et Cosmochimica Acta*, *67*, 1359-1374.

[1032]   Walter, B. P., Heimann, M. & Matthews, E. (2001). Modeling modern methane emissions from natural wetlands 2. Interannual variations 1982-1993. *J. Geophys. Res.*, *106(D24)*, 34207-34220.

[1033]   Walter, K. M., Zimov, S. A., Chanton, J. P., Verbyla, D. & Chapin, F. S. (2006). Methane bubbling from Siberian thaw lakes as a positive feedback to climate warming. *Nature*, *443*, 71-75.

[1034]   Wang, X. F., Auler, A. S., Edwards, R. L., Cheng, H. H., Cristalli, P. S., Smart, P. L., Richards, D. A. & Shen, C. C. (2004). Wet periods in northeastern Brazil over the past 210 kyr linked to distant climate anomalies. *Nature*, *432*, 740-744.

[1035] Wang, Z. P., Han, X. G., Wang, G. G., Song, Y. & Gulledge, J. (2008). Aerobic methane emission from plants in the Inner Mongolia Steppe. *Environ. Sci. Technol.*, *42*, 62-68, 10.1021/es0712241.

[1036] Washburn, L., Clark, J. F. & Kyriakidis, P. (2005). The spatial scales, distribution, and intensity of natural marine hydrocarbon seeps near Coal Oil Point, California. *Marine and Petroleum Geology*, *22(4)*, 569-578.

[1037] Weinberger, J. L., Brown, K. M. & Long, P. E. (2005). Painting a picture of gas hydrate distribution with thermal images. *Geophysical Research Letters*, *32(4)*.

[1038] Weitemeyer, K. A. & Buffett, B. A. (2006). Accumulation and release of methane from clathrates below the Laurentide and Cordilleran ice sheets. *Global & Planetary Change*, *53*, 176-187.

[1039] Wickland, K. P., Striegl, R. G., Neff, J. C. & Sachs, T. (2006). Effects of permafrost melting on CO2 and CH4 exchange of a poorly drained black spruce lowland. *Journal of Geophysical Research-Biogeosciences*, 111, G0201 1, doi: 10.1 029/2005JG000099.

[1040] Wood, W. T., Gettrust, J. F., Chapman, N. R., Spence, G. D. & Hyndman, R. D. (2002). Decreased stability of methane hydrates in marine sediments owing to phase-boundary roughness. *Nature*, *420(6916)*, 656-660.

[1041] Xu, W. Y., Lowell, R. P. & Peltzer, E. T. (2001). Effect of seafloor temperature and pressure variations on methane flux from a gas hydrate layer: Comparison between current and late Paleocene climate conditions. *Journal of Geophysical Research-Solid Earth*, *106(B1 1)*, 26413-26423.

[1042] Yuan, D. X., Cheng, H., Edwards, R. L., Dykoski, C., Kelly, M. J., Zhang, M. L., Qing, J. M., Lin, Y. S., Wang, Y. G., Dorale, J. A., An, Z. S. & Cai, Y. J. (2004). Timing, duration, and transitions of the Last Interglacial Asian Monsoon. *Science*, *304*, 575-578.

[1043] Zachos, J. C., Pagani, M., Sloan, L., Thomas, E. & Billups, K. (2001). Trends, rhythms, and abberations in global climate 65 Ma to present. *Science*, *292*, 686-693.

[1044] Zachos, J. C., Rohl, U., Schellenberg, S. A., Sluijs, A., Hodell, D. A., Kelly, D. C., Thomas, E., Nicolo, M., Raffi, I., Lourens, L. J., McCarren, H. & Kroon, D. (2005). Rapid acidification of the ocean during the Paleocene-Eocene thermal maximum. *Science*, *308(5728)*, 1611-1615.

[1045] Zachos, J. C., Wara, M. W., Bohaty, S., Delaney, M. L., Petrizzo, M. R., Brill, A., Bralower, T. J. & Premoli-Silva, I. (2003). A transient rise in tropical sea surface temperature during the Paleocene-Eocene thermal maximum. *Science*, *302(5650)*, 1551-1554.

[1046] Zeebe, R. E. & Westbroek, P. (2003). A simple model for the CaCO3 saturation state of the ocean: The "Strangelove", the "Neritan", and the "Cretan" Ocean *Geochemistry Geophysics Geosystems*, *4*.

[1047] Zhang, Y., Chen, W. J. & Cihlar, J. (2003). A process-based model for quantifying the impact of climate change on permafrost thermal regimes. *J. Geophys. Res.*, D22, doi: 10.1029/2002JD003354.

[1048] Zhang, Y., Chen, W. J. & Riseborough, D. W. (2007). Temporal and spatial changes of permafrost in Canada in the 21st century. *J. Geophys. Res.*

[1049] Zhuang, Q., Melillo, J. M., McGuire, A. D., Kicklighter, D. W., Prinn, R. G., Steudler, P. A., Felzer, B. S. & Hu, S. (2007). Net emissions of CH4 and CO2 in

Alaska: Implications for the region's greenhouse gas budget. *Ecological Applications, 17*, 203-212.

[1050] Zillmer, M., Flueh, E. R. & Petersen, J. (2005a). Seismic investigation of a bottom simulating reflector and quantification of gas hydrate in the Black Sea. *Geophysical Journal International, 161(3)*, 662-678.

[1051] Zillmer, M., Reston, T., Leythaeuser, T. & Flueh, E. R. (2005b). Imaging and quantification of gas hydrate and free gas at the Storegga slide offshore Norway. *Geophysical Research Letters, 32(4)*, L04308, doi: 10.1 029/2004GL02 1535.

[1052] Zimov, S., Schuur, E. A. G. & Chapin, F. S. (2006). Permafrost and the global carbon budget. *Science, 312*, 1612-1613.

[1053] Zuhlsdorff, L. & Spiess, V. (2004). Three-dimensional seismic characterization of a venting site reveals compelling indications of natural hydraulic fracturing. *Geology, 32(2)*, 101-104.

[1054] Zuhlsdorff, L., Spiess, V., Hubscher, C., Villinger, H. & Rosenberger, A. (2000). Implications for focused fluid transport at the northern Cascadia accretionary prism from a correlation between BSR occurrence and near-sea-floor reflectivity anomalies imaged in a multi-frequency seismic data set. *International Journal of Earth Sciences, 88(4)*, 655-667.

In: Rapid Climate Change Past Evidence and Future Prospects   ISBN: 978-1-60741-422-3
Editor: Alice E. Bennett   © 2011 Nova Science Publishers, Inc.

Chapter 2

CLIMATE CHANGE: SCIENCE HIGHLIGHTS

A. Jane Leggett, Specialist in Energy and Environmental Policy

SUMMARY

Scientific conclusions have become more compelling regarding the influence of human activities on the Earth's climate. In 2007, the Intergovernmental Panel on Climate Change declared that evidence of global warming was "unequivocal." It concluded that "[m]ost of the observed increase in globally averaged temperatures since the mid-20th century is very likely due to the observed increase in anthropogenic [human-related] greenhouse gas [GHG] concentrations."

The IPCC concluded that human activities have markedly increased atmospheric concentrations of "greenhouse gases" (GHG), including carbon dioxide (CO_2), methane (CH_4), nitrous oxide (N_2O), and gases (such as chlorofluorocarbons, CFC) that are controlled under the Montreal Protocol to protect the stratospheric ozone layer. From the beginning of the Industrial Revolution, CO_2 has risen from about 280 parts per million (ppm) to about 386 ppm today (up 38%). The concentration of CO_2 is higher now than in at least 800,000 years before present.

Additional human influences on the climate that are not easily compared to GHG emissions could, nonetheless, be managed to moderate regional and global climate change. These include tropospheric ozone pollution (i.e., smog), particulate and aerosol emissions, and land cover change. New chemicals, such as nitrogen trifluoride (NF_3), also may play a small role.

Without radical changes globally from current policies and economic trajectories, experts uniformly expect that GHG emissions will continue to grow and lead to continued warming of the Earth's climate. Experts disagree, however, on the timing, magnitude and patterns of future climate changes. In the absence of concerted climate change mitigation policies, for a wide range of plausible GHG scenarios to 2100, the IPCC projected "best guess" increases in global average temperatures from 1.8°C to 4.0°C (3.2°F to 7.2°F). Although these temperature changes may seem small, they compare to the current global, annual average temperature of around 14°C (57°F). While precipitation overall is expected to increase, its distribution may become more uneven: regions that now are dry are likely to get drier, while regions that now are wet, are likely to get wetter. Extreme precipitation and droughts are expected to become more frequent. Experts project that warming ocean waters will expand, and melting glaciers and ice sheets will further add to sea level rise. The Arctic Ocean could become ice free in summers within a few decades. Ocean salinity is expected to fall, and the Meridional Overturning Circulation in the Atlantic Ocean could slow, reducing ocean productivity and altering regional climates in both North America and Europe. The climate would continue changing for hundreds of years after GHG concentrations were stabilized, according to most models. There are also possibilities of abrupt changes in the state of the climate system, with unpredictable and potentially catastrophic consequences. Much concern is focused now, among scientists and economists, about the likelihoods and implications of exceeding such thresholds of abrupt change, sometimes called "tipping points."

This chapter summarizes highlights of scientific research and assessments related to human- induced climate change. For more extensive explanation of climate change science and analytical methods, see CRS Report RL33849, *Climate Change: Science and Policy Implications*.

INTRODUCTION

The focus of policy-makers on climate change science has shifted from debate over whether the Earth's climate has changed and whether human-related greenhouse gases are responsible for a major portion of it. Focus has turned to debate over the magnitude and patterns of future climate change, how adverse such changes may be, and how projections may inform mitigation and adaptation policy choices. There is also growing recognition of a wider variety of human-related "forcings" of climate change than the six "Kyoto gases" to include other kinds of pollution, newly developed chemicals, and land use patterns.

Understanding of potential impacts of climate change has deepened as well, but remains hindered by poor time and spatial resolution of predictions and, especially, by the wide divergence of projections across different climate models. This divergence reflects continuing uncertainties concerning clouds, oceans, and vegetation feedbacks to climate change, among other questions. It also reflects re-emerging recognition of the importance of socio-economic factors in technological change, climate change projections, potential damages (or opportunities) of those changes, and appropriate policy responses. Confidence has grown in some aspects of climate change science, allowing policy-makers to deliberate over appropriate risk management strategies, but needs for further research continue.

This chapter highlights major scientific observations, conclusions, and issues. The principal scientific findings from the 2007 Fourth Assessment Report (AR4) of the Intergovernmental Panel on Climate Change (IPCC)[1] continue to stand, and will be summarized in this chapter.[2] A fuller explanation of climate change processes, analytical methods, uncertainties, and controversies is provided in CRS Report RL33849, *Climate Change: Science and Policy Implications*, by Jane A. Leggett.

OBSERVED WARMING AND ADDITIONAL
METRICS OF CLIMATE CHANGE

The IPCC in 2007 declared that "[w]arming of the climate system is unequivocal.... Observational evidence from all continents and most oceans shows that many natural systems are being affected by regional climate changes." The Earth's climate has warmed by $0.6°$ to $0.9°$ Celsius (1.1 to $1.6°$ Fahrenheit) since the Industrial Revolution and approximately 0.5°C compared to the 20th century average. (See **Figure 1**.) Precipitation has increased over the past century, although some regions have become wetter and some have become drier, consistent with scientists' understanding of how heightened greenhouse gas concentrations affect climate regionally. Observed increases in ocean temperatures, altered wind patterns, extreme weather events, melting glaciers and sea ice, and timing of seasons are also attributed in part to greenhouse gas forcing. Although there is substantial natural variability in the climate system, a warming trend continued through 2008, with the year tied with 2001 as the eighth warmest globally since reliable measurements began in 1880.[3]

Figure 1 also shows that the global climate varies from year-to-year and over longer cycles, as well as showing a century-long trend to warming. Some influences are natural and some are very likely human-related. At regional and local scales, climate is generally even more variable. This illustrates the caution that should be exercised in trying to detect changes in trends in the context of variability. One or a series of hot or cool years, or a few extreme weather events, do not necessarily represent more than normal climate variability. For example, better scientific understanding of natural phenomena like La Nina, and solar variability have helped scientists to understand their cooling influence on global average temperatures since the year 2000, or the impact of El Nino on the extreme high temperature of 1998.[4] Detecting changes in the climate system requires measurements and assessment over decades or longer.

The U.S. Climate Change Science Program (CCSP) produced a new synthesis report in 2008, "reanalyzing" the U.S. climate from 1950 to the present,[5] evaluating observed and modeled changes, and attributing changes to different factors. The researchers estimated average warming over North America to have been 0.9 °C (1.6°F) from 1951 to 2006, with almost all of the warming later than 1970. The reanalysis concluded, among other findings, that:

- The largest yearly average regional temperature increases have occurred over northern and western North America, with up to 2.0°C (3.6°F) warming in 56 years over Alaska, the Yukon Territories, Alberta, and Saskatchewan. On the other hand, there have been no significant yearly average temperature changes in the southern United Sates and eastern Canada.
- There has not been a significant trend in North American precipitation since 1951, although there have been substantial changes from year to year and even decade to decade.

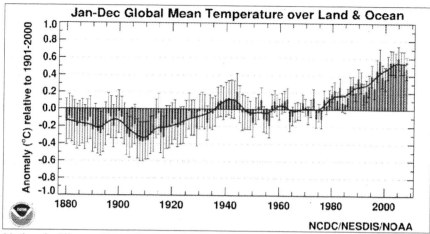

Source: National Climate Data Center, Climate of 2008, National Oceanic and Atmospheric Administration, December 16, 2008.

Notes: An anomaly is the difference between the value in a given year (or other time period) and the long-term average for a specified period, which in this case is one century. It describes how the annual value differs from a defined "normal."

The blue smoothed line in that graph is obtained by applying a "21-point binomial filter" to the time series plotted as red bars. The "whisker" (thin black vertical) lines represent confidence or possible error levels. Levels of confidence have improved sizably over the past century.

Figure 1. Global Annual Temperature Anomalies Compared to the 20th Century Average 1880 - 2008

The report did not find a significant trend in continental precipitation. Nor did it find systematic changes in how often or where severe droughts have occurred in the contiguous United States over the past decades. However, the demands and competition for water resources have increased in some regions, so that the vulnerability to drought is more severe.

ATTRIBUTION OF OBSERVED CHANGES
MOSTLY TO GREENHOUSE GASES

In 2007, the IPCC fourth assessment report concluded that "[m]ost of the observed increase in globally-averaged temperatures since the mid-20th century is very likely due to the observed increase in anthropogenic GHG concentrations."[6] According to the report, natural phenomena, such as volcanoes, solar variability and land cover change, have undoubtedly influenced the observed climate change, but the dominant driver of change since the 1970s is estimated to be the increase of greenhouse gases (GHG) in the Earth's atmosphere due to emissions from human- related activities.

The CCSP reanalysis of North American climate data[7] concluded that there is more than a 66% likelihood that more than half the average continental warming since 1951 has been due to human activities, but also that the regional differences in summer surface temperatures are unlikely to have been driven by human influences alone. The reanalysis found that sea surface temperatures likely have affected temperature trends, regional differences in temperature and regional and seasonal differences in precipitation. It also concluded that sea surface temperatures have likely contributed to multi-year droughts.[8]

Human-Related Influences on Climate Change

Although the most potent greenhouse gas in the Earth's atmosphere is water vapor, it is thought not to be directly influenced at a large scale by human activities.[9] Most policy attention has been given to the "basket" of six gases covered by the Kyoto Protocol: Carbon dioxide (CO_2), methane (CH_4), nitrous oxide (N_2O), sulfur hexafluoride (SF_6), hydrofluorocarbons (HFC), and perfluorocarbons (PFC). CO_2 is the most important human-influenced GHG globally.

Less policy attention has been give to other GHG and other human-related "forcings" of climate change, which might also offer climate change abatement opportunities:

- certain synthetic chlorinated and fluorinated chemicals (e.g., chlorofluorocarbons (CFC), hydrochlorofluorocarbons (HCFC), the production of which is controlled to reverse destruction of the ozone layer in the stratosphere (but which continue to be emitted from certain sources);
- tropospheric ozone (or "smog"), which is controlled in many countries as an air pollutant with adverse health and environmental effects, and is not emitted but is formed in the atmosphere due to emissions of nitrogen oxides (NOx), volatile organic compounds (VOC), and carbon monoxide (CO);
- regional scale air pollutants, such as sulfates, and tiny carbon-containing particles called *black carbon aerosols*.[10]

In some regions and over some periods, these air pollutants may dominate local climate changes, including how much precipitation falls and where. Aerosols also darken reflective surfaces, such as snow, and absorb more of the Sun's radiation. On snow, scientists have shown particles to increase melt, and hence create feedbacks that may accelerate climate changes. Some researchers argue that abatement of black carbon aerosols could help to slow atmospheric warming and melting in the Arctic and mountainous regions.[11]

New types of emissions are emerging that also have small, but potentially growing, effects on climate change. For example, nitrogen trifluoride (NF_3) is a gas used in certain manufacturing processes, introduced as a more benign alternative to CFC, but with a strong Global Warming Potential.[12] While the current influence of this new, substitute gas is very small (well less than half a percent of CO_2 forcing), its appearance points to the need for attention to Global Warming Potentials as new chemicals are developed, and for flexibility in incorporating new climate change influences into policy approaches over time.

Land use and land use changes have long been recognized for their influence on local climates, and increasingly on global climate. For example, the built environment is often constructed with dark materials, such as asphalt roads or roof shingling, that absorbs the Sun's radiation and heats the environment. These *urban heat islands* have a very small global effect. However, other changes, such as the warming influences of loss of snow cover, and the influences of changing vegetation (e.g. forest losses in the tropics), and possibly of agricultural irrigation, have been raised as meriting attention for their regional effects and potential global influence. While these influences do not reduce the importance of CO_2 emissions and other GHG forcing, they raise opportunities for enhancing mitigation and management of climate changes locally and globally.

Trends in Atmospheric Concentrations of Greenhouse Gases

Carbon dioxide (CO_2) concentrations have grown from a pre-industrial concentration of about 280 parts per million volume (ppm) to 386 ppm in 2008 (See **Figure 2**).[13] The IPCC had concluded in 2007 that "[a]tmospheric concentrations of CO_2 (379ppm) and methane (1774 parts per billion - ppb) in 2005 exceed[ed] by far the natural range over the last 650,000 years." The IPCC found that the increases in CO_2 concentrations since the Industrial Revolution were due primarily to human use of fossil fuels, with land-use changes (primarily deforestation) making a significant but smaller contribution. While over the past few decades, countries have trended towards using cleaner, lower carbon fuels (such as natural gas instead of coal), the IPCC noted that "the long-term trend of declining CO_2 emissions per unit of energy supplied reversed after 2000.[14]

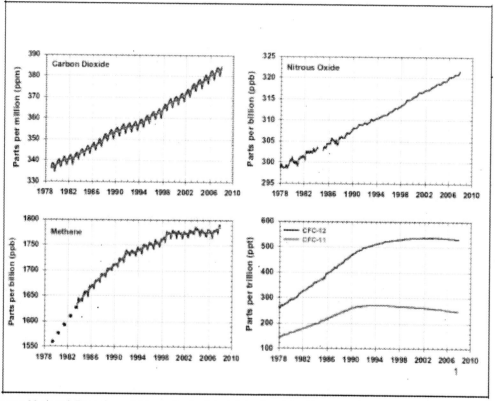

Source: National Oceanic and Atmospheric Administration, http://www.esrl.noaa.gov/gmd/aggi/.
Notes: Global averages of the concentrations of the major, well-mixed, long-lived greenhouse gases –
    carbon dioxide, methane, nitrous oxide, CFC-12 and CFC-11 from the NOAA global flask
    sampling network since 1978.
    These gases account for about 97% of the direct radiative forcing by long-lived greenhouse gases
    since 1750. The remaining 3% is contributed by an assortment of 10 minor halogen gases (see
    text). Methane data prior to 1983 are annual averages from Etheridge et al. (1998), adjusted to the
    NOAA calibration scale [Dlugokencky et al., 2005].

Figure 2. Global Average Atmospheric Concentrations of Five Major Greenhouse Gases 1978 to 2008

Methane concentrations also grew from a pre-industrial value of about 715 ppb to about1786 ppb in 2008.[15] (See **Figure 2**.)The rate of methane growth slowed and had been negative in several years since about 1992 for a variety of reasons, including economic restructuring, methane recovery for energy value (e.g., from landfills, animal wastes), etc. Methane concentrations grew only slightly since around 1999 but turned upwards again in 2007. Nitrous oxide emissions continue to grow at a roughly constant rate. In contrast, CFC and HCFC have level or declining concentrations since the early- to mid-1990s.

Greenhouse Gas Emissions and Growth Globally

The United States contributes almost one-fifth of net global greenhouse gas emissions and China contributes slightly more. China's CO_2 emissions exceeded those of the United States sometime around 2005, though changes in 2008 in economic growth rates, energy use and uncertainty preclude precise estimates for China. With its robust economic growth— dependent on industrialization fueled largely by coal—China is likely to remain the largest global emitter of CO_2 for the foreseeable future, although the government has established ambitious policies to improve energy efficiency, promote renewable energy and reduce polluting emissions.[16] Future greenhouse gas emissions will likely grow most rapidly in developing economies, as they strive to eliminate poverty and raise income levels towards those of the wealthier "Annex 1" countries. Future GHG trajectories are widely uncertain, depending largely on the rate and composition of economic growth, as well as technology and policy choices.

OBSERVED IMPACTS OF CLIMATE CHANGES

The IPCC concluded in 2007[17] that: " ... discernible human influences extend beyond average temperature to other aspects of climate." Human influences have:

- very likely contributed to sea level rise during the latter half of the 20th century
- likely contributed to changes in wind patterns, affecting extra-tropical storm tracks and temperature patterns
- likely increased temperatures of extreme hot nights, cold nights and cold days
- more likely than not increased risk of heat waves, area affected by drought since the 1970s and frequency of heavy precipitation events.

Anthropogenic warming over the last three decades has likely had a discernible influence at the global scale on observed changes in many physical and biological systems.

Extent of Arctic Sea Ice Near Lowest Recorded Levels

Sea ice at the poles is a vital component of the Earth's current climate system. Sea ice controls key aspects of Arctic atmospheric circulation, polar warming and other critical components of the Earth's climate system. Polar sea ice is of cultural and iconic value to some people. It also affects a number of human activities, such as shipping, fishing, resource accessibility, and tourism. Sea ice is important to current Arctic ecology, such as habitat for polar bear, seals, whales and others.

Arctic sea ice shrunk to its smallest extent in 2007, and nearly to the same extent in 2008, since satellite measurements began in 1979. Sea ice cover has reached perhaps 50% below the sea ice extent of the 1950s. Average sea ice extent in September 2008 was about 4.67 million square kilometers, compared to the record low of September 2007 of 4.28 million square kilometers (1.65 million square miles). Compared to the average extent of sea ice between 1979 and 2000, 2008 was 34% below and 2007 was 39% below. The rate of sea ice decline since 1979 has reached approximately 11% per decade, or 78,000 square kilometers (28,000 square miles) per • year.[18] While rapid Arctic ice loss appears in climate model runs, the loss of Arctic sea ice extent has been more rapid than produced by climate models.[19]

While the record melting of Arctic sea ice is associated with GHG-induced warming, the winds in 2007 pushed sea ice from the Arctic toward the Atlantic Ocean. Simultaneously, Arctic currents seemed to be reversing, returning to the pre-1990s direction. In addition, low cloudiness led to more solar warming than usual. According to NASA, "The results suggest not all the large changes seen in Arctic climate in recent years are a result of long-term trends associated with global warming." In 2008, there were near-record lows despite return of more "normal" wind and atmospheric conditions.

Earlier seasonal melting of sea ice triggers a positive feedback that increases ocean warming, further increasing sea ice melting, and so on.[20] Updated estimates now project that the Arctic Ocean could be ice-free in summer as early as 2040,[21] 2030[22] (or sooner) if recent accelerations in sea ice loss continue. Some scientists have expressed concern that recently observed sea ice loss may have passed a "threshold" or a spiral of warming feedbacks.[23] Melting ice in the Arctic would contribute very little to global sea level rise because it already floats, with its volume already displacing sea water; however, it contributes to concern because of the impact loss of Arctic sea ice would have in warming Arctic waters and atmosphere, and consequent effects on warming and melting of the Greenland Ice Sheet.

Melting of the Greenland Ice Sheet

Between 1979 and 2005, the area of Greenland that melted on at least one day per year grew by 42%, while the mean temperature rose by 2.4°C.[24] However, recent changes in rates of melting of the Greenland Ice Sheet point to variability in the climate system and the difficulties in discerning trends among the changes. Beginning in the late 1990s, ice flows from two of Greenland's biggest glaciers that flow into the ocean accelerated rapidly, surprising scientists at the speed of change. The lack of prediction of the phenomenon contributed to the IPCC's decision not to include the contribution of ice sheet melting in its projections of sea level rise due to global warming in the 21st Century. The high melting rates in 2005 startled many scientists and raised major concerns about the potential impacts on sea level rise. Some scientists have argued that policy goals to address climate change should avoid passing certain "tipping points" of the climate system that could have potentially catastrophic impacts, naming melting of the Greenland Ice Sheet on one of these thresholds. (See section "Projections of Future Climate" for further discussion of possible tipping points.)

However, as melting rates in 2006 returned closer to the average, they have exposed greater variability and complexity in ice dynamics than previously understood.[25] Now some scientists believe that warming waters near the glacier outlets accelerate ice flows and results in retreat of their floating leading margins. However, after a point, the glacier regains stability on its grounding rock and the retreat slows, though it will continue to melt with warm air temperatures.

Melting and Thickening of Ice in Antarctica

Over the past few decades, the atmosphere over Antarctica has warmed. Satellite observations analyzed in 2007 indicate that the Antarctic ice sheet is losing mass overall; the losses are mainly from the western Antarctic ice sheet. NASA satellites revealed that snow is melting farther inland, at higher altitudes than before and, increasingly, on the Ross Ice Shelf, which buffers land-based glaciers from the warmer ocean air.[26] Some high elevation regions of the Antarctic ice sheet do not show a significant rate of change or show less melting. Researchers identified a link between changes in temperatures and the duration and area of melting in Antarctica, suggesting a connection to global climate change. In another 2007 study, the British Antarctic survey found that 300 glaciers studied increased their average flow rate by 12% from 1993 to 2003. This was attributed to thinning of the lower glaciers at the edge of the sea, allowing the glaciers above them to flow faster, similar to phenomena observed in Greenland. Unlike Greenland, the Western Antarctic Ice Sheet is not well grounded like the outlet glaciers of Greenland, so that disintegration of the lead glacial margins could lead to persistently accelerated flows of ice to the sea. The researchers tied local warming on the Antarctic Peninsula—some of the fastest recent warming on Earth (nearly 3°C, or 4.4°F, over 50 years)—to retreat of 87% of its glaciers and the observed increase in their flow rates.[27]

In 2008, parts of the Antarctic Pennisula's Wilkins Ice Sheet disintegrated in three stages, which is especially significant because two of the stages occurred during the cold season. The pattern of breakup was smaller but similar to that of the Larsen A and B ice shelves, in1995 and 2002 respectively. According to the NSIDC, preliminary studies of the sea floor below the Larsen B ice shelf suggest that the 2002 disintegration was the first such in 12,000 years.[28] While the general warming of ocean waters in both the Arctic and Antarctica contributes to loss of ice sheets, two studies in 2008 indicate that other factors (winds and current changes) may circulate warm water in the vicinity of ice shelves.[29]

No Melting of Some Permanent Ice Fields

Not all glaciers and ice fields are experiencing increased melting. One study published in 2008 indicated that snow accumulation has doubled in the south-western Antarctica Peninsula since 1850, with rates accelerating in the past few decades. [30]In Europe, while glaciers between 2,000 and 4,000 meters in altitude have lost an average of 1-1.5 kilometers of length through the 20th Century, others at high altitude—above 4,200 meters—have changed very little in the same period. Some melting did occur, however, during the 2003 extreme heat wave.

Contributions of Melting Ice and Warming Oceans to Sea Level Rise

A 2008 assessment of satellite-based data suggests that most of the sea level rise observed in recent years can be explained by an increased mass of the oceans (i.e., more water). Of the global melting of ice contributing to observed sea level rise, about half has come from relatively small land-based glaciers, with the other half contributed by melting of the Greenland and Antarctic ice sheets.[31] One report published in 2007 concluded that the net amount of melting ice from glaciers and ice caps flowing to the oceans each year is about 100 cubic kilometers—or about the volume of Lake Erie.

With further warming, the acceleration of dynamic ice melt could raise the estimates of sea-level rise by an additional 4 to 10 inches by 2100. Recent articles have proposed a range of new estimates for sea level rise in the 21st Century that would include contributions of sea ice melt, particularly from Greenland. Pfeffer et al. conclude that physical constraints would preclude more than 2 meters (6.6 feet) of sea level rise over the coming century (with a range of 2.6 to 6.6 feet), and put forward a best guess, with low confidence, of about 0.8 meters rise by 2100.[32] Grinsted et al. suggest a range of 0.9 to 1.3 meters (3 to 4.3 feet) of sea level rise in 2090 to 2100, using a moderate climate change scenario.[33]

Hydrological Changes in the Western United States

A modeling study published in 2008 concluded that human factors may have induced as much as 60% of the changes observed between 1950 and 1999 in the hydrological cycle in the western United States. Climate changes were found to have influenced river flows, winter air temperatures and snow pack. The authors concluded that these changes, and their human influences, suggest an impending water supply crisis in the West.[34]

Observed Ecological Impacts of Climate Change

A growing number of studies are published each year investigating possible linkages between climate change and ecological changes. Results from a few released in 2008 are highlighted here.

One study concluded that warming of the Southern Ocean around Antarctica is threatening King penguin populations on the continent.[35]

A number of new studies continue to underscore threats to coral reefs globally by a variety of stressors that include heat stress from warm ocean events, and "ocean acidification" caused by absorption of CO_2 from the atmosphere by the oceans. One of these studies concluded that almost one-third of 704 reef-building coral species that could be assessed with data show enhanced risk of species extinction. It further concluded that the share of coral species at risk has risen in recent decades. The Caribbean was the region with the highest share of corals at high risk of extinction, while the Coral Triangle in the western Pacific had the greatest share of coral species in all categories at risk.[36] Another 2008 study concluded that throughout Australia's Great Barrier Reef, coral calcification (a measure of growth) has decreased by 14% since 1990. The authors further concluded that the severity and abruptness of the observed decline was unprecedented in at least the past 400 years.[37]

While risks to coral reefs from global warming and ocean acidification are increasingly studied, the associated risks to reef fish communities have been acknowledged but not documented. A 2008 study assessed the impacts of the major 2008 coral bleaching events across seven countries, 66 sites and 26 degrees of latitude in the Indian Ocean. The study concluded that, while impacts across sites were variable, ocean scale integrity of fish communities was lost, reflected in size structure, diversity, and food-chain composition of the reef fish. The authors also found that management regimes did not appear to affect the ecosystem responses to the bleaching event, suggesting a need to develop strategies for system-wide resilience to climate variability and change.[38] At least one study found evidence that some corals may be able to adjust to bleaching events by shifting the types of algae (zooxanthellae) with which they co-depend.[39]

In many ecological systems, climate is a primary—but not the sole—factor influencing the survival and behaviors of species. With the climate change experienced in recent decades, land- use, climate change and other factors have been associated with substantial range contractions, extinction of at least one species, and numerous changes in the timing of animal and plant behavior.

Polar bears are among the species that depend on sea ice for hunting and must fast during ice-free periods. The Western Hudson Bay of Canada has had ice-free summer periods for many years and, although the local polar bear population had previously appeared healthy, more recent observations have revealed lower survival rates among cubs and young bears.[40] Similar patterns have now emerged in Southern Hudson Bay and the Southern Beaufort Sea.[41]

Observations of several forest systems suggest that they are adapting to changes in climate more effectively than some scientists had expected. More specifically, NASA satellite imaging indicates that U.S. forests are adapting to the climate change experienced to date, and that the overall productivity response to weather and seasonal conditions has been closely linked to the number of different tree species in a forest area.[42] In Brazil, the productivity of Amazon forests has been resilient in spite of short but severe drought conditions in 2005, contrary to predictions of some ecosystem models, although whether the resistance will be sustained under longer drought—expected with climate change—is unknown.[43] Studies have shown increases in primary productivity in the Amazon as well as in above-ground biomass. They also show, however, changes in the composition of plant species, with fast-growing species faring better than slow- growing ones. The authors attributed these changes to global environmental changes, including elevated levels of CO_2 in the atmosphere.[44] Another study examined the influences of high temperatures on tropical forest uptake of CO2 from the atmosphere. It found that elevated temperatures initially raised CO_2 uptake, then CO_2 uptake declined. The authors concluded that, in the particular tract of forest studied, temperatures were approaching a threshold above which CO_2 uptake would drop sharply.[45]

WITHOUT FURTHER GHG MITIGATION POLICIES, GHGEMISSIONS WILL GROW

The U.S. Climate Change Science Program (US CCSP) released its second report in 2007, entitled "Scenarios of Greenhouse Gas Emissions and Atmospheric Concentrations and Review of Integrated Scenario Development and Application."[46] This research produced new scenarios of future GHG emissions and concluded that "In the reference scenarios,[47] economic and energy growth, combined with continued fossil fuel use, lead to changes in the Earth's radiation balance that are three to four times that already experienced since the beginning of the industrial age."[48]

This research also explored scenarios aimed at stabilizing the growth of GHG concentrations in the atmosphere at four increasingly stringent levels: roughly 750 parts per million (ppm), 650 ppm, 550 ppm, and 450 ppm (including multiple GHGs as CO_2-equivalents[49]). The analysis concluded, "The timing of GHG emissions reductions varies substantially across the four radiative forcing stabilization levels. Under the most stringent stabilization levels [450-550 ppm] emissions begin to decline immediately or within a matter of decades. Under the less stringent stabilization levels [750 ppm], CO_2 emissions do not peak until late in the century or beyond, and they are 1 1/2 to over 2 1/2 times today's levels in 2100."[50]

The results of the CCSP reference scenarios are similar to those of the 2000 Special Report on Emission Scenarios (SRES) of the IPCC, though the latter explored a wider range of uncertainty in its reference projections. The SRES projected global GHG emissions, without further climate change mitigation policies, to increase by 25-90% (CO_2-equivalent) from 2000 to 2030, with CO_2-equivalent concentrations growing in the atmosphere to 600-1550 ppm.

PROJECTIONS OF FUTURE CLIMATE

Scientists have found it very likely that rising greenhouse gas concentrations, if they continue unabated, will increase global average temperature above natural variability by at least 1.5° Celsius (2.7° Fahrenheit) during the 21st Century (above 1990 temperatures), with a small likelihood that the temperature rise may exceed 5 ° C (9 ° F). The projections thought most likely by many climate modelers are for greenhouse gas-induced temperature rise of approximately 2.5° to 3.5 ° C (4.5 to 6.3 ° F) by 2100.[51] Future climate change may advance smoothly or sporadically, with some regions experiencing more fluctuations in temperature, precipitation, and frequency or intensity of extreme events than others. Wet regions are expected to get more precipitation and dry regions are expected to become drier. Floods, droughts, storms and other extreme weather events are projected to increase, with impacts for ecological and human systems.

A report by the CCSP found that short-lived air pollutants, such as tropospheric ozone and black carbon aerosols could contribute as much as 20% of global warming by 2050. It found that one "climate model using projected changes in emissions and pollutant levels that occur primarily over Asia predicts significant increases in surface temperature and decreases in rainfall over the continental United States during the summertime throughout the second half of this century."[52] While most policy attention has been on the long-lived GHG, such as carbon dioxide and methane, including other climate forcings, such as short-lived air pollution, offers additional opportunities to abate near-term climate change, with the co-benefits of reducing health and environmental impacts.

With projected global warming, sea levels could rise by between 18 and 59 centimeters (between 7 and 23 inches) by 2100 due to expansion of oceans waters as they warm and additions of meltwater (at current rates of melting) from land-based glaciers and ice caps. The IPCC scientists were unable to include a quantitative estimate of the risks of accelerated melting or possible collapse of the Greenland or Antarctic ice sheets due to inadequacies of existing understanding of their dynamics.

CONCERN ABOUT ABRUPT "TIPPING POINTS" IN THE CLIMATE SYSTEM

Some people are concerned less about chronic damages due to slow and continuing climate change than they are about the potential for abrupt "tipping points" in the current climate system. Once the climate system reaches certain thresholds of change or tipping points, major aspects of the Earth's climate could change abruptly and in uncertain ways.[53] **Figure 3** shows one recent set of estimates of where dangerous "tipping points" may exist in the climate system relative to potential future global temperature increases. It also shows an estimate of the likelihoods, or probability density function (the black curve), that human-related GHG emissions since 1750 have already committed the planet to degrees of warming. In other words, according to the estimates represented in this figure, it is most likely that greenhouse gas emissions from 1750 to 2005 will lead to global average warming of 1°-$3^\circ C$, and potentially result in ice-free Arctic summers, major reduction of area and volume of the Hindu-Kush-Himalaya-Tibetan (HKHT) glaciers, major melting of the Greenland Ice Sheet, and so non. According to the estimates in the cited study, it is less likely that the committed warming is already greater than the "tipping points" for collapse of the Amazon rain forest and other identified components.

PROJECTIONS OF FUTURE IMPACTS

Some impacts of climate change are expected to be beneficial in some locations with a few degrees of warming (e.g., increased agricultural productivity in some regions, less need for space heating, opening of the Northwest Passage for shipping and resource exploitation). Most impacts are expected to be adverse (e.g., lower agricultural productivity in many regions, drought, rising sea levels,[54] spread of disease vectors, greater needs for cooling). Risks of abrupt, perhaps unidentified climate changes in aspects of the environment, with accompanying dislocations, are expected to increase as global average temperature increases, and could push natural and socioeconomic systems past key thresholds.[55] An example that recently occurred is the rapid and widespread infestation of beetles in western North America.[56] Such changes could precipitate major reorganization of ecosystems, and under some circumstances, stresses on nearby or dependent human systems.

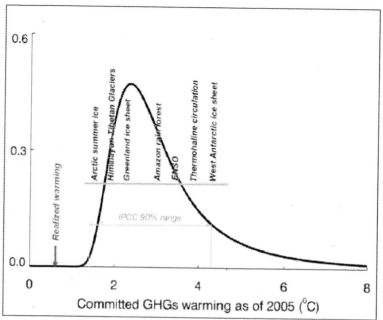

Source: 1. V. Ramanathan and Y. Feng, "On Avoiding Dangerous Anthropogenic Interference with the Climate System: Formidable Challenges Ahead," Proceedings of the National Academy of Sciences 105, no. 38 (September 23, 2008

Notes: Seven climate system components are identified that could undergo major, potentially abrupt and catastrophic, changes, as global temperatures increase. (ENSO includes El Nino/La Nina oscillations, and thermohaline circulation is the large-scale ocean overturning and circulation driven by temperature differences between the tropics and the poles and ocean salinity/freshwater). Each "climate-tipping" component is graphed above an estimated Increases in global average temperature (compared to pre-industrial levels) that could initiate the tipping, according to T.M. Lenton et al., Tipping Elements in the Earth's Climate System, Proceedings of the National Academy of Sciences, Vol.105:1786-1793 (2007). The estimated likelihoods (probability density function) that GHG emissions had already committed the climate system to different degrees of warming are represented by the black curve. See Ramanathan and Feng for further detail.

Figure 3. Possible "climate-tipping" elements and their possible likelihoods for committed global temperature change that may precipitate the tipping

Impacts on water resources are expected to be among the most serious due to climate change, combined with rising demand and management issues in many regions. Earlier melting of snowpack and ice in some regions (e.g., the Andes, the U.S. West) and loss of ice and snow would reduce water supply to settlements that now depend on the snowmelt, and change the seasonality of water supply. Snowfall in some regions is expected to increase, however, with greater atmospheric moisture.

One CCSP assessment examined the implications of projected sea level rise, with emphasis on the U.S. mid-Atlantic region. It concluded:

Today, rising sea levels are submerging low-lying lands, eroding beaches, converting wetlands to open water, exacerbating coastal flooding, and increasing the salinity of estuaries and freshwater aquifers. Other impacts of climate change, coastal development, and natural coastal processes also contribute to these impacts. In undeveloped or less developed coastal areas where human influence is minimal, ecosystems and geological systems can sometimes shift upward and landward with the rising water levels. Coastal development, including buildings, roads, and other infrastructure, are less mobile and more vulnerable. Vulnerability to an accelerating rate of sea-level rise is compounded by the high population density along the coast, the possibility of other effects of climate change, and the susceptibility of coastal regions to storms and environmental stressors, such as drought or invasive species.[57]

Agriculture in many regions, especially in the tropics, may experience losses of productivity, especially where higher temperatures coincide with increased dryness and limits on irrigation. The general shift upward in growing season temperatures could raise typical temperatures to or above those considered extreme today.[58] Agriculture may also confront challenges from invasive species, including pests. One 2008 study examined four key pests of maize in the United States with climate change. It projected increased winter survival and expanded ranges of all four pests, especially of the corn earworm, which may be resistant to insecticides. The authors concluded that even with pest management adaptations, the effects of climate change could increase seed and insecticide costs, decrease yields and alter crop yield variability.[59]

In 2008, a number of studies highlighted risks to human health with climate change, and at least one study found potential benefits of reduced cold weather mortality.[60] Raised risks and adaptation costs could be associated with increased diarrhoeal disease, renal disease, heat stress, malaria, cholera, malnutrition, and respiratory and other health effects associated with elevated ozone and particulate air pollution due to higher temperatures, among other concerns.[61]

Effects on, and concerns about, mental health have been raised by several studies, especially associated with projected increases in extreme weather events.[62]

One 2008 study found potential for higher temperatures to result in lower ratios of male to female births and reduced male longevity in humans, at least in Nordic countries.[63]

As the degree and distribution of climate changes continue, ranges of species are likely to change. Climate change is highly likely to create substantial changes in ecological systems and services in some locations, and may lead to ecological surprises. The disappearance of some types of climate also raises risks of extinctions of species, especially those with narrow geographic or climatic distributions, and where existing communities disintegrate. One study projected that, under the highest IPCC emissions scenario, 12% to 39% of the Earth's land areas may experience novel climates (i.e., climate conditions not existing now) while 10% to 48% of land areas' climates may disappear by 2100 AD. In the lowest IPCC climate change scenarios, 4% to 20% of land areas gain novel climates and 4% to 20% see existing climates disappear.[64]

Because climate change will occur with different magnitudes and characteristics in different regions, resulting dislocations and disparities across locations are expected to increase pressure on international aid and migration, with possible implications for political stability and security. Impacts may be alleviated with investments in adaptation. Adaptation as a strategy, however, is thought to be more challenging and potentially less effective the more widespread, uncertain and severe the climate changes.

APPENDIX. SUMMARY FOR POLICYMAKERS OF THE SYNTHESIS REPORT OF THE FOURTH ASSESSMENT REPORT OF THE INTERGOVERNMENTAL PANEL ON CLIMATE CHANGE

On November 16, 2007, government officials from most countries—including the United States—agreed on a Summary for Policymakers of the Synthesis Report of the IPCC Fourth Assessment Report. The Synthesis Report is derived from three technical reports: "The Physical Science Basis" (February 2007); "Impacts, Adaptation and Vulnerability" (April 2007); and "Mitigation of Climate Change" (May 2007). It represents a consensus among government officials and researchers, and will "constitute the core source of factual information about climate change [upon which policymakers will] base their political action... in the coming years" (IPCC Media Advisory, November 17, 2007). Key conclusions are excerpted (and slightly reordered) below:

"Warming of the climate system is unequivocal...." (p. 1) "Observational evidence from all continents and most oceans shows that many natural systems are being affected by regional climate changes." (p. 2)

"Global GHG emissions due to human activities have grown since pre-industrial times.... Carbon dioxide (CO2) is the most important anthropogenic [greenhouse gas] GHG. Its annual emissions grew by about 80% between 1970 and 2004. The long-term trend of declining CO2 emissions per unit of energy supplied reversed after 2000." (p. 4)

"Atmospheric concentrations of CO2 (379ppm) and CH4 (1774 ppb) in 2005 exceed by far the natural range over the last 650,000 years. Global increases in CO2 concentrations are due primarily to fossil fuel use, with land-use change providing another significant but smaller contribution." (p. 4)

"Most of the observed increase in globally-averaged temperatures since the mid-20th century is very likely due to the observed increase in anthropogenic GHG concentrations." (p. 5)

"Even if the concentrations of all greenhouse gases and aerosols had been kept constant at year 2000 levels, a further warming of about 0.1°C per decade would be expected. Afterwards, temperature projections increasingly depend on specific emission scenarios." (p. 6)

"Anthropogenic warming could lead to some impacts that are abrupt or irreversible, depending upon the rate and magnitude of the climate change." (p. 13)

"The uptake of anthropogenic carbon since 1750 has led to the ocean becoming more acidic.... Increasing atmospheric CO2 concentrations lead to further acidification.... [P]rogressive acidification of oceans is expected to have negative impacts on marine shell-forming organisms (e.g. corals) and their dependent species." (p. 11)

"Sea level rise under warming is inevitable.... The eventual contributions from Greenland ice sheet loss could be several metres ... should warming in excess of 1.9-4.6°C above pre-industrial be sustained over many centuries." (p. 21)

"Some systems, sectors and regions are likely to be especially affected by climate change.

terrestrial ecosystems: tundra, boreal forest and mountain regions because of sensitivity to warming; mediterranean-type ecosystems because of reduction in rainfall; and tropical rainforests where precipitation declines

coastal ecosystems: mangroves and salt marshes, due to multiple stresses

marine ecosystems: coral reefs due to multiple stresses; the sea ice biome because of sensitivity to warming

water resources in some dry regions at mid-latitudes and in the dry tropics, due to changes in rainfall and evapotranspiration, and in areas dependent on snow and ice melt

agriculture in low-latitudes , due to reduced water availability

low-lying coastal systems, due to threat of sea level rise and increased risk from extreme weather events

human health in populations with low adaptive capacity." (p. 11)

"[M]ore extensive adaptation than is currently occurring is required to reduce vulnerability to climate change. There are barriers, limits and costs, which are not fully understood." (p. 14)

"[International cooperation] will help to reduce global costs for achieving a given level of mitigation, or will improve environmental effectiveness. Efforts can include ... emissions targets; sectoral, local, sub-national and regional actions; RD&D programmes; adopting common policies; implementing development oriented actions; or expanding financing instruments." (p. 19)

"Decisions about macroeconomic and other non-climate policies can significantly affect emissions, adaptive capacity and vulnerability." (p. 19)

"Determining what constitutes 'dangerous anthropogenic interference with the climate system' in relation to Article 2 of the UNFCCC involves value judgements." (p. 19)

"Limited and early analytical results from integrated analyses of the costs and benefits of mitigation indicate that they are broadly comparable in magnitude, but do not as yet permit an unambiguous determination of an emissions pathway or stabilisation level where benefits exceed costs." (p. 23)

"Many impacts can be reduced, delayed or avoided by mitigation." (p. 20)

"There is high agreement and much evidence that all stabilisation levels assessed can be achieved by deployment of a portfolio of technologies that are either currently available or expected to be commercialised in coming decades, assuming appropriate and effective incentives are in place.... " (p. 22)

"An effective carbon-price signal could realise significant mitigation potential in all sectors. Modelling studies show global carbon prices rising to 20-80 US$/tCO2-eq by 2030 are consistent with stabilisation at around 550 ppm CO 2-eq by 2100. For the same stabilisation level, induced technological change may lower these price ranges to 5-65 US$/tCO2-eq in 2030." (p. 18)

"All assessed stabilisation scenarios indicate that 60-80% of the reductions would come from energy supply and use, and industrial processes, with energy efficiency playing a key role in

many scenarios. Including non- CO2 and CO2 land-use and forestry mitigation options provides greater flexibility and cost-effectiveness. Low stabilisation levels require early investments and substantially more rapid diffusion and commercialisation of advanced low emissions technologies. Without substantial investment flows and effective technology transfer, it may be difficult to achieve emission reduction at a significant scale. Mobilizing financing of incremental costs of low-carbon technologies is important." (p. 22)

"The macro-economic costs of mitigation generally rise with the stringency of the stabilisation target." (p. 22)

"Impacts of climate change are very likely to impose net annual costs which will increase over time as global temperatures increase. Peer-reviewed estimates of the social cost of carbon in 2005 average US$12 per tonne of CO2, but the range from 100 estimates is large (- $3 to $95/t CO2). This is due in large part to differences in assumptions regarding climate sensitivity, response lags, the treatment of risk and equity, economic and non-economic impacts, the inclusion of potentially catastrophic losses, and discount rates. Aggregate estimates of costs mask significant differences in impacts across sectors, regions and populations and very likely underestimate damage costs because they cannot include many non-quantifiable impacts." (p. 23)

"Choices about the scale and timing of GHG mitigation involve balancing the economic costs of more rapid emission reductions now against the corresponding medium-term and long-term climate risks of delay." (p. 23)

End Notes

[1] The IPCC is organized under the auspices of the United Nations and engages participation of more than 2000 scientists from around the world. According to its website, "The IPCC was established to provide the decision-makers and others interested in climate change with an objective source of information about climate change. The IPCC does not conduct any research nor does it monitor climate related data or parameters. Its role is to assess on a comprehensive, objective, open and transparent basis the latest scientific, technical and socio-economic literature produced worldwide relevant to the understanding of the risk of human-induced climate change, its observed and projected impacts and options for adaptation and mitigation. IPCC reports should be neutral with respect to policy, although they need to deal objectively with policy relevant scientific, technical

and socio-economic factors. They should be of high scientific and technical standards, and aim to reflect a range of views, expertise and wide geographical coverage," http://www.ipcc.ch/about/index.htm (extracted November 26, 2007). Previous assessment reports of the IPCC were published in 1990, 1995, and 2001.

[2] For the reader's convenience, key findings from *this Summary for Policy Makers* are provided in the **Appendix** of this chapter. CRS has not independently verified all the findings in the IPCC report.

[3] National Climate Data Center, *Climate of 2008*, National Oceanic and Atmospheric Administration, December 16, 2008, http://www.ncdc.noaa.gov/oa/climate. The NCDC is the federal government's official source for climate data.

[4] National Climate Data Center, *2008 Annual Climate Review: U.S. Summary*, National Oceanic and Atmospheric Administration, Asheville, NC, December 16, 2008, http://www.ncdc.noaa.gov/oa/climate/

[5] Randall Dole, Martin Hoerling, and Siegfried Schubert, *Reanalysis of Historical Climate Data for Key Atmospheric Features: Implications for Attribution of Causes of Observed Change*, Final Report, CCSP Synthesis and Assessment Product 1-3 (Asheville: NOAA/NCDC, 2008), http://www.climatescience.gov/Library/sap/sap1-3/finalreport/default.htm.

[6] IPCC, "Summary for Policymakers of the Synthesis Report of the IPCC Fourth Assessment Report" (Intergovernmental Panel on Climate Change, 2007), at http://www.ipcc.ch/index.htm (accessed November 27, 2007), p. 1.

[7] Dole, op.cit.

[8] Dole, op.cit.

[9] Water vapor in the atmosphere is indirectly affected by human activities, as greenhouse-gas induced global warming would lead to an increase in atmosphere moisture. Depending on how this occurs, this indirect effect is considered a "positive feedback" that reinforces initial warming.

[10] CCSP *Atmospheric Aerosol Properties and Impacts on Climate*, A Report by the U.S. Climate Change Science Program and the Subcommittee on Global Change Research. [Mian Chin, Ralph A. Kahn, and Stephen E. Schwartz (eds.)]. National Aeronautics and Space Administration, Washington, D.C., USA. (2009); CCSP, *Climate Projections Based on Emissions Scenarios for Long-Lived and Short-Lived Radiatively Active Gases and Aerosols*. A Report by the U.S. Climate Change Science Program and the Subcommittee on Global Change Research. H. Levy II, D.T. Shindell, A. Gilliland, M.D. Schwarzkopf, L.W. Horowitz, (eds.). Department of Commerce, NOAA's National Climatic Data Center, Washington, D.C., USA, 100 pp. (2008)

[11] See, for example, Kathy S Law and Andreas Stohl, "Arctic air pollution: origins and impacts," Science (New York, N.Y.) 315, no. 5818 (March 16, 2007): 1537-40.

[12] Global Warming Potential (GWP) is an index of different molecules' potential to influence climate, molecule for molecule, compared to carbon dioxide. CO_2 has a GWP of 1, but its far greater abundance in the atmosphere compared with other greenhouse gases makes it the most important human-related GHG.

[13] The 2008 value is the preliminary annual mean measured at Mauna Loa. ftp://ftp.cmdl.noaa.gov/ccg/co2/trends co2_annmean_mlo.txt.

[14] CRS Report RS22970, *Are Carbon Dioxide Emissions Rising More Rapidly Than Expected?*, by Jane A. Leggett and Jeffrey Logan.

[15] http://www.noaanews.noaa.gov/stories2008/20080423_methane.html.

[16] CRS Report RL34659, *China's Greenhouse Gas Emissions and Mitigation Policies*, by Jane A. Leggett, Jeffrey Logan, and Anna Mackey.

[17] IPCC, op.cit.

[18] Data from the National Snow and Ice Data Center (NSIDC), http://nsidc.org/sotc/sea_ice.html, based on calculations by Walt Meier, NSIDC.

[19] J. Stroeve, M.M. Holland, W. Meier, T. Scambos, and M. Serreze. 2007. Arctic sea ice decline: Faster than forecast. Geophysical Research Letters doi:10.1029/2007GL029703.

[20] Donald K. Perovich et al., "Increasing solar heating of the Arctic Ocean and adjacent seas, 1979-2005: Attribution and role in the ice-albedo feedback," Geophysical Research Letters 34 (October 11, 2007).

[21] Marika Holland, Cecilia M. Bitz, and Bruno Tremblay, "Future abrupt reductions in the summer Arctic sea ice," Geophysical Research Letters 33, no. L23503 (2006) http://www.cgd.ucar.edu/oce/mholland/abrupt_ice/holland_etal.pdf (accessed December 22, 2006).

[22] According to Mark Serreze, US National Snow and Ice Data Center, University of Colorado, as quoted in David Adam, "Ice-free Arctic could be here in 23 years," The Guardian, September 5, 2007.

[23] See, for example, http://www.reuters.com/article/environmentNews/idUSL2815198120070928?sp=true.

[24] Marco Tedesco, "A New Record in 2007 for Melting in Greenland," EOS Transactions 88 (September 1, 2007).

[25] Richard B. Alley, Mark Fahnestock, and Ian Joughin, "Understanding Glacier Flow in Changing Times," *Science* 322, no. 5904 (November 14, 2008): 1061-1062, doi:10.1126/science.1166366.

[26] NASA, "NASA Researchers Find Snowmelt in Antarctica Creeping Inland," September 20, 2007, at http://www.nasa.gov/centers/goddard/news/topstory/2007/antarctic_snowmelt.html (accessed November 30, 2007).

[27] H. D. Pritchard and D. G. Vaughan, "Widespread acceleration of tidewater glaciers on the Antarctic Peninsula," Journal of Geophysical Research 112 (June 6, 2007).

[28] http://nsidc.org/news/press/larsen_B/2002_seafloor.html.

[29] Rignot, E., J.L. Bamber, M.R. van den Broeke, C. Davis, Y. Li, W.J. van de Berg, and E. van Meijgaard. 2008. Recent Antarctic ice mass loss from radar interferometry and regional climate modelling. Nature Geoscience 1: 106- 110. And Stammerjohn, S.E., D.G. Martinson, R.C. Smith, and R.A. Iannuzzi. 2008. Sea ice in the western Antarctic Peninsula region: Spatio-temporal variability from ecological and climate change perspectives. Deep Sea Research Part II: Topical Studies in Oceanography doi:10.1016/j.dsr2.2008.04.026.; as described by NSIDC in the State of the Cryosphere, http://nsidc.org/sotc/iceshelves.html.

[30] Elizabeth R. Thomas, Gareth J. Marshall, and Joseph R. McConnell, "A doubling in snow accumulation in the western Antarctic Peninsula since 1850" (January 12, 2008). http://www.agu.org/pubs/crossref/2008/2007GL032529.shtml.

[31] A Cazenave et al., "Sea level budget over 2003–2008: A reevaluation from GRACE space gravimetry, satellite altimetry and Argo," Global and Planetary Change 65, no. 1-2 (January 2009): 83-88, doi:10.1016/j.gloplacha.2008.10.004.

[32] W. T. Pfeffer, J. T. Harper, and S. O'Neel, "Kinematic Constraints on Glacier Contributions to 21st-Century Sea-Level Rise," Science 321, no. 5894 (September 5, 2008): 1340-1343, doi:10.1126/science.1159099.

[33] Aslak Grinsted, J. Moore, and S. Jevrejeva, "Reconstructing sea level from paleo and projected temperatures 200 to 2100 A.D.," Climate Dynamics, doi:10.1007/s00382-008-0507-2, http://dx.doi.org/10.1007/s00382-008-0507-2., using the IPCC "A1B" scenario of GHG emissions.

[34] Tim P. Barnett et al., "Human-Induced Changes in the Hydrology of the Western United States," Science 319, no. 5866 (February 22, 2008): 1080-1083, doi:10.1126/science.1152538.

[35] Céline Le Bohec et al., "King Penguin Population Threatened by Southern Ocean Warming," Proceedings of the National Academy of Sciences of the United States of America 105, no. 7 (February 19, 2008): 2493–2497, doi:10. 1073/pnas.071203 1105.

[36] Kent E. Carpenter et al., "One-Third of Reef-Building Corals Face Elevated Extinction Risk from Climate Change and Local Impacts," Science (July 10, 2008): 1159196, doi:10.1126/science.1159196.

[37] Glenn De'ath, Janice M. Lough, and Katharina E. Fabricius, "Declining Coral Calcification on the Great Barrier Reef," Science 323, no. 5910 (January 2, 2009): 116-119, doi: 10.1126/science. 1165283.

[38] Nicholas A. J. Graham et al., "Climate Warming, Marine Protected Areas and the Ocean-Scale Integrity of Coral Reef Ecosystems," PLoS ONE 3, no. 8 (2008): e3039, doi:10.1371/journal.pone.0003039.

[39] 1. Nicholas A. J. Graham et al., "Climate Warming, Marine Protected Areas and the Ocean-Scale Integrity of Coral Reef Ecosystems," PLoS ONE 3, no. 8 (2008): e3039, doi:10.1371/journal.pone.0003039.

[40] Regehr, Eric et al. "Survival and Population Size of Polar Bears in Western Hudson Bay in Relation to Earlier Sea Ece Breakup," Journal of Wildlife Management, v. 71, no. 8 (2007), pp. 2673-2683. See also CRS Report RL33941, Polar Bears: Listing Under the Endangered Species Act, by Eugene H. Buck, M. Lynne Corn, and Kristina Alexander

[41] USGS, USGS Science to Inform U.S. Fish & Wildlife Service Decision Making on Polar Bears: Executive Summary (Reston, VA, 2007), http://www.usgs.gov/newsroom/special/polar%5Fbears/.

[42] NASA, "NASA Satellites Can See How Climate Change Affects Forests," http://www.nasa.gov/centers/goddard/news/topstory/2006/forest_changes.html, (accessed November 28, 2007).

[43] Scott R. Saleska et al., "Amazon Forests Green-Up During 2005 Drought," Science (September 20, 2007); Yadvinder Malhi et al., "Climate Change, Deforestation, and the Fate of the Amazon," Science (November 29, 2007).

[44] Jérôme Chave et al., "Assessing Evidence for a Pervasive Alteration in Tropical Tree Communities," PLoS Biology 6, no. 3 (March 2008): e45, doi:10.1371/journal.pbio.0060045.

[45] Catia M. Domingues et al., "Improved estimates of upper-ocean warming and multi-decadal sea-level rise," Nature 453, no. 7198 (June 19, 2008): 1090-1093, doi:10.1038/nature07080.

[46] See CCSP, Scenarios of Greenhouse Gas Emissions and Atmospheric Concentrations. Sub-report 2.1A of Synthesis and Assessment Product 2.1 by the U.S. Climate Change Science Program and the Subcommittee on Global Change Research.[Clarke, L., J. Edmonds, H. Jacoby, H. Pitcher, J. Reilly, R. Richels] Department of Energy, Office of Biological & Environmental Research, Washington, DC., USA. (2007) p. 3. http://www.climatescience.gov/ Library/sap/sap2-1/default.php.

[47] "Reference scenarios" typically represent the researchers' best estimates of future trajectories without significant policy changes. They are frequently used, as in this project, to compare with, estimate the impacts of, specific policy scenarios.

[48] CCSP, Scenarios of Greenhouse Gas Emissions and Atmospheric Concentrations. Sub-report 2.1A of Synthesis and Assessment Product 2.1 by the U.S. Climate Change Science Program and the Subcommittee on Global Change Research.[Clarke, L., J. Edmonds, H. Jacoby, H. Pitcher, J. Reilly, R. Richels] Department of Energy, Office of Biological & Environmental Research, Washington, DC., USA. (2007) p. 3. http://www.climatescience.gov/ Library/sap/sap2-1/default.php.

[49] In order to compare and aggregate different greenhouse gases, various techniques have been developed to index the effect each greenhouse gas to that of carbon dioxide, where the effect of CO2 equals one. When the various gases are indexed and aggregated, their combined quantity is described as the CO2-equivalent. In other

words, the CO2- equivalent quantity would have the same effect on, say, radiative forcing of the climate, as the same quantity of CO2.

[50] CCSP, *Scenarios of Greenhouse Gas Emissions and Atmospheric Concentrations*. Sub-report 2.1A of Synthesis and Assessment Product 2.1 by the U.S. Climate Change Science Program and the Subcommittee on Global Change Research.[Clarke, L., J. Edmonds, H. Jacoby, H. Pitcher, J. Reilly, R. Richels] Department of Energy, Office of Biological & Environmental Research, Washington, DC., USA. (2007) p. 3. http://www.climatescience.gov/Library/ sap/sap2-1/default.php. Scenarios *of Greenhouse Gas Emissions and Atmospheric Concentrations*. Sub-report 2.1A of Synthesis and Assessment Product 2.1 by the U.S. Climate Change Science Program and the Subcommittee on Global Change Research.[Clarke, L., J. Edmonds, H. Jacoby, H. Pitcher, J. Reilly, R. Richels] Department of Energy, Office of Biological & Environmental Research, Washington, DC., USA. (2007) p. 3. http://www.climatescience.gov/ Library/sap/sap2-1/default.php.

[51] To put the magnitude of these potential increases in context, the current global, annual mean temperature (GMT) of the Earth is approximately 14°C (57°F). The difference between the current GMT and the low point of the last Ice Age, about 21,000 years ago, was roughly 7-8°C (44-46oF).

[52] CCSP, *Climate Projections Based on Emissions Scenarios for Long-Lived and Short-Lived Radiatively Active Gases and Aerosols*. A Report by the U.S. Climate Change Science Program and the Subcommittee on Global Change Research. H. Levy II, D.T. Shindell, A. Gilliland, M.D. Schwarzkopf, L.W. Horowitz, (eds.). Department of Commerce, NOAA's National Climatic Data Center, Washington, D.C., USA, 100 pp. (2008)

[53] See, for example, CCSP, 2008: Abrupt Climate Change. A report by the U.S. Climate Change Science Program and the Subcommittee on Global Change Research [Clark, P.U., A.J. Weaver (coordinating lead authors), E. Brook, E.R. Cook, T.L. Delworth, and K. Steffen (chapter lead authors)]. U.S. Geological Survey, Reston, VA, 459 pp. (2008)

[54] See CCSP, *Coastal Sensitivity to Sea-Level Rise: A Focus on the Mid-Atlantic Region*. A report by the U.S. Climate Change Science Program and the Subcommittee on Global Change Research. [James G. Titus (Coordinating Lead Author), K. Eric Anderson, Donald R. Cahoon, Dean B. Gesch, Stephen K. Gill, Benjamin T. Gutierrez, E. Robert Thieler, and S. Jeffress Williams (Lead Authors)], U.S. Environmental Protection Agency, Washington D.C., USA. (2009).

[55] See, for example, CCSP, Thresholds of Climate Change in Ecosystems. A report by the U.S. Climate Change Science Program and the Subcommittee on Global Change Research. [Fagre D.B., Charles C.W., Allen C.D., Birkeland C., Chapin F.S. III, Groffman P.M., Guntenspergen G.R., Knapp A.K., McGuire A.D., Mulholland P.J., Peters D.P.C., Roby D.D., and Sugihara G.] (2009) U.S. Geological Survey, Department of the Interior, Washington D.C., USA.

[56] CRS Report R40203, *Mountain Pine Beetles and Forest Destruction: Effects, Responses, and Relationship to Climate Change*, by Ross W. Gorte.

[57] CCSP, Coastal Sensitivity to Sea-Level Rise: A Focus on the Mid-Atlantic Region. A report by the U.S. Climate Change Science Program and the Subcommittee on Global Change Research. [James G. Titus (Coordinating Lead Author), K. Eric Anderson, Donald R. Cahoon, Dean B. Gesch, Stephen K. Gill, Benjamin T. Gutierrez, E. Robert Thieler, and S. Jeffress Williams (Lead Authors)], U.S. Environmental Protection Agency, Washington D.C., USA. (2009).

[58] David. S. Battisti and Rosamond L. Naylor, "Historical Warnings of Future Food Insecurity with Unprecedented Seasonal Heat," Science 323, no. 5911 (January 9, 2009): 240-244, doi:10. 1126/science. 1164363.

[59] Noah S. Diffenbaugh et al., "Global warming presents new challenges for maize pest management," http://www.iop.org/EJ/article/1748-9326/3/4/044007/erl8_4_044007.html.

[60] A. Analitis et al., "Effects of Cold Weather on Mortality: Results From 15 European Cities Within the PHEWE Project," Am. J. Epidemiol. (October 24, 2008): kwn266, doi:10.1093/aje/kwn266.

[61] CCSP, *Analyses of the Effects of Global Change on Human Health and Welfare and Human Systems*. A Report by the U.S. Climate Change Science Program and the Subcommittee on Global Change Research. [Gamble, J.L. (ed.), K.L. Ebi, F.G. Sussman, T.J. Wilbanks, (Authors)]. U.S. Environmental Protection Agency, Washington, DC, USA (2008); Bernard Doyon, Diane Bélanger, and Pierre Gosselin, "The potential impact of climate change on annual and seasonal mortality for three cities in Québec, Canada," International Journal of Health Geographics 7 (2008): 23, doi: 10.11 86/1476-072X-7-23; Kristie L Ebi, "Adaptation costs for climate change-related cases of diarrhoeal disease, malnutrition, and malaria in 2030," Globalization and Health 4 (2008): 9, doi:10.1 186/1744-8603-4-9; Michael Emch et al., "Seasonality of cholera from 1974 to 2005: a review of global patterns," International Journal of Health Geographics 7 (2008): 31, doi: 10.1 186/1476-072X-7-3 1; Paola Michelozzi et al., "High Temperature and Hospitalizations for Cardiovascular and Respiratory Causes in 12 European Cities," Am. J. Respir. Crit. Care Med. (December 5, 2008): 200802-217OC, doi:10.1 164/rccm.200802-217OC; Paul Reiter, "Global warming and malaria: knowing the horse before hitching the cart," Malaria Journal 7, no. Suppl 1 (2008): S3, doi:10.1 186/1475-2875-7-S1- S3;.

[62] Jessica G Fritze et al., "Hope, despair and transformation: Climate change and the promotion of mental health and wellbeing," International Journal of Mental Health Systems 2 (2008): 13, doi: 10.1186/1752-4458-2-13; Alana

Hansen et al., "The Effect of Heat Waves on Mental Health in a Temperate Australian City," Environmental Health Perspectives 116, no. 10 (October 2008): 1369–1375, doi:10.1289/ehp.1 1339.

[63] Ralph Catalano, Tim Bruckner, and Kirk R. Smith, "Ambient temperature predicts sex ratios and male longevity," Proceedings of the National Academy of Sciences (February 4, 2008): 0710711104, doi: 10. 1073/pnas.071071 1104.

[64] John W. Williams, Stephen T. Jackson, and John E. Kutzbach, "Projected distributions of novel and disappearing climates by 2100 AD," Proceedings of the National Academy of Sciences of the United States of America 104, no. 14 (April 3, 2007).

INDEX

D

E

G

F

H

S

T

156,